How to Find a Habitable Planet

• • •

Books in the *SCIENCE ESSENTIALS* series bring cutting-edge science to a general audience. The series provides the foundation for a better understanding of the scientific and technical advances changing our world. In each volume, a prominent scientist—chosen by an advisory board of National Academy of Sciences members—conveys in clear prose the fundamental knowledge underlying a rapidly evolving field of scientific endeavor.

The Great Brain Debate: Nature or Nurture, by John Dowling

Memory: The Key to Consciousness, by Richard F. Thompson and Stephen Madigan

The Faces of Terrorism: Social and Psychological Dimensions, by Neil J. Smelser

The Mystery of the Missing Antimatter, by Helen R. Quinn and Yossi Nir

The Long Thaw: How Humans Are Changing the Next 100,000 Years of Earth's Climate, by David Archer

The Medea Hypothesis: Is Life on Earth Ultimately Self-Destructive? by Peter Ward

How to Find a Habitable Planet, by James Kasting

The Little Book of String Theory, by Steven S. Gubser

Enhancing Evolution: The Ethical Case for Making Better People, by John Harris

Nature's Compass: The Mystery of Animal Navigation, by James L. Gould and Carol Grant Gould

· James Kasting ·

How to Find a Habitable Planet

• • • • •

With a new afterword by the author

PRINCETON UNIVERSITY PRESS
PRINCETON AND OXFORD

Published by Princeton University Press, 41 William Street,
Princeton, New Jersey 08540

In the United Kingdom: Princeton University Press, 6 Oxford Street,
Woodstock, Oxfordshire OX20 1TW

press.princeton.edu

Second printing, and first paperback printing, with a new afterword by the author, 2012
Paperback ISBN 978-0-691-15627-9

The Library of Congress has cataloged the cloth edition of this book as follows

Kasting, James F.
How to find a habitable planet / James Kasting.
p. cm.
Includes bibliographical references and index.
ISBN 978-0-691-13805-3 (cloth : alk. paper)
1. Habitable planets. 2. Exobiology. I. Title.
QB820.K37 2010
576.8′39—dc22 2009034490

British Library Cataloging-in-Publication Data is available

This book has been composed in Minion family text, Minion display text,
and Lucinda Handwriting display

Printed on acid-free paper. ∞

Printed in the United States of America

10 9 8 7 6 5 4 3 2

To my wife, Sharon, for allowing me time to work on this manuscript and for being the hub of our wonderful family.

• • •

Contents

Preface

This book grew out of my experiences on a NASA committee on which I was privileged to serve during 2005–2006. The committee's assigned task was to perform a preliminary design study for a future space-based telescope called *Terrestrial Planet Finder—Coronagraph,* or *TPF-C,* for short. NASA hoped to use this telescope to find Earth-like planets around other stars and to study their atmospheres spectroscopically. Spectroscopy, for those unfamiliar with the term, is the technique of separating light (or other forms of electromagnetic radiation) into its component wavelengths, much as cloud particles do when they decompose sunlight to create a rainbow. *TPF-C,* had it flown, would have looked at visible and near-infrared light emitted by the parent star and reflected by the planet's atmosphere and surface. (The *near-infrared* region of the electromagnetic spectrum refers to wavelengths just longer than those of visible light.) If one were to look at Earth itself at these wavelengths, one should be able to detect oxygen (O_2), ozone, water vapor, and maybe even the presence of chlorophyll at the Earth's surface. Chlorophyll is the pigment used by green plants to collect light for photosynthesis, and O_2 is for the most part a by-product of photosynthesis; hence, seeing either one of these signals in a planet's spectrum would be a strong indication that the planet supported life. This was the goal that motivated NASA, along with many of us on the *TPF-C* committee. Everyone realized that this mission had the potential to make exciting discoveries—ones that could potentially revolutionize our perspective as to where terrestrial life, and humans, fit into the broader framework of life in the universe.

Unfortunately, NASA ran out of money and canceled the mission before even the preliminary design phase was completed. NASA has several major projects under way at the moment, the largest of which include finishing the Space Station and sending astronauts back to the Moon. A large infrared telescope, the *James Webb Space Telescope* (*JWST*), is also being constructed at the moment, and it will likely need to be finished before another large space telescope can be built. *JWST* will probably not be able to find Earth-like planets, but it will study the disks of material around stars from which planets are formed, and, so, it, too, is an exciting project. But *TPF-C* is such a potentially ground-breaking type of mission that most of us who have been involved with it are confident that it will reappear on NASA's mission list at some point in the not-too-distant future, albeit not necessarily in the same form as the mission that was canceled. It is likely to be several years, though, before work is restarted, and so I have taken advantage of this break in activity to write this book. One of my goals in doing so is to let people know about this and related planet-finding projects and perhaps to help shorten the downtime before *TPF-C* is flown.

The title of this book is derived from another book with a closely related title: *How to Build a Habitable Planet*, by Wallace (Wally) Broecker. Broecker's original book was published in 1985 and is due to reappear soon in a second edition, to be published by Princeton University Press. Broecker wrote his book for an undergraduate, non-science-major class that he taught for many years at Columbia University in New York. His book has received enthusiastic reviews from a generation of professors and students for the creative style by which he made complex concepts easy to comprehend. The present book goes into more detail on many topics that Broecker introduced, before going on to discuss methods for finding extrasolar planets. I have nonetheless tried to make it accessible to anyone with a strong interest in science, and some basic knowledge thereof. Much of the material in the book has been used in various courses that I have taught during the past 20 years at Penn State. In astrobiology classes, in particular, the students come from a variety of different physical and biological science backgrounds, and so I am forced to introduce each topic at a level that tries to include all of them.

I hope that this will also make the book readable by people to whom science is a hobby, rather than a profession.

My own teachers and students have contributed to this book in various ways. James C. G. Walker, now retired from the University of Michigan, was my guiding light (although not my actual thesis advisor) as a graduate student and for many years thereafter. Jim worked out the details of the CO_2/climate feedback mechanism that is now considered to be an essential contributer to Earth's long-term habitability. James Pollack (now deceased) was my postdoctoral advisor at NASA Ames Research Center. This Jim introduced me to the wonders of planetary science and to the technical details of runaway greenhouse atmospheres. Jim himself had been exposed to this subject as a graduate student of Carl Sagan. William (Bill) Schopf invited me to be part of his Precambrian Paleobiology Research Group at UCLA during the 1980s and thereby introduced me to the field of astrobiology, well before the term itself had been invented. And Heinrich (Dick) Holland, now retired from Harvard University, has been a scientific mentor throughout my career, ever since we met at a carbon cycle conference in Florida in 1984. Dick taught me about geochemistry, the rise of atmospheric oxygen, and many other issues related to Earth's long-term evolution, and he has been my sounding board for new ideas for the past 25 years.

This book was begun during a visit to the Laboratoire des Sciences du Climate et de l'Environment (LSCE) south of Paris in the summer of 2006. I am grateful to my host, Gilles Ramstein, and his colleagues at LSCE for inviting me to come and for many fruitful discussions of Earth's climate history. I am also indebted to Susan Brantley, Director of the Earth and Environmental Science Institute at Penn State University, for providing funds that allowed me to continue writing during fall of 2006.

The ideas in this book come indirectly from many others as well. I would like to acknowledge particularly Zlatan Tsevanov and Lia Lapiana at NASA and my colleagues on the *TPF*-C committee. My thanks also to the *TPF* Project at NASA's Jet Propulsion Laboratory in Pasadena, who co-funded my sabbatical there three years ago. The people I learned from there include Ginny Ford, Marie Levine, Stuart Shaklan,

Sarah Hundai, Chas Beichman, Steve Unwin, Steve Ridgeway, Dan Coulter, Jim Fanson, Wes Traub, and Mike Devirian. I also learned a lot from two "Pale Blue Dot" meetings organized at NASA Ames Research Center in the mid-1990s by Dave DesMarais and from my more recent experience as part of the joint NASA/NSF Exoplanet Task Force, chaired by Jonathan Lunine. This committee, which just issued its final report in late spring of 2008, delved deeply into different methods for searching for extrasolar planets and attempted to chart a coherent strategy for exoplanet exploration for the next 15–20 years. Thanks to Jonathan and to everyone else on this committee from whom I learned so much.

Finally, I'd like to thank my editor at Princeton University Press, Ingrid Gnerlich. Her careful readings and insightful comments have made the book much more readable. And thank you, dear Reader, for picking up this book and for being interested in the same topic that interests me! I hope you find some ideas here that intrigue you.

• Part I •

Introduction

• • •

Wherein we contemplate the factors that make life possible and pose the question of whether it might exist elsewhere in the universe . . .

· *Chapter One* ·

Past Thinking about Earth-Like Planets and Life

Are there other planets like Earth and, if so, do they support life? These two questions motivate NASA's proposed *Terrestrial Planet Finder* missions, which are at the heart of this book. These missions, and much of what is discussed in this book, are entirely new. But the questions themselves are fundamental in nature and have been discussed for a long time. The issue of whether or not the Earth was unique was debated in the 4th century BC by the Greek philosophers Aristotle and Epicurus. Aristotle, like his teacher Plato, thought that there could be only one Earth:

> Either, therefore, the initial assumptions must be rejected, or there must be one center and one circumference; and given this latter fact, it follows from the same evidence and by the same compulsion, that the world must be unique.[1]

Epicurus, who lived slightly later, disagreed. Epicurus believed that all matter was composed of microscopic atoms, and this idea led him to postulate the existence of many different worlds. Around 300 BC, in a letter to Herodotus, he wrote: "There are infinite worlds both like and unlike this world of ours" inhabited by "living creatures and plants and other things we see in this world."[2]

Unfortunately, as Mike Devirian of NASA's Jet Propulsion Laboratory in Pasadena is fond of pointing out, Epicurus died painfully (of a bladder infection) in 269 BC, and support for his views died with him. And so the Earth-centered picture of the Universe prevailed for more than 1800 years after Aristotle's death. Aristotle's view of astronomy was further elaborated by another Greek philosopher, Ptolemy, who lived from AD 90 to 168. Ptolemy developed a complicated scheme for predicting the orbits of the five known planets: Mercury, Venus, Mars, Jupiter, and Saturn. This scheme, which became known as the Ptolemaic system, was remarkably successful, even though it was based on physically incorrect principles. To explain the apparent retrograde motions of the three outermost planets—the fact that they appear to reverse their motion across the sky at some times during the year—Ptolemy postulated that the planets moved in small circular *epicycles* superimposed on larger circular orbits around the Earth. This Earth-centric view of the Solar System was later backed by the Roman Catholic Church, whose clergy found this way of thinking to be in accord with their own teachings about God and about man's place in the universe. So it was not surprising that it dominated astronomical thinking for more than 1500 years.

The Ptolemaic system remained unchallenged until the early 16th century, when the Polish mathematician/astronomer Nicolas Copernicus published a new theory in which he postulated that the Sun was at the center of the universe and that the Earth and all the other planets revolved around it. In his theory, the apparent retrograde motion of the planets was explained by the changing vantage point of the Earth as it moved around the Sun and the fact that the Earth moves more quickly than do the planets beyond it. It was not until early in the next century that the critical test of the two theories was made. The Italian mathematician Galileo, using a telescope that he improved upon, but did not invent, discovered that Venus and Mercury exhibited phases, like the Moon. This observation was consistent with the Copernican view of the universe, but not with the older Ptolemaic view. Tycho Brahe had an intermediate theory, in which the other planets went around the Sun, while the Sun went around the Earth, and this could not be disproved by Galileo's observations. The Catholic Church supported the modified

Ptolemaic system, and Galileo was threatened with execution if he did not renounce his support for the Copernican model. As every graduate of a first-year general astronomy class knows, Galileo recanted and was spared, but he spent the remainder of his life under house arrest. So things continued to go poorly for astronomers who believed that the Earth was not unique.

The patron saint of planet-finders, though, is Giordano Bruno. Bruno was an Italian philosopher who lived in the late 16th century. Bruno was a gadfly who held all sorts of beliefs that conflicted with Catholic Church doctrine. He was also a believer in the existence of other Earths and in extraterrestrial life. In his book, *De L'infinito Universo E Mondi*, published in 1584, Bruno wrote:

> There are countless suns and countless earths all rotating around their suns in exactly the same way as the seven planets of our system . . . The countless worlds in the universe are no worse and no less inhabited than our Earth.

For this and other heretical statements, Bruno was sentenced to be burned at the stake. Unlike Galileo, Bruno refused to recant, and the sentence was carried out in the year 1600 in Campo dei Fiore, Rome, where a statue of Bruno can be found today (figure 1.1). Today, as Devirian notes when consoling frustrated would-be planet-finders, scientists may lose their funding, but the other consequences of their activities are fortunately much less severe.

The Habitable Zone and the Importance of Liquid Water

In modern times, the question of whether habitable planets exist has been taken up by mainstream scientists. The idea that stars were other suns took hold, and researchers, including the French mathematician Laplace, developed theories for how planets might form around them. It did not take too long for some astronomers to again start speculating about whether any of them might be habitable. In a book written in 1953,

Figure 1.1 Statue of Giordano Bruno in Campo dei Fiore, Rome. Bruno was an early advocate of the existence of other worlds who was executed for his beliefs.[3]

the famous astronomer Harlow Shapley defined what he termed the "liquid water belt" as being that region in a planetary system in which liquid water could exist at a planet's surface.[4] In making this distinction, Shapley was acknowledging something that biologists had been aware of for many years, namely, that liquid water is essential for all known forms of life. Although some organisms, notably those that form spores, can survive for long time periods in the absence of water, none of them can metabolize or reproduce unless liquid water is available.

Biology depends on liquid water for reasons that make good sense from a chemical standpoint. The water molecule has a bent shape, as shown in figure 1.2. Electrons are more attracted to the oxygen atom in the middle than to the hydrogen atoms at the two ends. This makes water a highly *polar molecule*, and that in turn makes it a powerful solvent for other polar molecules. (A polar molecule is one, like the water molecule, in which the electrons are concentrated toward one end.) Terrestrial life consists of various complex, carbon-containing compounds, most of which are polar and thus soluble in water. One can, of course, imagine an alien life form that might utilize some other polar solvent, or

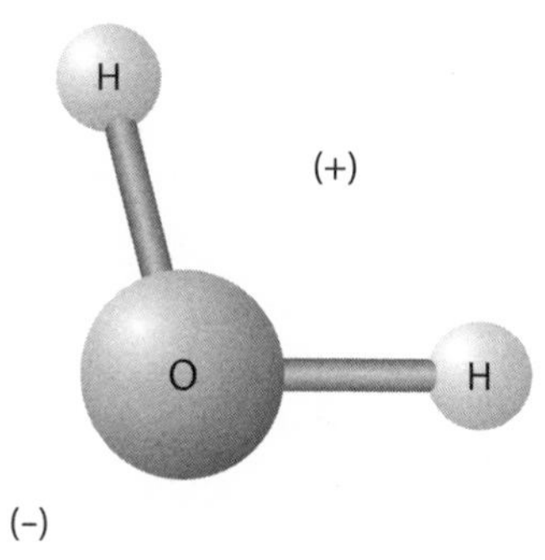

Figure 1.2 The geometry of the water molecule. The side closest to the hydrogen atoms is positively charged. The opposite side is negatively charged. The separation of charge in this molecule has important implications, both for planetary habitability and for life.

even one that is not carbon-based, like us. But for most biologists, and for astronomers like Shapley as well, the idea of looking for life on planets that contain water seems like a good place to begin.

Astronomers are not at all dismayed that liquid water is needed for life, because water is thought to be one of the most abundant chemical compounds in the universe. The two atoms from which it is formed, hydrogen and oxygen, are the first and third most abundant elements—helium is second—and they are present in high concentrations in nearly all stars. Hence, planets around most stars might be expected to contain H_2O. We shall see later on that the amount of H_2O may differ greatly from one planet to another, and so we should not expect that all planets should have as much water as does Earth. But this is not a reason to be pessimistic. There could still be many planets with enough water to support life.

Liquid water has other properties that may contribute to creating stable, life-supporting planetary environments. The highly polar nature of the H_2O molecule also gives liquid water an extremely high *heat capacity*. The heat capacity of a substance is a measure of how much heat, or thermal energy, must be added to it to raise its temperature by a given amount. The high heat capacity of water arises because the positively charged ends of some water molecules are attracted to the negatively charged ends of other molecules, forming what are referred to as *hydrogen bonds*. When water is heated, much of the added energy goes into loosening the hydrogen bonds between the individual molecules, and thus less of it is available to make the molecules move faster, and thereby raise its temperature.

The fact that the temperature of liquid water changes relatively slowly has important implications for Earth's climate. Coastal regions, for example, are more temperate than continental interiors. The reason is that the high heat capacity of the ocean greatly reduces its seasonal cycle in temperature. Land surfaces, by contrast, have relatively low heat capacities and more pronounced seasonal cycles. This difference between land and sea may become even more important on planets whose orbits are less circular (or more *eccentric*) than Earth's orbit, or whose *obliquities* are higher than Earth's. (A planet's obliquity is the tilt of its spin axis with respect to its orbital plane. Earth's obliquity is 23.5 degrees.) On such planets, land surfaces could become either extremely hot or extremely cold at different times during the year, rendering them uninhabitable. But ocean temperatures would fluctuate much less, and so marine life would be less likely to be affected by these factors. So liquid water is important both for the chemistry of life itself and for the habitability of planets on which life may evolve.

We should also note, right from the start, that planets that lack liquid water at their surfaces, and so are outside of Shapley's "liquid water belt," could still be habitable if they have subsurface liquid water. Mars is one example of such a planet. Although its surface is entirely frozen, computer models predict that liquid water might be present at a depth of several kilometers, as a consequence of the continuing flow of heat from the planet's interior. If so, then microbes similar to terrestrial methanogens, single-celled organisms that make a living converting carbon dioxide and hydrogen (and assorted organic substrates) into methane, could conceivably be present. Reported observations of methane in Mars' atmosphere, discussed further in chapter 8, could be evidence for such subsurface life. Jupiter's moon, Europa, may also harbor subsurface water, and so could be an abode for carbon-based life. (This idea was anticipated many years ago by science fiction author Arthur C. Clarke in his novel *2010: Odyssey Two*. Clarke envisioned squid-like creatures swimming in Europa's subsurface seas.) Both Mars and Europa are high on NASA's priority list as destinations for unmanned, and eventually manned, planetary probes. These priorities are well justified—after all, nothing would be of greater interest to astrobiologists than the chance to actually find and study an extraterrestrial organism.

For astronomers interested in the possibility of life on planets around other stars, however, Shapley's restriction of the liquid water belt to planets possessing surface liquid water makes sense. For the time being, at least, the stars are too distant for us to explore directly. Any observations that we may make of planets orbiting around them will be obtained using telescopes based either on Earth or, more likely, in space. We may indeed be able to study such planets by using spectroscopy, as pointed out in the preface to this book. But we will be looking at the planet's atmosphere and surface, not at the (putative) organisms themselves. For life to be able to modify a planet's atmosphere in a detectable way, it needs to be present at the planet's surface.* That way, it can take advantage of the abundant energy from the planet's parent star to colonize the planet's surface and to modify its atmosphere, just as photosynthetic organisms have done on Earth. We will return to the topic of the liquid water belt, also called the *habitable zone* or *ecosphere*, in chapter 10, as it is a key concept in the search for Earth-like extrasolar planets.

Carl Sagan and the Drake Equation

In the last 40 years, the question of whether life might exist elsewhere in the universe has been popularized by the famous astronomer Carl Sagan. Sagan was interested not just in whether life exists elsewhere, but also in whether intelligent beings might exist, with whom we might one day make radio contact. In his 1966 book, *Intelligent Life in the Universe*,[5] coauthored with Soviet astrophysicist I. S. Shklovskii, Sagan expressed the odds of making contact in the form of a mathematical relationship

*The possible detection of methane in Mars' atmosphere, which *may* be biogenic, could negate this statement. The upper limit on martian methane, however, is 10–100 parts per billion, which is just barely detectable from nearby Earth, and far too low to detect from interstellar distances. Thus, from a practical standpoint, life does need to be present at a planet's surface in order to create a detectable atmospheric signature.

that is usually called the Drake equation, although Sagan himself had a hand in crafting it and is said to have preferred the name "Sagan-Drake equation." Frank Drake is a radio astronomer who headed the SETI (Search for Extraterrestrial Intelligence) Institute for many years. Drake and Sagan developed their equation to help guide the discussion at a meeting on intelligent life in the universe held at the Green Bank Radio Observatory in 1961. The equation estimates the number of advanced, communicating civilizations in the galaxy, N, to be equal to the product of seven parameters:

$$N = N_g f_p n_e f_l f_i f_c f_L$$

Here, N_g is the number of stars in our galaxy; f_p is the fraction of stars that have planets; n_e is the number of Earth-like planets per planetary system; f_l is the fraction of habitable planets on which life evolves; f_i is the probability that life will evolve to an intelligent state; f_c is the probability that intelligent life will develop the capacity to communicate over long distances (for example, by radio telescope); and f_L is the fraction of a planet's lifetime during which it supports a technological civilization.

The Drake equation cannot actually be solved, as it involves several terms—the last four in particular—that no one knows how to evaluate. We will nonetheless make an attempt to do so in chapter 15, using information gleaned from the studies described in this book. For now, let us stick to what was already known before the modern extrasolar planet-finding era began. Based on detailed star counts in representative areas of our galaxy, the leading term, N_g, is about 4×10^{11}, or 400 billion. Hence, if the other factors are anywhere close to one, then N should be a large number, and there might well be intelligent civilizations that would be close enough to converse with, albeit slowly, using radio telescopes like the giant Arecibo telescope shown in figure 1.3 (see color section). Sagan himself was an unapologetic optimist on this question. His estimated value for N from his later book *The Cosmic Connection*[7] was 1 million. If that number is correct, then the nearest intelligent civilization is probably no further than a few hundred light-years from Earth.

Whether other intelligent beings exist is, as Sagan realized, the question that we would all ultimately like to answer. And the SETI Institute exists to try to do just that. Many scientists, however, would be happy for now if we could just evaluate the first four terms of the equation, which together would give us the probability of the existence of habitable planets and life. We will focus on those terms in this book, bearing in mind that even if we are successful in estimating them, the ultimate question of whether extraterrestrial intelligence exists remains to be addressed.

As a footnote to this discussion, I should note that at the time of this writing (early 2008), the Arecibo radio telescope is in danger of being shut down because of lack of funding (most of which comes from the U.S. National Science Foundation). The reasons for its financial troubles have nothing to do with SETI, which does not even use the telescope at the present time, and which only "piggybacked" on other, unrelated telescopic searches even at the best of times. The closure of Arecibo would eliminate a powerful tool that could potentially be used to search the galaxy for evidence of extraterrestrial life. But there may be other ways to do this (see chapter 15), and so I will leave it to the astronomers to decide how best to deal with the future of Arecibo.

Other Perspectives on Planetary Habitability: Rare Earth and Gaia

Not all modern authors are as optimistic as Carl Sagan about the possibilities for life elsewhere. In their book *Rare Earth*, published in 2000, Peter Ward and Donald Brownlee offered up a direct challenge to his ideas.[8] Ward is a paleontologist who has studied impacts and mass extinctions and who has written several other popular books. Brownlee is an astronomer (and a distinguished member of the National Academy of Sciences) who is famous for his measurements of the interplanetary dust particles that fall into Earth's atmosphere from space. Indeed, these tiny bits of material are sometimes referred to as "Brownlee particles" in his honor.

The thesis of *Rare Earth* is that complex life—by which the authors mean animals and higher plants—is rare in our galaxy, and presumably throughout the universe. Humans and hypothetical aliens are both animals, of course, and so this would imply that intelligent life is also rare. Simple, unicellular life may be widespread, according to this hypothesis; the authors do not suggest that life itself is uncommon. But, from their point of view, Earth offers a uniquely stable environment for the development of higher plants and animals, for a variety of reasons. High on their list are the operation of plate tectonics (which Ward and Brownlee consider to be an unusual geological process) and the stabilization of Earth's spin axis by the Moon (which, as virtually everyone agrees, formed as the result of a statistically unlikely glancing collision with a body the size of Mars). In their view, these and other cosmic accidents permitted complex life to evolve on Earth, but it is highly unlikely that this same evolutionary pathway would have been followed on any other rocky planet. Humans are therefore probably alone in the galaxy.

The predictions of *Rare Earth* are not likely to be directly tested in the near future. The planet-finding missions discussed in this book will most likely be incapable of distinguishing between microbial life and complex, multicellular life. We shall return to *Rare Earth* in chapter 9, however, because many of the issues that Ward and Brownlee raise are relevant to more general issues of planetary habitability and thus bear directly on the question of whether other Earth-like planets exist. Also, we will need to consider the question of complex life as well if we hope to make an estimate for the number N in the Drake equation.

An alternative, completely different approach to planetary habitability is the *Gaia hypothesis* put forward by James Lovelock and Lynn Margulis. The name is taken from Greek mythology, in which Gaia was the goddess of Mother Earth. Lovelock is a British scientist and inventor who has written a series of books on his hypothesis.[10-13] Margulis is an American biologist, previously married to Carl Sagan, who is famous for her contributions to evolutionary theory, particularly her development of the concept of *endosymbiosis*—the idea that certain organelles within eukaryotic organisms (including plants and animals) are the result of the past incorporation of other free-living organisms. In their

Figure 1.4 Painting of Gaia, the Greek goddess of Mother Earth.[9]

papers and books, Lovelock and Margulis have argued that life itself has played a critical role in keeping the Earth habitable throughout its 4.5-billion-year history. The theory stemmed originally from Lovelock's analysis of the *faint young Sun problem*, which we discuss in chapter 3. Lovelock suggested that photosynthetic organisms drew CO_2 out of the atmosphere at just the right rate to compensate for steadily increasing solar luminosity over time. Although the biota in Lovelock's model exerted no conscious control over this process, he nevertheless suggested that they act as a cybernetic control system that counteracts such perturbations and that has moderated Earth's climate throughout its history. If this hypothesis is valid, a planet might need to be inhabited in order to remain habitable. And if this in turn is true, then the question of whether other habitable planets exist is inextricably linked to the question of whether life has originated elsewhere than on Earth. We will return to Gaia, too, in chapter 9 to see whether or not we agree with Lovelock's theory.

The foregoing brief history of thinking about habitable planets is by no means complete. Many other scientists and philosophers have speculated about whether habitable planets and life might exist elsewhere. As we have seen, the debate has been carried out sporadically for literally thousands of years and remains unresolved at the present time. But this question is now timely and exciting because, as we shall see, astronomers are on the verge of being able to answer these questions observationally. If they can manage to do so, and especially if evidence for Earth-like planets and life is found, the philosophical implications would be profound. Indeed, such a discovery would be no less world-shaking than Galileo's proof that the Earth goes around the Sun.

• Part II •

Our Habitable Planet Earth

• • •

Wherein we describe in more detail how our own planet formed and why it has remained habitable throughout its history despite the gradual brightening of the Sun, and despite being occasionally beset by nearly unimaginable climate catastrophes . . .

· Chapter Two ·

Critical Updates on How Planets Are Built

Before we get too far along in talking about how to find habitable planets, it might be wise to step backward for a moment and talk briefly about how planets are built. As one might expect, how planets are formed turns out to be critical for determining whether they might be habitable. And, even though there are many papers and books on this topic, including the book by Wally Broecker that was mentioned in the preface, most readers may not be familiar with all of the arguments, especially those that have appeared only within the last few years. Furthermore, although Broecker provides an excellent introduction to the topic, his perspective is primarily that of a geochemist, and so his discussion is focused on explaining chemical abundances in planets and meteorites and on the extremely important concepts of radiometric age dating and geologic time. (That said, Broecker's discussion overlaps with some of the material presented in the first two sections of this book, because both of us are interested in the question of what *keeps* planets habitable). Astronomers have also thought a lot about planet formation, and their approach is typically more mechanistic in nature. Let's therefore take a brief tour of recent astronomical thinking on this topic, making notes as we do so about which aspects of this process are most important for planetary habitability.

Chapter Two

The Conventional Wisdom regarding Planet Formation

At the time that Broecker's book was written (1985), astronomers thought they already had a pretty good theory for how planets were formed. It was, and still is, generally accepted that both stars and planets form from the collapse of interstellar clouds of gas and dust. In the case of our own Solar System, the central parts of the cloud collapsed to form the Sun, and the remainder of the material was spun out by rotation into a disk, often termed the *solar nebula.* We know now that such disks exist because we can observe them telescopically around many young stars. Figure 2.1 shows a *Hubble Space Telescope* picture of a disk around the star Beta Pictoris—the first such disk to be directly imaged at visible wavelengths.

The next step of planet formation is still somewhat controversial. Most astronomers believe that planets form from *accretion* of solid materials that condensed from the disk. The term "accretion" refers to the process by which orbiting particles collide with each other to form larger *planetesimals*, and eventually planets. Rocky, *terrestrial planets* like Venus and Earth are formed almost entirely from these chunks of solid material. *Gas giants* such as Jupiter and Saturn are thought to have solid cores that formed by accretion. Once these cores grew larger than about 10–15 Earth masses, though, they were able to capture gaseous hydrogen and helium from the surrounding solar nebula. The largest gas giant, Jupiter, grew to over 300 Earth masses and has a composition close to that of the Sun, indicating that most of its material was captured gravitationally from the nebula. Some astronomers, notably Alan Boss of the Carnegie Institute in Washington, DC, have argued that gas giant planets may form as stars themselves do, from gravitational collapse of the disk itself. The issue might potentially be resolved if we can send another spacecraft to Jupiter and determine (through detailed study of its gravitational field) whether it really does have a rocky/icy core. But this analysis is considered technically challenging and might or might not yield a definitive answer. So, for now, I will follow the standard *core accretion* model for forming both terrestrial and giant planets, bearing in mind that some disagreement still exists.

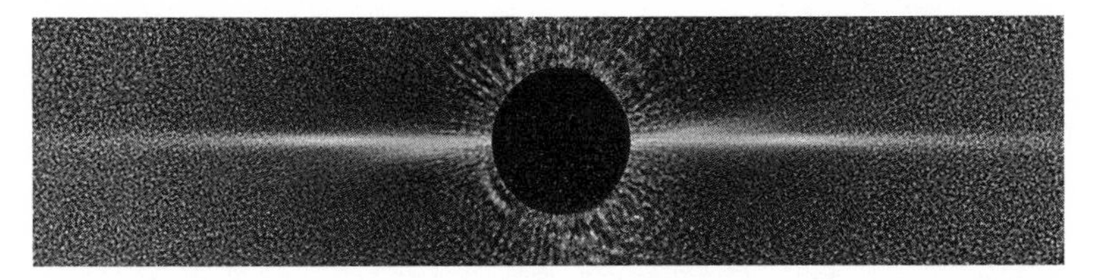

Figure 2.1 Photograph of the disk surrounding Beta Pictoris. Note the large scale of the photograph. By comparison, Neptune orbits the Sun at about 30 AU; thus, the Beta Pictoris disk extends well beyond the boundaries of our own Solar System. The central star itself is blocked out in the photo. (Photo courtesy of NASA; obtained from http://www.centauri-dreams.org/?p=1495.)

With the exception of the disk around Beta Pic, most of this story was already known 25 years ago. Astronomers also had a pretty good idea for *why* the Solar System was structured as it is, with rocky, terrestrial planets on the inside and gas giants further out. The inner planets are rocky because the inner parts of the solar nebula were hotter, and rocky materials have the highest condensation temperatures (or, equivalently, the highest vaporization temperatures); hence, they were the only materials that could condense in the hot, inner parts of the Solar System. John Lewis, an astronomer from the University of Arizona, expressed this idea elegantly in his *equilibrium condensation model* for planetary formation.[1] In this model, Lewis assumed that the growing *protosun* was surrounded by a solar nebula composed of gas and dust that had the same overall composition as the Sun itself.* According to Lewis, the order in which different materials should condense from such a nebula as it cools is as follows:

1. Refractory oxides of calcium, titanium, aluminum, and magnesium ($CaTiO_3$, $Ca_2Al_2SiO_7$, $MgAl_2O_4$)
2. Metallic iron-nickel (Fe–Ni) alloy
3. $MgSiO_3$ (enstatite)—a silicate mineral that is abundant in Earth's mantle
4. Alkali aluminosilicates

*Lewis also assumed that the solar nebula contained the same amount of mass as the Sun—an estimate that is probably too high by at least a factor of 10—but we can ignore this aspect of his model, as it has only a weak effect on his predictions.

5. FeS (troilite)—an iron-sulfide mineral
6. FeO silicates (iron-bearing silicates)
7. Hydrated silicates—silicates containing chemically bound water
8. H_2O (water)
9. NH_3 (ammonia)
10. CH_4 (methane)
11. H_2 (molecular hydrogen)
12. He (helium)

The *refractory* oxides at the top of the list are compounds that vaporize only at extremely high temperatures. The species making up the lower half of the list are *volatile* compounds with relatively low vaporization temperatures. Or, to put it another way, they are species that are gases at room temperature. The last two compounds on the list, H_2 (molecular hydrogen) and He (helium), condense only at extremely low temperatures, and hence were probably always present in the solar nebula as gases.

A careful look at Lewis's list explains much of what we observe about the composition of planets in our own Solar System. The terrestrial planets formed closer to the Sun, from metal oxides and iron, or iron–nickel alloy, as these were the only compounds that could condense out in the warm inner parts of the solar nebula. In Lewis's model, the nebula became cool enough at around 5 AU to allow H_2O to condense out as water ice. (1 AU = 1 astronomical unit = mean Earth-Sun distance.) The presence of Jupiter at 5.2 AU from the Sun is thus nicely explained. As noted in chapter 1, oxygen is the third most abundant element in the Sun, and therefore in the solar nebula as well, and so when water was able to condense out as ice the abundance of solid particles in the disk increased enormously. This allowed Jupiter's core to accrete very rapidly—so fast that it grew quickly to more than 10 Earth masses, and hence was able to capture gas from the solar nebula before the nebula disappeared. Observations made over the last 20 years of disks around other young stars indicate that protostellar disks exist for only a few million years before they are dissipated by planet formation or other processes.

Lewis's model can also explain another observation that we have not yet mentioned. The four outer planets—excluding Pluto, which has been reclassified as a "planetoid"—do not share the same composition. As mentioned earlier, Jupiter has nearly the same elemental composition as the Sun. Saturn, by comparison, is deficient in hydrogen and helium by about a factor of 3. And Uranus and Neptune, the two outermost planets, contain relatively little hydrogen and helium. Instead, they are composed mostly of compounds 8–10 on Lewis's list: water (H_2O), ammonia (NH_3), and methane (CH_4). In the cool outer regions of the solar nebula, these compounds should all have condensed out as ices. But because planets that are farther out from the Sun take longer to orbit, their icy cores would not have accreted as fast as did Jupiter's core, and so the gas and dust should have been largely gone from the nebula by the time they formed. Indeed, Uranus (14.5 Earth masses) and Neptune (17 Earth masses) are basically preserved icy cores and are now classified as *ice giants* rather than as gas giants. We shall see later on (chapter 8) that the story of their formation may be even more complicated and that they may have moved, or *migrated*, outward from their original locations. Other giant planets, including Jupiter and Saturn in our Solar System, may also migrate during or shortly after their formation as a consequence of gravitational interactions with the gaseous solar (or stellar) nebula. But that is a relatively recent idea which is not part of the "conventional wisdom," and hence we will postpone this discussion to later chapters.

Where Did Earth's Water Come From?

A key problem for Lewis's equilibrium condensation model had to do with how Earth obtained its water. Lewis assumed that Earth got its water from hydrated silicate minerals such as serpentine, $(Mg, Fe)_3Si_2O_5(OH)_4$, and chlorite, $(Mg, Fe, Al)_6(Si, Al)_4O_{10}(OH)_8$. As one can tell from their lengthy chemical formulas, these are complex silicate molecules with water molecules incorporated into their structures. Lewis's list indicates that these compounds should have condensed at intermediate temperatures, somewhere between iron silicates and water ice. Hence, Earth and Mars could plausibly have obtained hydrated silicates, whereas Venus, in his model, did not. Based on this reasoning, Lewis argued for many years that Venus started off dry. Alternatively, Venus could have started off with lots of water, like Earth, but then lost it because of a runaway greenhouse effect. We discuss this theory later on in chapter 6.

The idea that hydrated silicates might have existed in the solar nebula is supported by observational evidence. Certain types of meteorites, especially the *carbonaceous chondrites*, are chock full of them. Carbonaceous chondrites, like the Allende meteorite shown in figure 2.2, are primitive, *undifferentiated* meteorites whose less volatile components, excluding hydrogen and helium, are thought to be similar in composition to the original solar nebula. (By contrast, *iron meteorites* are thought to come from the fragmentation of planetesimals that had already grown large enough to separate, or differentiate, iron cores.) Carbonaceous chondrites are rich in both organic matter and water. As we shall see below, they may well be the ultimate source of Earth's water, along with its carbon and nitrogen. The ones we find today, though, are thought to have originated from the outer part of the asteroid belt, between about 2.5 and 3.5 AU. The material that condensed from the nebula near 1 AU was probably much drier than this. That is a good thing, actually, as an Earth formed entirely from carbonaceous chondrites would have had an embarrassingly large amount of water—some 300 times as much as is now present! Lewis did not believe that the bulk of the Earth could have been formed from such material (although an Australian geochemist, Ted Ringwood, actually proposed this in the scientific literature[2]). Earth might, however, have formed from material similar to *ordinary chondrites*, which are much less water-rich than carbonaceous chondrites, but which could still have provided the rather meager amount of water present in Earth's oceans (about 1.4×10^{21} kg). Earth itself has a mass of about 6×20^{24} kg.

The problem for Lewis's model was to explain how the hydrated silicate minerals formed in the first place. Even though his model predicted that these compounds should have been thermodynamically stable near 1 AU, the timescale required to form them by direct interactions between condensed silicate minerals and gaseous H_2O in the nebula turned out to be greater than the age of the universe![4] So this type of chemical reaction could not have occurred near 1 AU. Luckily, the hydrated sili-

Figure 2.2 The Allende carbonaceous chondrite. These meteorites are considered to be the most similar in composition to the original solar nebula.[3]

cates found in carbonaceous chondrites can be explained in another way. In the outer asteroid belt, water ice was able to condense, albeit at a somewhat later time than did the ice that formed Jupiter's core. Once the ice had condensed, it became incorporated into growing planetesimals. Those planetesimals also contained radioactive nuclides, including short-lived ones like aluminum-26, which would have generated heat as they decayed. That heat melted the water ice in the planetesimals, forming liquid water that then reacted with the surrounding silicates.

The conclusion from all of this is that the process of planetary formation must have been more complex than had been originally assumed. Some components of the Earth, including most of its water and other volatiles, must have condensed elsewhere in the solar nebula and then somehow become incorporated into the planet. And, to be fair, Lewis himself realized as much by the time his 1984 book was written. The details of how this happened, though, remained to be explained.

New Models for Planetary Accretion and Delivery of Water

This brings us to the modern era. Astronomers have made great strides over the last two decades in understanding planetary formation from a theoretical standpoint. Much of this progress has been a consequence of the advent of powerful computers. Computer power, as many readers will know, has increased exponentially for the last several decades, following a relationship known as Moore's law. Moore's law says that computing power doubles roughly every two years. This is not something that can continue forever, but it has proven accurate during the recent past. Astronomers have taken advantage of this increase in computer power, and this has led to many new insights into how planets formed.

The first important steps in modeling the accretion of planets had been taken in the 1960s and 1970s, before the advent of computers. The Russian astrophysicist Victor Safronov developed a detailed, *analytical model* of the accretion process.[5] (An analytical model is one in which the equations can be solved with a pencil and paper, without the need for a computer.) Safronov assumed that the planets grew by capturing small bodies from an accretion zone close to where the planet itself orbited. Each planet cleared out a ring of material near its own circular orbit. When combined with Lewis's model described above, Safronov's theory yielded a detailed description of what materials should be incorporated into each planet. But, as we have already seen, that prediction still did not explain the origin of Earth's water.

This view of accretion was changed dramatically by work[6–8] performed in the 1980s by George Wetherill, an astronomer at the Carnegie Institution of Washington, DC. Wetherill was already famous among geochemists for his work on determining the age of the Earth and, later, of meteorites. But he created an equally successful second career for himself as an astronomer by using computers to refine Safronov's model of planetary accretion. Computers at that time were not nearly as fast as those of today, and so by today's standards Wetherill's models were not very elaborate. Wetherill, though, was a clever mathematician. He modified a mathematical technique developed originally by the Estonian astronomer Ernst Öpik to compute meteorite orbits, and he used it to study the accretion process. Rather than trying to simulate the entire process, which is still considered to be difficult or impossible, he started from a time in which the planetesimals had already grown to about the size of Earth's Moon. (The Moon is approximately 1/80 of an Earth mass.) To keep his model computationally fast, he considered only the inner parts of the Solar System—the region containing the terrestrial planets. In Wetherill's model, unlike Safronov's, the planetesimals interacted with each other gravitationally as they grew. He found, unsurprisingly in retrospect, that they perturbed each other's orbits in a big way. Consequently, planetesimals that started out in one location in the nebula often ended up in another. This implied that planets are composed of material that started from a variety of different orbital distances. Although Wetherill's calculations did not always yield an Earth-like planet at 1 AU, the planets that formed at or near that orbital distance typically included material from as far out as the asteroid belt region. This is just what was needed to explain the origin of Earth's water! According to Wetherill's calculations, the impact of just one modest-sized planetesimal from the water-rich asteroid belt region could have provided more than enough water to fill Earth's oceans.

Wetherill's model made another important prediction. It suggested that the final stages of the accretion process should have been dominated by large impacts. By "large," I mean really large. As discussed at greater length in chapter 9, the impact that is thought to have formed Earth's Moon probably involved a planetesimal that was Mars-sized or larger. Mars is about 1/10 of an Earth mass, so this would indeed have been a spectacular collision. Before Wetherill's model was published, collisions of this magnitude were rarely postulated or, if they were, they were considered to be extremely speculative. We know now that they are to be expected, although this does *not* necessarily imply that Moon-forming impacts are common, because such impacts must also occur at the right relative velocity and the right angle.

Could Earth's Water Have Come from Comets?

Another potential source of water and other volatiles is comets. Comets are small bodies with diameters ranging from kilometers to tens of kilometers that are composed of roughly equal mixtures of ice and rock. They originate from outside the Solar System proper, that is, from beyond

Neptune's orbit (~30 AU). Most of the time, they are invisible to even the most powerful telescopes, but occasionally they come close enough to the Sun that they heat up and develop tails composed of material from vaporized ices, allowing them to be easily observed. The brightest ones, such as comets Halley and Hyakatake, can be seen at night with the naked eye, if one can find a dark place from which to look, away from city lights that spoil the viewing in most urban areas.

The comets that we observe today come from either of two regions: the Oort Cloud or the Kuiper Belt (see figure 2.3). The Oort Cloud is a spherical shell of comets surrounding the Solar System that extends outward to roughly 50,000 AU, or approximately 1 light-year. (This is about 1/4 the distance to the nearest star, Proxima Centauri.) The Oort Cloud is the source of most *long-period comets*, many of which are observed only once. The orbits of these comets are randomly distributed in space and are nearly parabolic, indicating that their source region must be spherical and extremely distant. The Kuiper Belt is reservoir of comets that lies within the plane of the Solar System, but beyond the orbit of Neptune. It is the source of most *short-period comets*, most of which have orbits that are prograde (i.e., in the same direction as those of the planets) and whose orbital planes are more or less aligned with those of the planets. Some long-period comets experience close encounters with planets and are perturbed into short-period comets. Such is the case for Comet Halley, for example. Halley is in a 76-year orbit, which makes it a short-period comet. But its orbit is inclined at 18° to the ecliptic plane (the plane of Earth's orbit), and it is retrograde, indicating that Halley probably came originally from the Oort Cloud.

Because they contain large amounts of water ice, comets could, in principle, have been the source of Earth's water. This idea was popular among planetary scientists for many years,[10] especially as it might also explain the abundance pattern of noble gases in Earth's atmosphere.* Comets also contain other volatile materials, including complex organic carbon compounds, and so some scientists have suggested that they may have contributed directly to the origin of life.[12, 13]

*The cometary delivery model of Owen et al.[11] can successfully account for the relative atmospheric ratios of the heavy noble gases, argon, krypton, and xenon. By contrast, models in which these noble gases are supplied by asteroids tend to predict too much xenon.

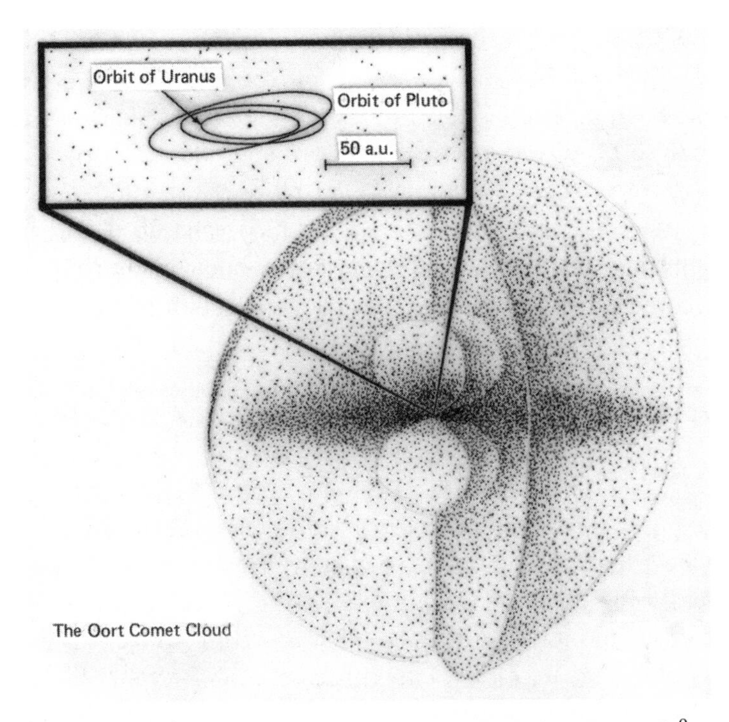

Figure 2.3 Diagram illustrating the Oort Cloud and the Kuiper Belt.[9] (Courtesy of Donald Yeoman, NASA Jet Propulsion Laboratory.)

Over the last two decades, however, the idea that comets supplied most of Earth's water has fallen out of favor. The main reason is that researchers have been able to measure the deuterium to hydrogen ratio of three different comets: Halley, Hayakatake, and Hale-Bopp. Deuterium (D) is an isotope of hydrogen that contains both a proton and a neutron in its nucleus. Normal hydrogen (H) has only a single proton. The D/H ratio of seawater is well known and has a value of 1.56×10^{-4}. The D/H ratio in all three comets that have been studied is about twice that value.[14]* If this is true for all comets, then they could not have been the

*A skeptic might say that three comets are not enough from which to draw firm conclusions. Furthermore, comets Halley, Hyakatake, and Hale-Bopp are all Oort Cloud comets. Might comets from the Kuiper Belt have lower D/H ratios? Attractive as this idea may sound, it is not likely to be correct. The D/H ratio of material in the original solar nebula is thought to have increased with distance from the Sun because the temperatures were lower and the dust grains, which were rich with organics, would have been more like the high D/H interstellar dust from which they originated. And, paradoxically, even though the Oort Cloud is further from the Sun than is the Kuiper

source of the bulk of Earth's water. As we discuss in more detail in chapter 6, the D/H ratio in a planet's atmosphere can increase with time if the planet loses hydrogen faster than it loses deuterium, but it cannot go back in the other direction. By contrast, carbonaceous chondrites, which come from the asteroid belt, have D/H ratios that scatter widely, but which on average are close to that of Earth's oceans. So these data are consistent with the idea that asteroids were the major source of Earth's water.

An Up-to-Date Simulation of Planetary Accretion

In more recent times, and with the development of faster computers, the simulations of the accretion process have gotten more detailed. Astronomers now make use of so-called *n-body codes*, in which a specified number of bodies, *n*, are given initial positions and velocities and are then allowed to interact with each other according to Newton's law of gravitation. When necessary, Einstein's corrections to Newton's law, for example, his theory of general relativity, are thrown in. (This is required, for example, to make accurate calculations of the orbit of Mercury in today's Solar System.) One such recent simulation is shown in figure 2.4 (see color section). This particular calculation extended from the Sun out to 5 AU. That was enough to include the four innermost planets in our own Solar System, but not the four giant planets. The calculation started from a swarm of 1885 planetesimals of various sizes, the average mass being about half the mass of the Moon. These planetesimals were initially assumed to be orbiting at various distances from the Sun, randomly picked between 0.4 and 5 AU (top left panel). The initial *eccentricities* and *inclinations* of the orbits were assumed to be zero. (The eccentricity is a measure of how elliptical a planet's orbit is. An orbit with zero eccentricity is circular. The inclination is the angle of the planet's orbital plane with

---

Belt, the comets that formed it are thought to have come originally from much closer in—many of them from the Uranus–Neptune region.[15] By contrast, Kuiper Belt comets formed right where they are today: beyond the orbit of Neptune. Hence, the prediction is that Kuiper Belt comets, when they are eventually studied, will be found to have D/H ratios that are even higher than those that have already been observed.

respect to the average, or *invariant,* plane of the system.) A Jupiter-mass planet, not shown, was assumed to be on a circular orbit just outside the calculation, at 5.5 AU. The colors of the dots represent the water content of the planetesimals, with blue colors showing water-rich bodies and red colors representing dry ones. The planetesimals shade from red to blue as one moves from a few tenths of an AU out to 5 AU. Water-rich planetesimals containing 5 percent water by mass are present beyond 2.5 AU.

A number of interesting things happened in this simulation, only a few of which will be mentioned here. Within a few hundred thousand years following the start of the simulation, the planetary embryos began to drift both inward and outward from their initial positions, and they were excited to higher eccentricities and inclinations. All of this happened because of their mutual gravitational interactions, or the way they interact and perturb each other gravitationally. Most importantly, water-rich planetesimals from beyond 2.5 AU were scattered inward toward the inner parts of the planetary system. Some of these ended up being incorporated into planets that remain close to the Sun. In this particular simulation, a 2-Earth-mass planet formed at 0.98 AU, very close to Earth's actual orbital distance. Besides being considerably larger than the real Earth, this planet was also much more water-rich. The fraction of Earth's total mass that is water (including water in Earth's mantle) is only about 10^{-3}, or 0.1 percent, which would make it yellowish-green in this figure. By contrast, the blue planet in the figure has a water mass fraction close to 10^{-2}. Such a planet, if it existed, would have oceans that were 30–40 kilometers deep, as compared to only about 3–4 km on Earth. Astronomers term such (currently hypothetical) planets "ocean planets."

The simulation shown in figure 2.4 also produced a 1.5-Earth-mass planet at 0.55 AU and a 1-Earth-mass planet at 1.9 AU. Both of these planets also have lots of water. The innermost one, though, is well inside the inner edge of the habitable zone, as we will see later on. If such a planet formed in a real planetary system, it would probably lose its water by the runaway greenhouse mechanism described in chapter 6. The outermost planet is close to, or beyond, the outer edge of the habitable zone, and so any water on that planet's surface would likely be frozen.

The calculation described above represents just one possible set of initial conditions in the solar nebula and one particular outcome. Astronomers have done many more such simulations, and they have yielded a variety of different results. The various outcomes can be quite different both from each other and from our own Solar System. In some cases, 3 or 4 Earth-sized planets form within the terrestrial planet region. In others, one ends up with only a single large planet. The differences in the results are caused partly by differences in the assumed initial size distribution of planetesimals—something that is not very well known because modelers have been unable to realistically simulate the earlier stages of the accretion process. The results are also strongly influenced by whether or not a Jupiter-like planet is assumed to lie just outside the region being studied. Jupiter is so massive that its gravity affects the orbits of all of the planetesimals and planets whose orbits lie inside it. Indeed, that is thought to be why the asteroid belt exists in our Solar System between 2 and 3.5 AU. Because of its proximity, Jupiter's gravitational influence was so strong in this region that the planetesimals there were never able to accrete to form a large planet. Jupiter still perturbs the asteroid belt region, and it occasionally flings asteroids out of their quasi-stable orbits, sending some of them on paths that have a chance of intersecting that of Earth. We shall return to this topic in chapter 9, as this is one of the factors that affect planetary habitability in Ward and Brownlee's rare Earth hypothesis.

Other differences in accretionary outcomes are *stochastic*, or probabilistic, in nature. Even if one starts from the same initial size distribution of planetesimals and assumes the same Jupiter-mass planet lying outside, one can still generate different results just by changing the initial velocities and positions of the planetesimals by a small amount. In other words, building planets by accretion is a bit like rolling dice. The outcomes are likely to be different each time simply because the odds of getting everything to be exactly the same twice in a row are very small. Thus, these simulations suggest that planetary systems are likely to be quite diverse, simply as a consequence of small differences in the initial conditions and in the way the gravitational interactions (including collisions) play out.

The other take-home point from this simulation and others like it is that the planets formed within the inner parts of planetary systems can be either much wetter or much drier than Earth. That is because some planets just happen to incorporate large, water-rich planetesimals from outside 2.5 AU, whereas other planets do not. Once again, the results are stochastic. If one does many simulations, though, and counts the terrestrial planets that are formed, one finds many more that are water-rich than water-poor.[17] Hence, if these simulations are at all representative of what actually happens during planetary formation, there should be lots of rocky planets with at least as much water as Earth. And that, of course, bodes well for the prospects for looking for extraterrestrial life.

· Chapter Three ·

Long-Term Climate Stability

In the previous chapter, we saw that there are good reasons to believe that Earth-like planets may form around other stars. But do they *remain* Earth-like as time goes on? The answer to this question is not obvious, as it brings up an issue that was mentioned briefly in chapter 1: the so-called *faint young Sun paradox*. Although we think of the Sun as providing a steady source of energy today—we even use the term "solar constant" to describe the flux at the top of Earth's atmosphere—on long timescales its output is thought to vary considerably. In particular, the Sun is predicted to have been roughly 30 percent less luminous when it formed, some 4.5 billion years ago, and to have brightened with time more or less linearly since then.[1] All other things being equal, this implies that the Earth's surface should have been completely frozen during the first half of its history. We know, however, that it wasn't completely frozen, because liquid water and life have both been around for much longer than that. This seeming mismatch between solar evolution theory and geology is often referred to as the "faint young Sun paradox." It's not really a paradox, though—it's simply a problem. And, as we shall see, it has a pretty good solution, or possibly several of them. Before we discuss these solutions, though, let's make sure that we understand why this problem comes about.

Solar Evolution Theory

The theory of stellar evolution is one of the triumphs of modern astrophysics. Its foundations were laid down in the 1950s and 1960s by astrophysicists such as Martin Schwarzschild and Frederick Hoyle.* Hoyle, Schwarzschild, and other astrophysicists that followed were able to estimate from first principles how stars form and evolve.

The physics that drives solar evolution is straightforward, even though the details are complex. The key point on which everyone agrees—the last prominent dissenter being Lord Kelvin in the 19th century—is that the Sun derives its energy from nuclear fusion. Fusion is the process of combining two or more light atomic nuclei to make one heavier nucleus. For the Sun, the net reaction can be written $4\,{}^{1}\mathrm{H} \rightarrow {}^{4}\mathrm{He} + \text{energy}$. Here, ${}^{1}\mathrm{H}$ is a hydrogen atom with a nucleus that consists of a single proton, giving it a mass number of 1. ${}^{4}\mathrm{He}$ is a helium atom with 2 protons and 2 neutrons, giving it a mass number of 4. One atom of helium-4, as it can also be written, is slightly less massive than 4 atoms of hydrogen; hence, energy is released during the reaction, as Einstein so famously demonstrated almost 100 years ago. The fusion reactions take place in the Sun's core (see figure 3.1). The energy is transferred upward by radiation at depth and by convection nearer the surface. It is eventually emitted from the *photosphere* (the visible surface of the Sun) as visible and near-infrared radiation.

The reason that nuclear fusion occurs in the Sun's core is that the temperature there is extremely high (~15 million kelvins), and the density is high as well. These are the conditions that are necessary for fusion reactions to occur. To understand why, one must think about the mechanics of the fusion process. The temperatures in the Sun's core are so high that the atoms are fully ionized, that is, their electrons are stripped off. Thus, the ${}^{1}\mathrm{H}$ is really just a proton, which carries a positive charge. For it to fuse with another proton, both particles must be moving toward each other fast enough to overcome the inherent electrostatic repulsion between the two positive charges. Once they get close enough, the strong attractive nuclear force can overwhelm the electrostatic repulsion and bind them together into a single atomic nucleus. This is where the high temperature and high density helps. The high density means that the

*Science fiction buffs—old ones like me, at least—may recognize Fred Hoyle as the author of a classic 1950s book entitled *The Black Cloud.* In the book, an interstellar cloud (which turns out to be alive) visits the Solar System to feed off of radiant energy from our Sun and, in doing so, cuts off much of the sunlight hitting the Earth. Hoyle is also noted for his advocacy of *panspermia*—the idea that life came to Earth from elsewhere in the cosmos. So he was clearly imaginative, in addition to being a good physicist.

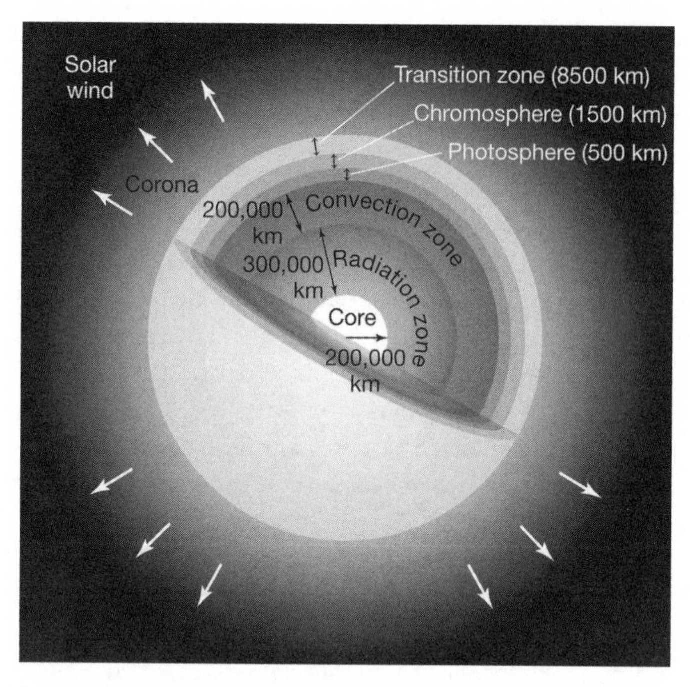

Figure 3.1 Diagram illustrating the structure of the Sun. The fusion reactions occur inside the Sun's core. The energy is carried upward by radiation at depth and by convection higher up.[2] (From *Astronomy Today*, 6th ed., E. Chaisson and S. McMillan (2008). Reproduced by permission of Pearson Publishing.)

protons are already close together, and the high temperature ensures that they are moving extremely fast. The temperature of a material is just a measure of the average speed of the particles that make it up.

Once one understands the underlying physics, the rest of the theory of solar evolution falls naturally into place. I am exaggerating slightly here, because the details are quite complex, but for our purposes the rest of the story is straightforward. As the Sun fuses hydrogen into helium, the density of its core increases, because one ^{4}He nucleus takes up less volume than do 4 individual protons. The Sun holds itself together by its own gravity, and so as its core density increases, the gravitational attraction within the Sun increases along with it. (Outside the Sun, its gravitational attraction remains approximately constant, because its mass is changing by only minute amounts.) This causes the Sun's core to contract and heat up. Or, to think of it another way, the core is forced to heat up so that the central pressure can stay high enough to keep the Sun's interior stable. A stable, hydrogen-burning star is one that is said to be on the stellar *main sequence*. Our Sun, which formed ~4.5 billion years ago, is about halfway through its own main sequence lifetime. We'll talk more about main sequence stars when we discuss habitable zones in chapter 10.

The Sun's core temperature and density thus increase as the Sun ages. This causes the fusion reactions to go faster, for the reasons just described. Because the Sun produces more energy as it gets older, it must radiate more energy as well, and so its luminosity (brightness) increases. This increase in brightness happens, it turns out, mostly by way of a gradual increase in the Sun's radius, rather than by an increase in its surface temperature. Hence, though the Sun is getting ever brighter, it is nearly the same color today as it was 4 billion years ago.*

Roughly 5 billion years from now, too little hydrogen will remain in the Sun's core to supply the energy needed to stabilize it, and the core will start to contract much more rapidly. At that point, all hell will break loose, from the standpoint of the planets orbiting around it! The outer layers of the Sun will expand and cool as the core contracts and heats up, causing the Sun to evolve into a *red giant*. Its luminosity will increase to as much as 10,000 times its present value, and its atmosphere will expand to envelope Mercury, and possibly Venus as well. The Sun will remain in this state for roughly a billion years before contracting once again to form an Earth-sized *white dwarf*. The Earth itself should still be present during the red giant stage, but that won't matter much because it will have been fried to a crisp. For stars slightly more massive than the Sun, the end is much more violent, because they explode as *supernovae* instead of contracting to the white dwarf stage. So astrophysicists have known since the 1950s that Earth-like planets, including Earth itself, cannot remain habitable forever. At best, they can support life only while their parent stars remain on the main sequence.

*Theoretically, the Sun could also have brightened by increasing its surface temperature, in which case it would now be bluer than it was originally, but any such change is thought to have been minimal. This last prediction—that the Sun is the same color as it was 4 billion years ago—is based on detailed calculations of how energy is transmitted through the Sun's "atmosphere" (its convective upper layers), and so it is not as robust as our understanding of the luminosity increase itself. The luminosity change depends only on the physics of the Sun's core which, as we have seen, is well understood.

Solar Mass Loss?

Even though we think we understand it, standard solar evolution theory could be wrong for a variety of reasons. The most plausible of these involves the possibility of solar mass loss. The luminosity of a star is strongly related to its mass. More massive stars are brighter for the reasons just discussed: their core pressures and temperatures are higher, making their fusion reactions go faster. So, if the young Sun were a few percent more massive than it is today, its luminosity could have been substantially higher than predicted by the standard model. Papers exploring this solar mass-loss hypothesis have appeared sporadically in the scientific literature.[3,4]

The reason why the young Sun could potentially have been more massive is that the Sun is always losing mass by way of the *solar wind*. The solar wind is a flux of charged particles, mostly H^+ and He^{2+}, that emanates outward from the Sun's extremely hot *corona*. The current solar wind mass flux is small—roughly 10,000 times too small to have appreciably affected the Sun's mass over the past 4.5 billion years. We know, however, that many young stars are considerably more "active" than our Sun: their coronae are hotter, and they have larger and more frequent flares. The evidence for this comes from X-ray observations of other stars made by rockets and satellites. This increase in activity is directly linked to the fact that young stars rotate rapidly compared to older ones. The Sun's rotation period is a leisurely one month, for example, whereas young stars have rotation periods ranging from a few hours to a few days. This fast rotation, coupled with their strong intrinsic magnetic fields, creates instabilities that heat up their outer atmospheres and coronae and that could conceivably drive strong stellar winds. If the Sun had lost several percent of its own mass in this way, this effect could have offset most of the change in luminosity predicted by the standard solar model.

Within the past few years, however, new data have become available that place strict upper limits on the amount of mass loss that takes place in young stars. Measuring stellar winds directly is essentially impossible because they consist primarily of fully ionized hydrogen and helium, and so do not produce detectable electromagnetic signatures. But stellar winds eventually collide with the surrounding interstellar medium, which itself consists of a tenuous mixture of ionized hydrogen and helium. At the shock front that sets them apart, significant concentrations of neutral hydrogen can build up, and this neutral hydrogen can be observed by using instruments like the STIS (Space Telescope Imaging Spectrograph) instrument on the *Hubble Space Telescope*. Based on such measurements, Brian Wood and his colleagues at the University of Colorado have been able to place bounds on how much mass loss occurs from young stars.[5,6] They find that Sun-like stars might lose as much as 3 percent of their mass in this manner. This would be enough to cancel out much of the predicted 30 percent luminosity decrease in the standard solar model.* However, by correlating their measurements with estimates of stellar ages, they also find that nearly all of this mass loss occurs within the first 200 million years of the star's history. If this same time evolution applies to our Sun, its luminosity should have been back in agreement with the standard solar model by about 4.3 billion years ago, which is well before the geologic record begins. So, from the standpoint of both geology and climate, the standard solar model appears to be essentially correct.

Electromagnetic Radiation and the Greenhouse Effect

The fact that the Sun was 20–30 percent less bright early in its history must clearly have had a big effect on the climate of the early Earth. All other things being equal, the early Earth should have been considerably colder than today. We shall see, however, that this simple prediction turns out to be incorrect. As discussed in more detail in the next chapter, the early Earth actually appears to have been somewhat warmer than at present. This is most likely because all other things were *not* equal: the

*The luminosity of a star like the Sun is roughly proportional to its mass to the 4.5 power. Planetary orbital distances are inversely proportional to stellar mass, and the flux at a planet's orbit is inversely proportional to the square of its distance from the star, so the flux received by a planet varies as the 6.5 power of stellar mass. An increase of 3 percent in the mass of the young Sun should thus change the flux incident upon the Earth by a factor of $1.03^{6.5} = 1.21$, for example, a 21 percent increase.

Earth was evolving along with the Sun, albeit for very different reasons. Importantly, Earth's surface temperature depends not only on how much sunlight it receives, but also on the *greenhouse effect* of its atmosphere. Most readers will already be familiar with the term "greenhouse effect" because one hears a lot about it these days in relation to global warming. Earth's greenhouse effect is currently increasing because humans are emitting large quantities of CO_2 and other *greenhouse gases* into the atmosphere from various activities, primarily the burning of fossil fuels. What, if anything, to do about this problem, is a topic that is discussed frequently these days in the popular media. As an aside, pay close attention to these discussions! Al Gore is right—global warming is real, and we need to begin taking action to curb it, sooner rather than later.

On long timescales, and in a more general sense, the greenhouse effect is a good thing because it keeps Earth's climate within the regime where liquid water is stable. Without it, Earth would be a frozen iceball, uninhabitable for surface life. Hence, we need to understand it if we wish to understand planetary habitability. To begin, let's briefly review the concept of *electromagnetic radiation*. We will need this concept later on as well, because electromagnetic radiation is the tool that we hope to use to look for Earth-like planets around other stars. So bear with me for a page or two, if you will. Readers who are already familiar with this topic can move on to the next section.

Electromagnetic radiation is the general term for various forms of propagating energy, many of which are familiar (e.g., visible light, radio waves, X-rays) because we make use of them in our day-to-day lives. Technically, electromagnetic radiation consists of crossed electric and magnetic fields that oscillate perpendicular to the direction of motion of the wave. Like other waves, they can be broken down into different wavelengths, forming a *spectrum*. The wavelengths of visible light form a spectrum within this spectrum, which we perceive as the colors of the rainbow. The electromagnetic spectrum is shown in figure 3.2 (see color section). Short-wavelength (high-frequency) radiation is toward the left; long-wavelength (low-frequency) radiation is toward the right.* The por-

*The frequency, ν, and wavelength, λ, of an electromagnetic wave are related by the formula $\lambda\nu = c$, where c is the speed of light.

tion of the spectrum that is most important for climate is the visible and the neighboring, slightly longer wavelength, infrared.

Electromagnetic radiation has another important property. Like all other forms of energy, it is quantized: it comes in distinct units called *photons*. An electromagnetic wave can also be thought of as a stream of photons. Physicists refer to this concept as *wave–particle duality*. For many purposes, including climate, this doesn't matter. There are typically so many photons that counting them is irrelevant. But for atmospheric chemistry, photons are important. As the famous German physicist Max Planck demonstrated back in 1900, the shorter the wavelength of a photon, the higher its energy.* Shorter-wavelength ultraviolet photons are capable of causing chemical reactions, such as splitting O_2, which cannot be triggered by longer-wavelength visible photons, no matter how many of them are present. Because we are discussing climate at the moment, we shall ignore photons for now. But keep them in mind, because we will come back to them later on when we talk about looking for Earth-like planets.

The greenhouse effect depends on the fact that objects at different temperatures emit radiation at different wavelengths: the hotter the temperature of the object, the shorter the wavelength of the radiation that it emits.** The Sun, which has a surface temperature of just under 6000 K, emits most of its energy in the visible, although a substantial fraction of its emitted energy extends into the longer-wavelength infrared region, and a smaller amount is given off in the shorter wavelength ultraviolet region. Visible radiation spans the wavelength region from 400 to 700 nm, or 0.4 to 0.7 μm.*** The Earth, which has a mean surface temperature of just under 300 K, emits most of its energy in the infrared.

The infrared energy emitted by the Sun and by the Earth is actually quite different. The Sun's emitted infrared energy is mostly between 0.7 and 5 μm. It is close to the visible portion of the spectrum, so it is often termed *near-infrared* radiation. Earth emits most of its infrared radiation

*Technically, the energy, E, of a photon is given by $E = h\nu = hc/\lambda$, where h is Planck's constant.

**For a blackbody (an object that emits and absorbs equally well at all wavelengths), this relationship can be expressed mathematically as Wien's law. The peak wavelength of emission of a blackbody is given by $\lambda_{max} \cong 2898/T$, where λ_{max} is wavelength in μm and T is temperature in kelvins.

***1 nm (nanometer) = 10^{-9} m; 1 μm (micrometer or micron) = 10^{-6} m.

between 5 and 500 μm. This radiation, which we perceive as radiant heat, is called *thermal-infrared* radiation. We will make more use of these definitions in later chapters when we discuss the remote detection of Earth-like planets around other stars.

What makes the greenhouse effect work is that different wavelengths of electromagnetic radiation interact differently with Earth's atmosphere. Earth's atmosphere is relatively transparent to visible radiation, so it allows most of the incident sunlight to pass through it. A small amount of solar near-infrared radiation is absorbed in the atmosphere. The sunlight that passes through is absorbed at Earth's surface and is then reemitted as thermal-infrared radiation. But this thermal-infrared radiation cannot pass easily through the atmosphere back into space, because it is absorbed by various greenhouse gases. The two most important ones in Earth's atmosphere are carbon dioxide (CO_2) and water vapor (H_2O). Other minor greenhouse gases include methane (CH_4), nitrous oxide (N_2O), ozone (O_3), and various man-made chlorofluorocarbons (CFCs).

Of the two main greenhouse gases (CO_2 and H_2O), H_2O is the more powerful, because it absorbs efficiently over a larger portion of the thermal infrared spectrum. But it affects climate in a different way than does CO_2, because H_2O is always near its condensation temperature. When the surface temperature gets colder, H_2O condenses out of the atmosphere; this reduces the greenhouse effect and makes the temperature even colder. When the surface temperature gets warmer, more H_2O evaporates from the oceans, increasing the greenhouse effect and making the temperature still warmer. Because it behaves in this manner, H_2O participates in a *positive feedback loop* that tends to amplify temperature changes caused by other factors. For example, if atmospheric CO_2 were to double in concentration (as it may do within the next century), the H_2O feedback would approximately double the resulting warming, from 1.1 to about 2.5°C.*

*The water vapor feedback is only one of many feedbacks in the climate system, but it is one of the most important and best understood. Sophisticated, 3-dimensional climate models predict a range of 2–4.5°C for CO_2 doubling because they include feedbacks from clouds and other processes in addition to the H_2O feedback.

CO_2 does not condense at typical Earth temperatures, and so its effect on climate is quite different from that of H_2O. Increases in CO_2 cause the surface temperature to increase, as we have just discussed; hence, on short timescales, CO_2 is a *climate forcer*. On longer timescales, however, atmospheric CO_2 concentrations can and do respond to changing surface temperatures. For example, on timescales of 10^4 to 10^5 years, atmospheric CO_2 levels vary by about 30 percent as Earth goes through glacial-to-interglacial cycles (see chapter 5). Atmospheric CO_2 is higher during the warmer interglacial periods and lower during the colder glacial periods, so this is another positive feedback on climate. For long-term planetary evolution, though, other processes dominate in the carbon cycle, and so the relationship between CO_2 and climate changes. Indeed, CO_2 is part of a strong *negative feedback loop*, as we shall see later in this chapter.

Planetary Energy Balance

We need one more piece of climate theory in order to understand the effect of changing solar luminosity on climate: the concept of *planetary energy balance*. The basic idea is simple: the energy being absorbed by the Earth must be equal to the energy given off.* The Earth is heated by visible radiation from the Sun, and it is cooled by emission of thermal-infrared radiation from its surface and atmosphere. Considerably more sunlight is received near the equator than at the poles, but the temperature differences are smoothed out to some extent by winds and ocean currents. And the nightside is roughly the same temperature as the dayside because the Earth is rotating rapidly. For planets that satisfy these constraints, a simple mathematical formula can be used to express the balance between incoming and outgoing radiation. It is often called the *planetary energy balance equation*. Readers who are averse to math can skip the next two paragraphs, but for those who can appreciate it, this

*Actually, the balance between energy absorbed and energy given off by the Earth is not quite exact. If it were, Earth's surface temperature would not change with time. But the imbalances are so small that for most purposes they can be ignored.

equation is a useful way of quantifying how a planet's temperature depends on the amount of sunlight it receives.

It is easiest to start by first considering the effect of solar flux changes on an airless body, like Earth's Moon. The *effective temperature,* T_e, of such an object (which for an airless body is just the average surface temperature) is given by the formula

$$\sigma T_e^4 = \frac{S}{4}(1 - A) \tag{1}$$

Here, S is the solar flux at the top of the atmosphere (1365 W/m^2 for modern Earth), A is the planetary *albedo*, or reflectivity (0.3 for modern Earth), and σ (sigma) is the Stefan-Boltzmann constant (5.67×10^{-8} $W/m^2/K^4$). (W/m^2 stands for "Watts per square meter.") If one plugs in the numbers, one can calculate that the effective temperature of the modern Earth is about 255 K, which is equivalent to −18°C, or 0°F.*

Now, you might notice here that this calculated temperature of −18°C is much colder than Earth's actual average surface temperature. Earth's actual average surface temperature, T_s, is really about 288 K, which is equivalent to 15°C, or 59°F. T_s is thus higher than T_e by some 33 K. This temperature difference is caused by the greenhouse effect of the atmosphere, which we discussed in the previous section. Thus, as stated there, the greenhouse effect itself is a very good thing: If Earth didn't have one, we would all be frozen solid. The bad news about the greenhouse effect, as I have already noted, is that we are making it stronger very quickly, and this is almost certainly not a good thing.

The Faint Young Sun Problem

Let's return now to the faint young Sun problem that we introduced at the beginning of the chapter. As mentioned there, the basic outlines of stellar evolution theory were developed in the 1950s and 1960s. It was

*The Kelvin scale is related to the Celsius scale by the relationship $T(K) = T(°C) + 273.15$. The Fahrenheit scale is related to the Celsius scale by the relationship $T(°F) = [T(°C) \times 1.8] + 32$.

not until much later, however, that anyone pointed out the consequences of this theory for planetary evolution. In 1972, Carl Sagan and George Mullen wrote a classic paper[7] that for the first time explored the effect of solar luminosity change on planetary climates.

Recall that the Sun is thought to have been dimmer by about 30 percent at the time when it formed. If one plugs that number into the planetary energy balance equation, and if one assumes for the sake of argument that the planetary albedo has remained unchanged, one finds that the earlier Earth's effective temperature should have been lower by some 22 K, yielding $T_e = 233$ K. If the greenhouse effect was the same then as it is now, 33 K, then the mean surface temperature at that time should have been 266 K, which is equivalent to -7°C, or about 19°F. That is already well below the freezing point of water. This calculation almost certainly underestimates the actual problem, though, because if surface temperatures were lower, atmospheric H_2O concentrations should have been lower as well, and the greenhouse effect should have been even smaller.

In their paper, Sagan and Mullen described a simple climate model calculation that took all of these factors into account. They assumed, once again for the sake of argument, that the atmospheric CO_2 concentration has remained constant over time. (In reality, atmospheric CO_2 concentrations were probably higher on the early Earth, but we will get to that a little later.) For H_2O, Sagan and Mullen made a different assumption: they fixed the atmospheric *relative humidity*. Relative humidity is the concentration of water vapor in air compared to the maximum amount of water vapor that air can hold at that temperature. By doing so, Sagan and Mullen included the positive feedback loop mentioned earlier: higher surface temperatures → more water vapor → stronger greenhouse effect → still higher surface temperatures.

The results of Sagan and Mullen's calculation were illuminating. Under these assumptions, the average surface temperature of the Earth should have been below the freezing point of water during the entire first half of its history. In the 1980s, I worked at NASA Ames Research Center with James Pollack (Sagan's former graduate student), and we did a similar calculation using our own climate model. The results of that calculation were essentially the same as Sagan and Mullen's results (see figure 3.3). I've used our figure here because our climate model was somewhat more elaborate and included improved absorption coefficients for CO_2 and H_2O. As one can see from the figure, the greenhouse effect increases with time, as it should, because of the positive feedback provided by water vapor. And, as in Sagan and Mullen's calculation, the mean surface temperature is predicted to be below the freezing point of water prior to about 2 billion years ago.

So what's the problem with this calculation? The problem, as Sagan and Mullen pointed out, is that the surface temperature evolution shown in figure 3.3 is completely at odds with the actual climate record. Amazingly, evidence for liquid water at Earth's surface now goes back as far as 4.4 billion years ago, or only about 150 million years after the planet formed. (I will henceforth follow geological terminology and use "Ga," or "giga-annum," as shorthand for "billions of years ago.") This evidence comes from oxygen isotopes in *zircons*[9] (zirconium silicate, $ZrSiO_4$), but it does not actually tell us that the ocean surface was liquid. The liquid water could have existed beneath a layer of ice. By 3.8 Ga, however, sedimentary rocks were being formed,[10] and by 3.5 Ga there is evidence for photosynthetic life (see the next chapter). Both of these observations suggest that Earth's climate was warm enough to support liquid water at its surface.

Another way to arrive at this same conclusion is to examine the Earth's glacial record. Evidence for continental-scale glaciation is absent prior to ~2.9 Ga (although, admittedly, few rocks are preserved from this time period, and so this observation could be misleading). The Earth also appears to have been ice-free between 2.8 and 2.4 Ga and again between 2.3 and 0.8 Ga (see figure 4.2 in the following chapter). By contrast, the present Earth has large polar ice caps on Antarctica and Greenland, and so in that sense we are in a "glacial" climate today. (More precisely, we are in an interglacial interval in the midst of a long-term glacial period.) Hence, the early Earth appears to have been, if anything, warmer than today, despite the fact that the Sun was less bright. This is the essence of the faint young Sun problem. Evidently, the simple model of

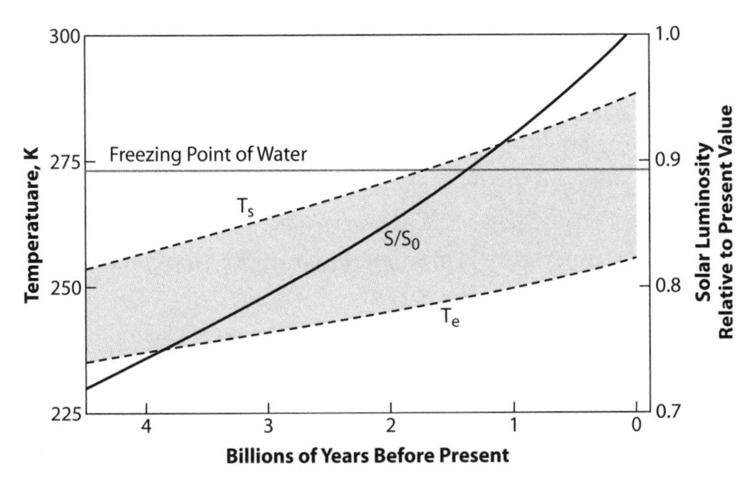

Figure 3.3 Diagram illustrating the faint young Sun problem. The solid curve represents solar luminosity relative to its present value.[1] The dashed curves represent the effective radiating temperature of the Earth, T_e, and the average surface temperature, *Ts*. Constant atmospheric CO_2 and fixed relative humidity are assumed in the calculation. The shaded region between T_e and T_s represents the greenhouse effect of Earth's atmosphere. (Figure taken from Kasting et al. 1988.[8])

climate evolution depicted in figure 3.3 cannot be right. The early Earth must have been warmed either by additional greenhouse gases or by some other mechanism.

Possible Solutions to the Problem

The faint young Sun problem has a number of possible solutions, some of which are more plausible than others. The most likely solution, as I will argue below, is higher concentrations of greenhouse gases, especially CO_2, and possibly CH_4 as well. But other possibilities should be mentioned, if only to demonstrate that they have not been overlooked. The most obvious is a drastic reduction in cloud cover. As we saw earlier, Earth's current albedo is roughly 0.3, meaning that it reflects 30 percent of the sunlight that hits it. Of this, about 0.25 (5/6 of the total albedo) is caused by clouds. If cloudiness were near zero on the early Earth, then the faint young Sun problem would be almost solved because the planet would absorb nearly as much sunlight as today. I say "almost" because clouds, high cirrus clouds in particular, also contribute to the greenhouse effect, and this contribution would be lost if clouds were not present.

The basic argument against relying on clouds to solve the faint young Sun problem is straightforward. If the oceans formed early, as we suggested in chapter 2, and if surface temperatures were above freezing, how could clouds not have been present? Clouds, like water vapor, are usually considered to act as a feedback on climate, rather than as a climate forcer.

That said, some researchers continue to support this idea. Recently, Henrik Svensmark of the Niels Bohr Institute in Copenhagen has suggested that cloudiness might have been lower on the early Earth because the flux of cosmic rays, which create ions that act as *cloud condensation nuclei* (CCNs), would have been smaller.[11] Cosmic rays are high-energy charged particles, mostly protons, which come from elsewhere in the galaxy. They are partially deflected by the Sun's magnetic field, which is carried throughout the Solar System by the solar wind. If the Sun was more active in the distant past, the interplanetary magnetic field should have been stronger, cosmic ray shielding could have been more efficient, and nucleation of cloud droplets might have been inhibited.

Although Svensmark's hypothesis cannot be disproved, it suffers from the same weakness as the solar mass loss theory discussed earlier. The Sun was most active during the first few hundred million years of its existence. So the flux of cosmic rays would have been high, and cloudiness would have been low, only during the very earliest part of the Earth's history. To be sure, other mechanisms for reducing cloud cover can be imagined, some of which might conceivably have acted over a longer period of time. Windblown dust can also act as CCNs, so smaller continents in the distant past might have helped to keep cloud cover low. CCN production is also modulated today by the biogenic trace gas dimethyl sulfide (DMS), which oxidizes to form CCN-like sulfate particles. This influence would have been absent prior to life's origin. But, in my view, all of these mechanisms for reducing cloud cover on the early Earth are highly speculative. Furthermore, they are unnecessary, as other plausible mechanisms for solving the faint young Sun problem exist.

What about geothermal heat? Geothermal heat is the heat coming from Earth's interior. Some of it, roughly half, comes from the energy released by the decay of radioactive elements such as uranium, potassium, and thorium in the crust and mantle, while the other half is left over from the Earth's formation. Both heat sources should have been stronger when the Earth was younger, and so the geothermal heat flux was almost certainly higher in the past. The numbers, though, just don't add up. The current global average geothermal heat flux is only 0.09 W/m^2, which should be compared with an average absorbed solar flux of 240 W/m^2.* To offset a 30-percent decrease in solar luminosity, the geothermal heat flux would have to have been about 800 times greater in the distant past. Models, however, suggest that at 4 Ga geothermal heat flow was only 2–4 times higher than today. The increased geothermal heat flux may indeed have been important, but its effects were indirect. It could have affected the carbon cycle and atmospheric CO_2 levels, as discussed further below.

Sagan and Mullen themselves suggested that the atmospheric greenhouse effect was larger in the past, and that this was the main factor countering reduced solar luminosity. This idea does not require any great stretch of one's imagination—there is absolutely no reason why the greenhouse effect should have always been the same. In Sagan and Mullen's model, the important additional greenhouse gases were ammonia (NH_3) and methane (CH_4). Both gases could conceivably have been more abundant in the distant past because they are both *reduced gases* that react with O_2. Methane, for example, can be converted to carbon dioxide by the reaction $CH_4 + 2O_2 \rightarrow CO_2 + 2H_2O$. This is what happens when one combusts natural gas (which is mostly CH_4). This reaction doesn't happen directly at room temperature—if it did there would be no methane in today's atmosphere—but it happens indirectly through atmospheric photochemistry. Hydrogen, H_2, is another reduced gas because it can react with oxygen to form water vapor: $2H_2 + O_2 \rightarrow 2H_2O$.

*The average absorbed solar flux is just the right-hand side of the planetary energy balance equation, $(S/4) \times (1 - A)$.

But hydrogen could not have solved the faint young Sun problem because it is not a good greenhouse gas.

As we shall see in the next chapter (and as Sagan and Mullen knew as well), atmospheric O_2 concentrations were very low during the first half of the Earth's history. Hence, the atmospheric lifetimes of gases that react with O_2 should have been longer than they are today, and so the concentrations of such gases should have been higher. We will see shortly that this argument works better for methane than it does for ammonia. But the basic idea of using reduced gases to bolster the greenhouse effect on the early Earth is a good one.

In their paper, Sagan and Mullen placed the primary emphasis on ammonia. Ammonia is an excellent greenhouse gas that absorbs across the thermal-infrared spectrum, like H_2O. Methane, as we shall see, is much less effective. According to their calculations, roughly 10–100 ppm (parts per million) of NH_3 would have been enough to keep the early Earth warm. However, ammonia has a problem that was pointed out later by photochemists.[12] In an atmosphere containing only small amounts of oxygen and ozone, ammonia is rapidly broken apart, or *photolyzed*, by solar ultraviolet radiation. Hence, a large source of ammonia would have been needed to maintain its concentration, and it is not obvious what this source could have been. To keep the ammonia concentration at appreciable levels, it would have been necessary to shield it from photolysis. In a paper with former graduate student Chris Chyba that appeared after his death,[13] Sagan suggested a mechanism for doing so. Organic haze formed from methane photolysis might have provided just such a shield. My students and I have been pursuing this hypothesis for the past several years, and I'll discuss it further in the next chapter. Although we don't think the shielding would have been as effective as Sagan and Chyba suggested, it is one more example of Sagan's scientific ingenuity.

Thus far, we have ignored another obvious possibility. CO_2 is the second most important greenhouse gas today, after H_2O. Could the atmospheric CO_2 concentration have been higher in the past? Figure 3.3 assumes that it was not, but this figure was constructed simply to illustrate a point. To answer this question, we need to consider the factors that affect the atmospheric CO_2 concentration over long timescales.

The Carbonate-Silicate Cycle and Controls on Atmospheric CO_2

The question of what controls atmospheric CO_2 levels has an obvious answer: the carbon cycle. As many people already know, CO_2 is continually being recycled between the atmosphere, plants, animals, and soils. Plants take up CO_2 from the atmosphere during *photosynthesis* and, in doing so, they release O_2. Animals, including humans, breathe in O_2 and exhale CO_2. Plants actually do this as well, as they also derive their metabolic energy from *respiration*. Hence, plants can be either a source or a sink for CO_2, depending on what season it is and whether the sun is shining. Bacteria decompose dead plants and animals, releasing CO_2 through the process of *decay*. This part of the carbon cycle is widely understood, as it is a familiar part of our daily environment.

Fewer people are aware that the carbon cycle has several different parts. The cycle just described is just one aspect of what is termed the *organic carbon cycle*. The organic carbon cycle itself has additional levels of complexity. Photosynthesis is largely balanced by respiration and decay. If this balance was exact, then no O_2 would be produced. But some dead organic matter—about 0.5 percent of the total amount that is produced by photosynthesis—is buried in soils and in ocean sediments where it is protected from O_2. It is this burial of organic carbon in sediments that generates the O_2 in Earth's atmosphere.[14] Atmospheric O_2 is eventually consumed by reaction with dispersed organic carbon and other reduced material (iron and sulfide minerals) in rocks during *weathering* (see further discussion below). This gives atmospheric O_2 an effective lifetime, or *residence time*, of about 4 million years.

Burial of organic matter is also a long-term loss process for atmospheric CO_2. James Lovelock invoked this CO_2 removal process in his "Gaian" mechanism for solving the faint young Sun problem (see chapter 1). He proposed that photosynthetic organisms, including plants, algae, and *cyanobacteria*,* pulled CO_2 out of the atmosphere at just the

*Cyanobacteria are photosynthetic bacteria that, unlike plants and algae, lack cell nuclei. We discuss them in some detail in the next chapter, as they are thought to have played a critical role in the rise of atmospheric O_2.

right rate to compensate for increasing solar luminosity.[15] But trying to resolve the faint young Sun problem in this manner creates a dilemma. The (long-term) organic carbon cycle controls atmospheric O_2 levels. If the cycle gets out of balance, then O_2 levels change in such a way that balance is restored. For example, if atmospheric O_2 starts to increase, then the amount of dissolved O_2 in the oceans also increases, and this reduces the rate of organic carbon burial, thereby cutting back on O_2 production. Or, at least, that is how we think this cycle operates.[14] More than one book has been written trying to explain the details of the O_2 feedback mechanism. Fortunately, we don't need to understand these details here. Rather, we can ask another question: If the organic carbon cycle regulates atmospheric O_2, how can it simultaneously regulate climate? The answer is that it probably can't, at least not in any long-term, sustainable way. It can affect climate by temporarily drawing down (or increasing) atmospheric CO_2, but it can't maintain CO_2 near some particular level because it is tied so strongly to atmospheric O_2. There just aren't enough degrees of freedom in the system for the organic carbon cycle to simultaneously regulate both gases.

Fortunately, the carbon cycle has another, completely independent part that may be able to do the job. It is sometimes called the *inorganic carbon cycle* or, more often, the *carbonate-silicate cycle*. To understand the difference from the organic carbon cycle, it helps to think about the chemistry that is involved. Organic carbon is carbon that is bonded mostly to hydrogen atoms or to other carbon atoms. Glucose, $C_6H_{12}O_6$, is a good example. Here the carbon atoms are connected in a 6-atom chain. Inorganic carbon, by contrast, is carbon that is bonded mainly to oxygen. CO_2 is a good example. Calcium carbonate, $CaCO_3$, is another. Calcium carbonate (limestone) is a common rock type that is found in many locations. Central Pennsylvania, where I live, is underlain by limestone. That is why we have lots of caves in the area. Calcium carbonate dissolves more readily than do the silicate minerals that make up the bulk of Earth's crust, and so it gets hollowed out by flowing groundwater that contains dissolved CO_2.

The carbonate rocks that one finds in Pennsylvania and elsewhere were formed originally by tiny organisms that live in the ocean. Promi-

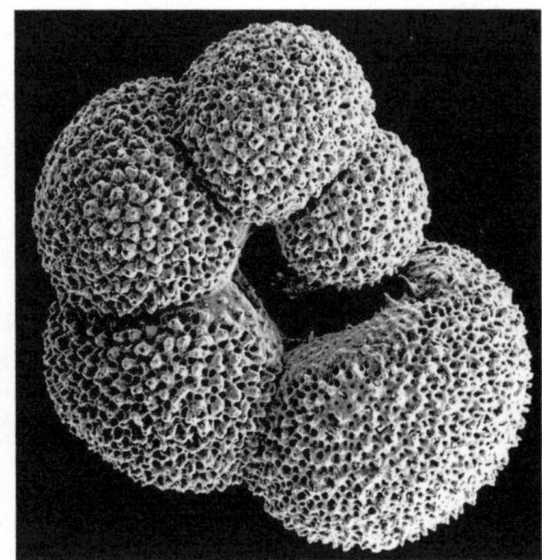

Figure 3.4 Two examples of marine plankton that make shells out of calcium carbonate. Left panel: a *coccolithophore* (about 10 μm in diameter). Right panel: a *foraminifer* (600 μm in diameter). (Courtesy of R. Bernstein, University of South Florida and Pearson Publishing.)

nent among these are the *coccolithophorids* and *foraminifera*—tiny phytoplankton and zooplankton, respectively, that live in the surface ocean (figure 3.4). Phytoplankton are single-celled plants that carry out photosynthesis; zooplankton are single-celled animals that eat the phytoplankton. Coral reefs and clam shells are also made out of calcium carbonate.

The carbonate-silicate cycle is the process by which CO_2 moves from the atmosphere into carbonate shells and eventually into carbonate rocks, and then is later returned to the atmosphere. Throughout this entire cycle, carbon remains bonded to oxygen.* Consequently, no O_2 is produced. This cycle thus operates independently from the organic carbon cycle and atmospheric O_2; hence, it is free to respond to, and to influence, climate without responding to changes in O_2.

Figure 3.5 illustrates how the carbonate-silicate cycle works. First, CO_2 from the atmosphere dissolves in rainwater to produce carbonic acid, H_2CO_3—the same weak acid that is found in soda pop. Some of this slightly acidic rainwater falls onto the continents, where it dissolves both carbonate and silicate rocks. Real silicate minerals have complex

*A chemist would say that the carbon never changes its valence state, or oxidation state. The valence state of organic carbon is zero. The valence state of carbon in CO_2 or carbonates is +4.

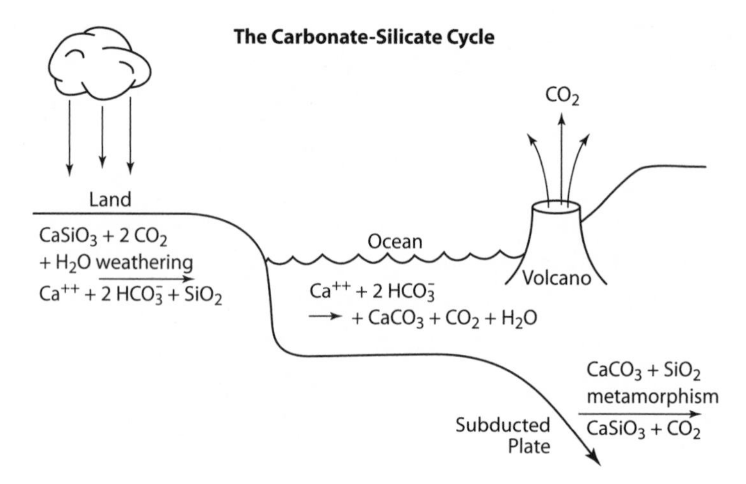

Figure 3.5 Diagram illustrating the carbonate–silicate cycle.

chemical formulas; however, for our purposes we can represent them by the simplest silicate mineral, *wollastonite* ($CaSiO_3$). Other silicate minerals contain magnesium, Mg, in addition to, or in place of calcium, Ca. The process by which rocks react with and are altered by rainwater is termed "weathering," and so the dissolution of silicate minerals is called *silicate weathering*. Carbonate rocks weather even more rapidly, as we have seen; however, the net effect of this process on the levels of atmospheric CO_2 is zero when averaged over long timescales, and so we can simply ignore it.

The rest of the carbonate-silicate cycle proceeds as follows: The products of silicate weathering, which include Ca ions (Ca^{2+}), bicarbonate ions (HCO_3^-), and dissolved silica (SiO_2), accumulate in groundwater and are carried by streams and rivers down to the ocean. There, organisms like the foraminifera and coccolithophorids use the Ca and bicarbonate to make shells of calcium carbonate. This process is termed *carbonate precipitation* because the carbonate minerals are precipitated out of seawater. When these organisms die, they fall down into the deep ocean, where most of the calcium carbonate redissolves. Some of the shells are preserved, however, and they accumulate on the seafloor to form sediments of calcium carbonate. The seafloor itself is part of a dynamic crustal system that we call *plate tectonics*. Seafloor is created at midocean ridges, and then slowly spreads away laterally. When an oceanic plate encounters a continental plate, the oceanic plate is *subducted* downward into the mantle. Some of the carbonate sediment is scraped off in the process, but some of it is carried down to depths where the pressures and temperatures are high. Under these conditions, carbonate rocks react with silica, which at this depth is in the form of the mineral quartz, re-forming calcium and magnesium silicates, and releasing gaseous CO_2. This last process is termed *carbonate metamorphism*. The CO_2 then comes back into the atmosphere through volcanism, thereby completing the cycle.

To understand how this cycle affects climate, we must know how fast it operates. The simple answer to this question is "slowly." If one compares the relative fluxes, the carbonate-silicate cycle is about 1000 times slower than the rapid photosynthesis–respiration-decay part of the organic carbon cycle. Hence, it is slow compared with human timescales and will therefore be of little help in taking up fossil fuel CO_2 over the next few decades to centuries (although it will ultimately remove all of this additional CO_2 on a timescale of a few million years[16,17]). It is fast enough, though, to recycle all of the carbon in the combined atmosphere-ocean system on a timescale of about half a million years. To a geologist, that is not such a long time. It is also short compared to the timescale for solar evolution, which is measured in hundreds of millions to billions of years. Hence, the carbonate-silicate cycle should respond quite effectively to changes in solar forcing. Let's consider now what that response might be.

The CO_2-Climate Feedback Loop

Earlier in this chapter, I mentioned that H_2O provided a feedback on climate, whereas CO_2 was a climate forcer. That is true on short timescales. But on long timescales, CO_2 itself is part of a feedback loop. A careful look at figure 3.5 reveals why. To make things simple, suppose that the oceans actually did freeze over early in Earth's history, as figure

3.3 would imply. (This is not a hypothetical question—this may have actually happened during Snowball Earth episodes, as we will see in chapter 5.) What would have happened to the carbonate-silicate cycle? If the oceans were frozen, no water would have evaporated from them, and no rain would have fallen on the continents. Some ice should have sublimated in the tropics, and some snow should have fallen at higher latitudes. But liquid water would have been entirely absent at the surface, and so silicate weathering would have proceeded slowly, or not at all. As in the case of solar nebula chemistry, described in the previous chapter, direct reactions between gases and silicates do not occur at any appreciable rate.

Consider what would have happened, then, if the early Earth had been frozen at its surface. Silicate weathering would have been inhibited, or stopped entirely. Volcanoes, however, would have continued to release CO_2 into the atmosphere. This is not obvious, perhaps, from figure 3.5, because it looks there as if the entire carbonate-silicate cycle should have ground to a halt. In reality, though, a huge amount of carbon is tied up in existing carbonate rocks, and CO_2 is released volcanically in multiple geologic environments, including subduction zones, midocean ridges, and mantle "hot spots" like Hawaii and Sicily. So volcanic release of CO_2 should have continued, even on a totally frozen Earth. Although we have not demonstrated this here, once CO_2 accumulated to a few tenths of a bar of pressure (about 1000 times its present concentration), its greenhouse effect should have been strong enough to melt the ice and to restart silicate weathering,* thereby allowing the carbonate-silicate cycle to come back into steady state.

The scenario just described is extreme. If the oceans ever did freeze over, this happened only a handful of times during Earth's history. How

*Actually, the issue of whether a Snowball Earth could have been deglaciated by CO_2 increases is controversial. We shall argue in chapter 5 that this *did* happen at least twice in the Late Proterozoic, around 600–700 million years ago. Some geologists don't accept the Snowball Earth model,[18] though, and some climate models suggest that deglaciation is difficult.[19] It might also have been more difficult to recover from Snowball Earth early in Earth's history when the Sun was less bright because CO_2 itself might have condensed from the atmosphere.[20] We will revisit the greenhouse effect of dense, cold CO_2-rich atmospheres in chapter 8 when we discuss the climate of early Mars.

did the carbonate-silicate cycle behave the rest of the time? It is not too hard to figure out how the system might work. We made a simple model of this system in a paper[21] that grew out of my own Ph.D. work at the University of Michigan. When the climate gets colder, evaporation and rainfall rates decrease, silicate weathering slows, and CO_2 builds up in the atmosphere. This increases the greenhouse effect, thereby making the climate warmer. Conversely, when the climate gets warmer, evaporation and rainfall increase, silicate weathering speeds up, and atmospheric CO_2 decreases, thus making the climate cooler. The response of the cycle is always such as to oppose the initial perturbation to surface temperature, and so it is a powerful negative feedback loop that helps to stabilize Earth's climate over long timescales. Indeed, it is probably one of the main reasons that Earth has remained habitable throughout its history. And, as we shall see, it is relevant as well to the question of how wide the habitable zone might be around the Sun and other stars. But we will save that story for chapter 10.

Before leaving the current chapter, we should consider what the implications of these ideas might be for the Gaia hypothesis, introduced in chapter 1. In Lovelock's original model, photosynthesis by plants and algae, followed by burial of organic carbon in sediments, caused atmospheric CO_2 to decrease at just the right rate to offset increasing solar luminosity. That idea probably doesn't work, for reasons given earlier: the organic carbon cycle cannot simultaneously control both O_2 and climate, and so atmospheric CO_2 levels are mostly controlled by the inorganic carbon cycle. "Inorganic" does not mean the same thing as "abiotic," though, even though these terms are sometimes used interchangeably. As we have already seen, the inorganic carbon cycle clearly involves organisms: nearly all of the calcium carbonate that forms today is precipitated by marine plankton and by corals. This particular process, it turns out, is not a strong argument in favor of Gaian control, because formation of carbonates is not the process that limits the rate of removal of CO_2 from the atmosphere. If organisms were not present, dissolved calcium and bicarbonate would simply build up in the oceans until $CaCO_3$ precipitated out on its own.* But, as Lovelock pointed out in one of his later books,[22] the process of silicate weathering (which *is* the rate-limiting step in CO_2 removal) is also influenced by biology. Trees and grasses pump up the CO_2 concentrations in soils to 20–30 times the atmospheric value, primarily through *root respiration* and from decay of root systems after the plants die. This enhances the silicate weathering rate and helps draw down atmospheric CO_2. Neither trees nor grasses existed before about 450 million years ago, so this particular "Gaian" influence is relatively recent. Even microorganisms in soils, though, can release *humic acids* (complex organic acids) that are more effective in dissolving silicate minerals than is carbonic acid. So silicate weathering rates may been influenced by the biota ever since life first arose on the planet.

Does this mean that the Gaia hypothesis is correct? That depends on what one means by the question, or by the hypothesis itself. There is little doubt that life can, and does, influence climate, along with many other aspects of the environment. We will see a prime example of this in the next chapter when we talk about methane and oxygen. But the carbonate-silicate cycle would still operate, albeit somewhat differently, even if life were not present. A lifeless Earth would likely be somewhat warmer, because silicate weathering would be less efficient and so CO_2 levels would be higher, but the climate should still stabilize well within the regime where liquid water is present. So, if liquid water is indeed the key to habitability, as suggested earlier, then a lifeless planet could still be habitable. That's a pleasing thought for former Trekkies like me because it implies that we might eventually find habitable planets that we could colonize without having to deal with the ethical issue of displacing an indigenous biota. That, after all, might be a violation of the Prime Directive.

*Indeed, plankton that form shells out of $CaCO_3$ are a relatively recent biological invention, having evolved only some 150 million years ago. And shell-forming animals like clams originated only 540 million years ago. Before that time, the oceans probably *were* more saturated with calcium and bicarbonate.

· Chapter Four ·

More Wrinkles in Earth's Climate History

In the last chapter, we saw that variations in atmospheric CO_2 have likely played a significant role in maintaining climate stability over time and in countering changing solar luminosity. CO_2 concentrations have probably varied for a number of different reasons, and this can explain much of what we observe about the climate record, especially during the second half of the Earth's history. But when one delves even further back, one finds that changes in CO_2 and solar luminosity are not enough by themselves to explain all of the patterns that we observe. Instead, other greenhouse gases may have supplied much of the required warming during the first half of Earth's history when atmospheric O_2 concentrations were low, just as Sagan and Mullen suggested over 30 years ago. Ammonia was probably photochemically unstable, as we have seen, but methane should have been long-lived in such an atmosphere. What role might methane have played in early climate history, and how does it compare with the effect of CO_2?

These questions are important for understanding Earth's history, and they may be important for understanding extrasolar planets as well, once we find them. My students and I think of the problem this way: If we were to look at the Earth's atmosphere at some arbitrary time, what would we see? During the last half of Earth's history, we would probably see both oxygen and ozone (O_2 and O_3). But if we looked at the Earth during the first half of its history, these gases would likely be absent. In their place, we might well see high concentrations of methane. If we saw methane, though, would this tell us that the planet was inhabited? This is a more difficult question—one that is still not answered satisfactorily. We will return to this and other related issues in chapter 14. In the meantime, we need to study the early Earth if we hope to understand extrasolar planets. On this topic, geologists, biologists, and astronomers share a mutual interest.

The Phanerozoic Climate Record

To begin, let's look in more detail at the geologic record of biological and climatic evolution. A geologic timescale showing the most recent 540 million years of planetary history is shown in figure 4.1. This period of Earth history is termed the Phanerozoic, which means "age of life." The name is actually a misnomer, as we now know that the history of life goes back much further in time—to 3.5 Ga or even earlier. But until the discovery of microfossils in the 1940s, paleontologists believed that the first organisms arose at the beginning of the Cambrian period, around 540 Ma. ("Ma" is shorthand for "mega-annum," or "millions of years ago.") This was the time, often termed the "Cambrian explosion," when multicellular organisms first learned to make shells out of calcium carbonate and silica. Consequently, the fossil record becomes much more detailed after this time.

As can be seen in figure 4.1, the Phanerozoic eon includes three different glacial periods. We are in one of these today (termed the "Late Cenozoic glaciation"), even though most of us don't think of the present climate as being glacial. That's because we're in the midst of a relatively warm interglacial period, the Holocene epoch, which is embedded within a series of deeper glaciations. (More on this in chapter 5.) From our long-term perspective, though, the Late Cenozoic glaciation has been going on for the past 35 million years, as that is when Antarctica started to become glaciated. The cold climatic conditions became more intense starting at about 2.8 Ma during the late Pliocene and Pleistocene

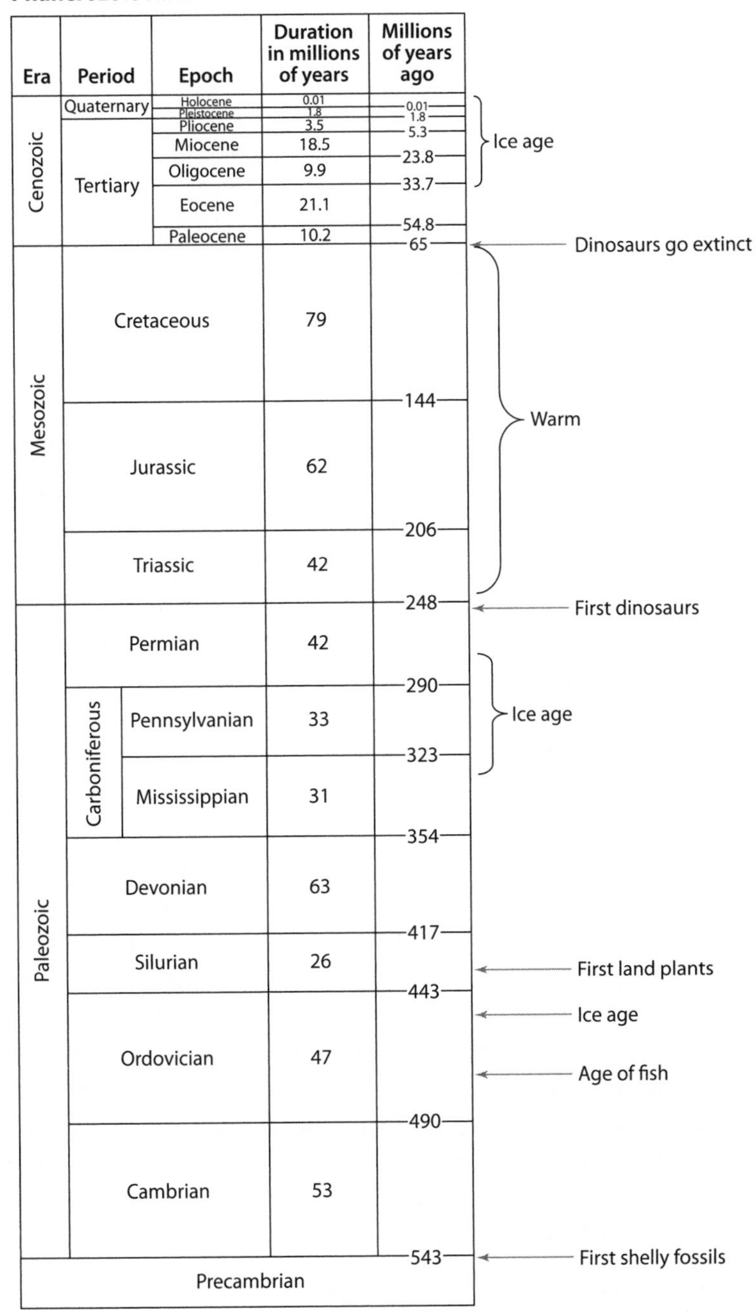

Figure 4.1 The Phanerozoic timescale, with major biological events indicated. Timings of the three different Phanerozoic ice ages are also shown.

epochs. Prior to 35 Ma, during most of the Mesozoic and early Cenozoic eras, the Earth was warm. The Mesozoic was the age of dinosaurs, some of which were living in semi-tropical climate conditions on the North Slope of Alaska. Before this, the climate was cold, as evidenced by the 80-million-year-long Permo-Carboniferous glaciation, which lasted from about 340 Ma until 260 Ma. Prior to that, the Paleozoic era was mostly warm, except for a brief glaciation during the late Ordovician, around 445 Ma, which appears to have lasted for less than one million years.

The causes of Phanerozoic climate change have been discussed extensively in the geological literature, notably by Yale University's Robert Berner.[1] Most of the variation is attributed to two main factors: changing atmospheric CO_2 concentrations and changing continental positions. The continued increase in solar luminosity with time is also important, as the Sun was still about 5 percent dimmer than today at the beginning of the Phanerozoic. The changes in atmospheric CO_2 and in continental positions go hand-in-hand, in a sense, because both are driven by plate tectonics. When carbonate-laden seafloor is subducted, or when continents containing carbonate rocks collide, extra CO_2 is added to the atmosphere through volcanism and *metamorphism*.* Conversely, when high mountains are created in regions where there is a lot of rainfall, silicate weathering is enhanced, and CO_2 concentrations may decrease.

One of the most widely circulated ideas about Phanerozoic climate change, termed the "Raymo-Ruddiman hypothesis," after its coauthors, Maureen Raymo of MIT and William Ruddiman of the University of Virginia, concerns the changes brought about by the uplift of the Himalayas.[2] The Himalayas and the Tibetan Plateau were created by the collision between India and Asia, which began around 40 Ma. The Himalayas generate their own rainfall by helping to cause the monsoons that soak India during northern hemisphere summer. The high Tibetan Plateau behind the Himalayan mountain chain heats up in the summer, causing the air above it to rise, and drawing in moist air from over the Indian ocean. This air drops its moisture as rain when it is forced up-

*Metamorphism refers to alteration of rocks at temperatures lower than those required to melt them. Metamorphism can occur in mountain-building regions, as well as subduction zones.

ward by the Himalayas, thereby accelerating chemical weathering over a wide region. According to Raymo and Ruddiman, rapid weathering of this freshly exposed rock may account for the drawdown of CO_2 and accompanying climatic cooling that occurred during the Late Cenozoic. So this may be at least part of the reason why we are in an ice age at the present time. Another, equally important tectonic change was the opening of the Drake Passage between South America and Antarctica around 35 Ma. This allowed the Antarctic Circumpolar Current to begin flowing around the continent, thereby reducing the delivery of heat from warm ocean currents and making it easier for the continent to become glaciated.

The Permo-Carboniferous glaciation, 340–260 Ma, is also thought to have been triggered by decreases in atmospheric CO_2, but in this case the suggested mechanism is quite different. As Berner has pointed out in his model simulations,[1] land plants evolved some tens of millions of years before this, during the Late Silurian period, and this should have affected CO_2 concentrations in several ways, some of which were discussed earlier in chapter 3. Land plants accelerate silicate weathering by pumping up CO_2 in soils, so the colonization of the land by plants should have helped draw down atmospheric CO_2. Furthermore, under the right conditions, plant material can become trapped in bogs, forming coal. Indeed, the Carboniferous period is so named because of the massive coal deposits that formed during this time. Coal formation removes CO_2 from the atmosphere-ocean system, and so the organic carbon cycle may also have helped lower CO_2 levels during the Carboniferous, thereby setting the stage for the glaciation that occurred toward the end of that period.

In contrast to these two most recent glaciations, both of which are relatively well explained, the Late Ordovician glaciation at 445 Ma is a complete mystery. This glaciation was so brief that it is contained within a single stage of the geologic record, called the Hirnantian. It also coincides with the second largest mass extinction of species in the Earth's history.* The Late Ordovician mass extinction may well have been caused by the glaciation, which caused sea level to drop by as much as 70 m

*The largest mass extinction was the End Permian event, at 251 Ma. The Cretaceous-Tertiary (K-T) mass extinction that eliminated the dinosaurs at 65 Ma was only the third largest such event.

(230 ft). But what caused the glaciation itself? Curiously, Earth's climate appears to have been warm and ice-free just prior to the Hirnantian, and it was warm again just afterward. This stands in stark contrast to the two later Phanerozoic glaciations, both of which involved tens of millions of years of gradually cooling climate.

One possibility, which is speculative, but intriguing, is that the Late Ordovician glaciation was triggered by some sort of extraterrestrial impulse. Alex Pavlov (my former Ph.D. student) and his colleagues at the University of Colorado have suggested that glaciations can be triggered when the Solar System drifts through dense interstellar dust clouds.[3] Dust from the cloud would be captured by the Earth and could form a layer in the stratosphere that would be thick enough to block an appreciable amount of sunlight. Glaciation could result if the perturbation lasted long enough for the oceans to cool (tens to hundreds of years). The expected frequency with which the Solar System encounters dense interstellar clouds is about once every billion years, so having one such event during the Phanerozoic is not implausible.

Alternatively, the Late Ordovician glaciation and mass extinction could conceivably have been caused by a nearby gamma-ray burst[4] or supernova. The high-energy photons emitted by these powerful explosions would split N_2 and O_2 molecules in Earth's stratosphere, forming nitrogen oxides that might destroy the ozone layer and allow harmful solar ultraviolet radiation to pass through the atmosphere. Nitrogen dioxide (NO_2) could also have absorbed incoming visible sunlight, thereby cooling the surface. This mechanism appears capable of explaining mass extinctions, through increased UV radiation. It is less clear whether it could have triggered a glaciation, as this would have been a one-time, nearly instantaneous event, and the nitrogen oxides that were formed should have been removed from the atmosphere within a few years. So the interstellar dust cloud hypothesis sounds more feasible. Nevertheless, some sort of extraterrestrial triggering mechanism would fit nicely with the short-lived nature of the Late Ordovician glaciation.*

*Since these last two paragraphs were written, a new paper has appeared that may make an extraterrestrial cause of the Late Ordovician glaciation seem less plausible. Julie Trotter and colleagues have used oxygen isotopes in phosphate minerals to show that ocean temperatures declined to

Precambrian Climate

Let's now turn our attention to earlier times, which are less well understood, but which constitute almost 90 percent of the history of the planet. The entire time span between the Earth's formation at 4.6 Ga and the dawn of the Cambrian is termed (not surprisingly) the Precambrian eon (figure 4.2). Before it was known to have harbored life, the Precambrian was treated as a single time unit. Since then, it has been subdivided into three eras: the Hadean (4.5–3.8 Ga), the Archean (3.8–2.5 Ga), and the Proterozoic (2.5–0.54 Ga). The Precambrian, like the Phanerozoic, appears to have been mostly warm. Indeed, if one takes the evidence at face value, the Earth has been ice-free for over 90 percent of its history. This is especially remarkable in light of the dimmer young Sun. All other things being equal, one might expect that the early Earth should have been cold.

Having said this, we should probably inject a word of caution. The geologic record becomes increasingly sparse as one goes further back in time. The observation that the Mid-Proterozoic Era was ice-free between about 2.2 and 0.8 Ga is probably statistically meaningful, as there are numerous exposed rocks of this age, so we should expect to see evidence for glaciation had it occurred. The earlier Archean era has fewer well-preserved rock outcrops, however, and so the absence of glacial deposits during most of this time could be purely an artifact. That is why we have placed a question mark beside the "warm" label for the early and middle Archean in figure 4.2.

The Precambrian eon does contain evidence for several glaciations. The best documented ones occurred during the Paleoproterozoic, around 2.4 Ga, and during the Neoproterozoic, around 0.6–0.75 Ga. The glaciation at 2.9 Ga is labeled with a question mark because it has been

---

modern values before dipping more sharply into the Hirnantian glaciation. If the cause of the glaciation was extraterrestrial, as argued here, then why should surface temperatures have dropped beforehand? We have also shown, in a paper about early Mars that is currently under review, that NO_2 is more likely to warm a planet's climate than to cool it. (It does so by reducing the planet's albedo.) Both of these new papers suggest that we may need to look for a more Earth-bound cause for the Late Ordovician glaciation.

Geologic Time

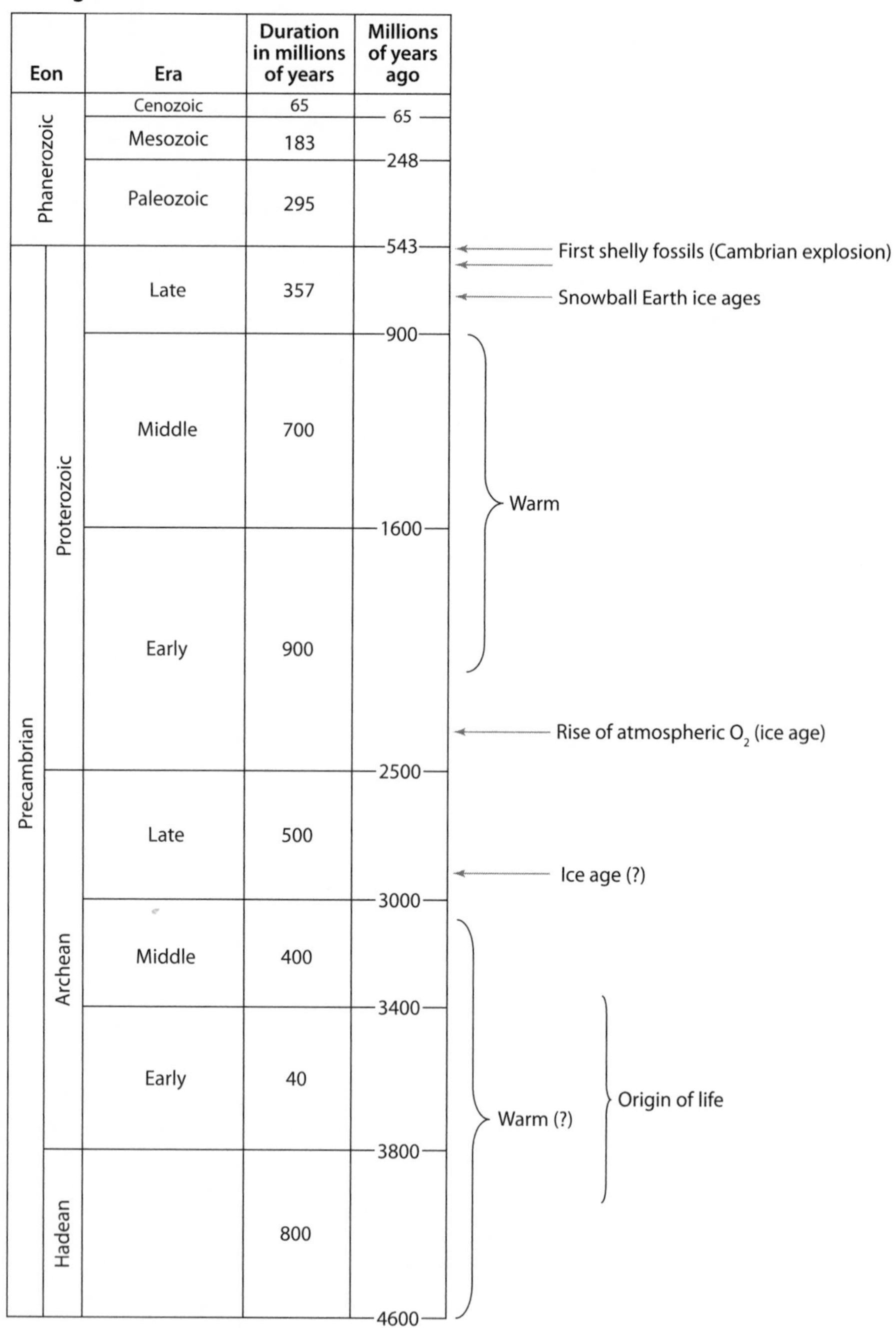

Figure 4.2 The geologic timescale, showing major events in Earth's history back to its origin at 4.6 Ga. Known periods of glaciation are indicated.

less thoroughly studied than the ones that came later. Firm evidence for glaciation is found on only one continent, Africa,[6,7] so it could have been a localized occurrence. (We discuss glacial indicators in more detail below.) By contrast, both the Paleoproterozoic and Neoproterozoic glaciations occurred on multiple continents at close to the same times.

The 2.4-Ga glaciation is particularly important for the story in this chapter because, as one can see from the figure, it occurred right when atmospheric O_2 concentrations first rose. That is unlikely to be a coincidence; hence, one may deduce that either the rise of O_2 triggered the glaciation, or else the glaciation triggered the rise of O_2.[7] For reasons discussed below, I favor the first interpretation. The two Neoproterozoic glaciations, at roughly 0.6 and 0.75 Ga, are so-called "Snowball Earth" episodes in which Earth's oceans may have frozen over completely. The evidence for this is quite strong, although many geologists still dispute this interpretation. The Paleoproterozoic glaciation at 2.4 Ga may have been a Snowball Earth episode as well,[8] but the evidence is less conclusive. The idea that such catastrophes can occur—and yet life somehow survives them—is sufficiently important for the issue of planetary habitability that I have devoted much of the following chapter to it.

Geologic Evidence for the Rise of Atmospheric O_2

I've mentioned several times now that atmospheric O_2 concentrations were low on the early Earth, but I have not yet said why we believe that to be true. So let's take a few moments to discuss the geologic evidence for the rise of atmospheric oxygen. The discussion here will necessarily be incomplete, as the literature on this topic goes back at least 50 years. The good news, though, is that one doesn't need to read all of it because there is now a pretty good consensus about how the story goes, even if geologists are still quibbling about the details.

Back in the late 1960's Preston Cloud from the University of California, Santa Barbara, noticed a clear pattern of oxidized and reduced minerals in ancient rocks.[9] Recall that reduced substances are those that can react with oxygen, whereas oxidized substances are those that have already reacted with oxygen. Sediments older than about 2.0–2.2 Ga contain reduced minerals, such as pyrite (FeS_2) and uraninite (UO_2), which are *detrital* in origin. Detrital minerals are grains of material that were eroded from rocks, carried by streams, and deposited in sediments without ever dissolving. Detrital uraninite and pyrite do not form today, except under unusual circumstances, because they oxidize and dissolve when exposed to O_2. The oxidation process evidently did not occur efficiently before ~2.2 Ga, however, because these minerals are common in sediments formed prior to this time. This is consistent with the idea that atmospheric O_2 levels were much lower. More recent age dating puts the oxic–anoxic transition closer to 2.4 Ga, but Cloud's reasoning remains unchanged. The retired Harvard geochemist Heinrich (Dick) Holland has carried on and extended many of Cloud's original ideas. Figure 4.3 shows his summary of the geologic evidence bearing on the rise of O_2, including that from detrital reduced minerals.

Additional geochemical clues supported the notion that O_2 concentrations were low during the first half of the Earth's history. In rocks younger than 2.4 Ga (to use the modern age dates), Cloud identified *redbeds*. These are reddish sandstones, such as the rocks that one sees in exposed cliff faces in the Arizona desert. The reddish color comes from tiny grains of *hematite* (Fe_2O_3) that coat larger sand grains. Hematite consists of oxidized, or *ferric*, iron, so this suggests that O_2 must have been present by this time.* Sandstones formed before 2.4 Ga are grayish in color because the iron in them is in the reduced, or *ferrous*, form. As an aside, one needs to be careful in using a mineral's color as evidence for its composition. Large chunks of hematite actually resemble pewter, as one can learn by examining the hematite offered for sale at gift shops or by looking at pictures of the dark, hematite-rich soils of Meridiani Planum, where the Mars *Opportunity* rover landed in early 2004. But when hematite occurs in tiny particles coating sand grains, it gives them a reddish hue. The smaller amounts of hematite found elsewhere on Mars give it its nickname, the Red Planet.

*One needs to be cautious here, because iron can sometimes be oxidized in the absence of free O_2. Some photosynthetic bacteria, for example, oxidize ferrous iron to ferric in the process of reducing CO_2 to organic matter. But this type of process cannot account for the ferric iron in redbeds.

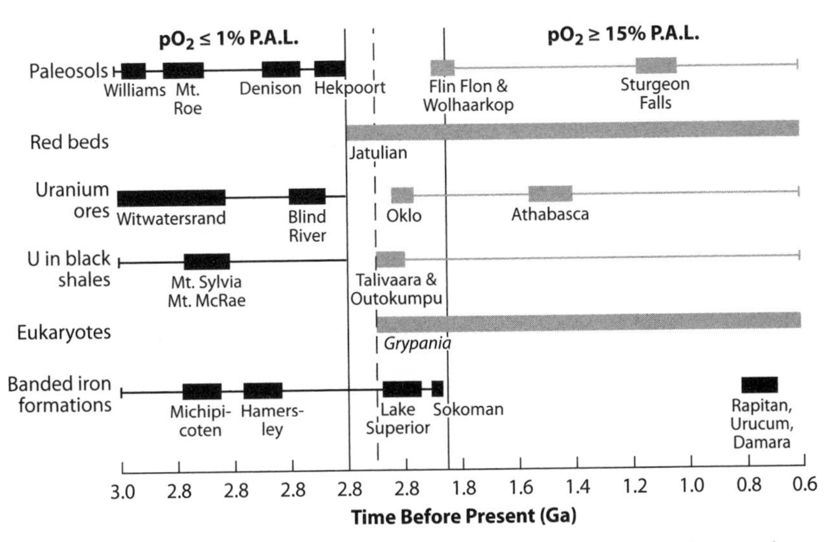

Figure 4.3 Diagram illustrating the "conventional" geologic evidence for the rise of atmospheric O_2. Black boxes indicate evidence for low O_2; gray boxes indicate evidence for high O_2. (Adapted from Holland 1994.[10])

Another rock type that helped form the basis of Cloud's theory was banded iron formations, or BIFs. BIFs are large, iron-rich deposits in which the iron (which occurs in a variety of chemical forms) is found in layers that alternate with chert, SiO_2 (figure 4.4, in color section). Chert is actually present in all the layers, but some layers are iron-rich and some are iron-poor. In figure 4.4, the dark layers are iron-rich and the lighter red layers are iron-poor. (The red color comes from hematite, which, as noted above, looks reddest when present in small quantities.) BIFs have been studied for many decades by economic geologists because they are the source of most of the iron used to make steel and automobiles today. The Hamersley BIF in Western Australia is a famous example. BIFs formed throughout the first half of the Earth's history, then disappeared for good at ~1.8 Ga, except for a few small formations that are associated with the Late Proterozoic Snowball Earth episodes. BIFs thus outlasted other low-O_2 indicators, probably because they reflect the oxidation state of the deep oceans, rather than that of the atmosphere.[10] If the deep oceans were oxygen-free, then iron would have been in its soluble, ferrous state, and could thus have been transported long distances to the sites where BIFs formed. The banding may have been caused when iron-rich deep water upwelled seasonally, as it does today in various locations such as the Grand Banks off the coast of Newfoundland, or in the great fisheries along the coast of Peru.

Figure 4.3 shows several other O_2 indicators as well. *Paleosols* (ancient soils) can be used for this purpose. Those that formed under a high-O_2 atmosphere tend to have retained their iron (because it was oxidized to insoluble ferric iron), whereas those that formed under a low-O_2 atmosphere are iron-deficient. *Eukaryotes* (organisms that have cell nuclei) can be used as O_2 indicators because almost all eukaryotes depend on O_2 for respiration. (More on this below.) Eukaryotes first appear in the fossil record around 2.1 Ga, consistent with the geological O_2 indicators. There is also additional evidence from sulfur isotopes (not shown here) that provides an even more conclusive argument that atmospheric O_2 first rose close to 2.4 Ga.[12] Because that story is tangential to the main topic of this book, and because it is consistent with the story told above, we will not go through it in detail. Skeptics of the Cloud/Holland model, though, have so far been unable to explain the sulfur isotope data. So we will assume for the rest of this chapter that Cloud and Holland were right.

Cause of the O_2 Rise: Cyanobacteria

The geologic record thus indicates quite clearly *when* atmospheric O_2 first rose. But *why* did it do so? Here one finds all sorts of disagreement. Everyone agrees on at least one point, though: atmospheric O_2 concentrations rose as a consequence of photosynthesis. As mentioned in the previous chapter, photosynthesis is the process by which plants and algae extract CO_2 from the atmosphere and convert it to organic matter, using the energy of sunlight to drive the reaction.

The story is actually more complicated than this, however, because there are different types of photosynthesis. The process that we have been talking about is *oxygenic photosynthesis.* It can be represented chemically by the following reaction:

$$CO_2 + H_2O \rightarrow CH_2O + O_2.$$

Here, "CH_2O" is geochemists' shorthand for organic matter. (The sugar, glucose ($C_6H_{12}O_6$), mentioned in the previous chapter can be thought of as just CH_2O times 6.) In this process electrons extracted from H_2O are used to chemically reduce CO_2 to organic matter. This process is termed "oxygenic" photosynthesis because it produces molecular oxygen, O_2, as a by-product.

Other forms of photosynthesis are also found in nature. An example is the process carried out by certain types of sulfur-dependent bacteria:

$$CO_2 + 2\,H_2S \rightarrow CH_2O + H_2O + 2\,S.$$

These bacteria use hydrogen sulfide, H_2S, to reduce CO_2 to organic matter. Instead of producing O_2, this reaction generates elemental sulfur. Still other photosynthetic bacteria use molecular hydrogen (H_2) or ferrous iron to reduce CO_2 to organic matter. Both hydrogen and ferrous iron should have been abundant on the Archean Earth, and so these bacteria were probably widespread in the early oceans. (A corollary to this observation is that the presence of ferric iron in BIFs does *not* necessarily indicate the presence of free O_2.[13]) These alternative forms of photosynthesis are collectively termed *anoxygenic photosynthesis* because they do not generate O_2.

When exactly did these different forms of photosynthesis arise? This is one area where arguments are still raging. Photosynthesis itself is considered to be quite ancient. Some of the oldest rocks known, the 3.5-Ga sediments from the Pilbara Group in Australia, contain layered structures called *stromatolites*, which are thought to represent the fossilized remains of photosynthetic microbial mats. But it is not certain whether or not they were generating O_2.

From a biochemical standpoint, it is clear that anoxygenic photosynthesis must have preceded oxygenic photosynthesis. Anoxygenic photosynthesis is carried out by different organisms using two completely independent sets of biochemical reactions, termed photosystem I and photosystem II. Oxygenic photosynthesis requires that both photosystems be present and that they be hooked up in series. That's because it is energetically more expensive to use H_2O to reduce CO_2 than it is to use reduced compounds like H_2S, H_2, and ferrous iron. But the advantages of doing so are enormous because H_2O is available in virtually unlimited quantities, in the oceans at least, whereas reduced compounds are relatively scarce. So the invention of oxygenic photosynthesis allowed biological productivity to increase dramatically,[14] while also paving the way for the development of higher organisms that depend on O_2 for respiration.

Another point on which virtually everyone agrees concerns which organisms were the first to perform oxygenic photosynthesis. Biologists are quite certain about the answer to this question, for reasons that will become apparent shortly. The key organisms were the *cyanobacteria*, some modern examples of which are shown in figure 4.5 (see color section). Cyanobacteria are *prokaryotic* organisms that lack cell nuclei. (Recall that *eukaryotic* organisms have cell nuclei.) Cyanobacteria were formerly called "blue-green algae," but this was actually a misnomer because algae are eukaryotes. Cyanobacteria also have the ability—which higher plants do not—of carrying out photosynthesis either anoxygenically or oxygenically, depending on how much H_2S or H_2 is present in their environment. For this reason, biologists long ago surmised that they must have been the first organisms to produce O_2.

One puzzle that is not yet fully understood concerns the timing of the rise in atmospheric O_2 as compared to the origin of cyanobacteria. Various geochemical indicators found in rocks suggest that both cyanobacteria and eukaryotes were already present by 2.8 Ga or earlier.[16,17]* Cyanobacteria—modern ones, at least—can produce O_2, and eukaryotes require O_2 to live, so it would appear that oxygenic photosynthesis had already been invented by this time. But, as we have seen, atmospheric O_2 did not become abundant until about 2.4 Ga. What caused this delay of 300 million years or more between the production of O_2 and its first appearance in the atmosphere? This question is not yet resolved, and some authors question whether the indicators of cyanobacteria and eukaryotes at 2.8 Ga are reliable.[19] The story would be much simpler, in their view, if cyanobacteria arose just shortly before atmospheric O_2 concentrations went up. This is a technical argument that we need not get into

*A recent paper by Sky Rashby and colleagues showed, however, that the 2-methylhopanes used as indicators of cyanobacteria are also found in other types of bacteria.[18] And an even more recent paper suggests that these compounds were added to the rocks at a later time by oil that migrated into them.[35] So the inference that cyanobacteria were present at 2.8 Ga is not as strong as it once was.

here. It is relevant to our broader story, though, because if we want to be able to predict whether O_2 should be present in the atmospheres of other planets, we need to understand exactly why it is present in Earth's atmosphere. Would any planet on which oxygenic photosynthesis evolved develop an O_2-rich atmosphere? That is a question worth pondering.

Methane, Methanogens, and the Universal Tree of Life

What would Earth's atmosphere have been like prior to the rise of O_2 at 2.4 Ga? In the last chapter we argued that it should have been richer in CO_2 as a consequence of the faint young Sun and the feedback loop between CO_2 and climate. And N_2, being largely unreactive, should also have been present in roughly the same concentration as today. What about other gases, though? Should any of them have been abundant? We are especially interested in *biogenic gases*—ones that, like O_2, are produced by organisms. We will focus on one gas in particular, methane (CH_4). Methane has already been mentioned as a possible additional greenhouse gas that might have helped keep the early Earth warm. Should methane have been abundant in the early atmosphere?

To answer this question, we need to think first about where CH_4 comes from today. Unlike CO_2, CH_4 is not given off in appreciable quantities by volcanoes, at least not by those that vent directly into the atmosphere. Small amounts of CH_4 are emitted by submarine volcanoes, notably the relatively cool (~60°C water temperature) vents located along the flanks of the mid-Atlantic ridge.[20] This methane may be produced by *serpentinization* reactions, in which hot water reacts with ferrous iron-rich rocks (peridotites) deep within the ridge systems to form greenish serpentine minerals. Iron, which was a major component of peridotite, tends to be excluded from the serpentine minerals that are formed. Instead, it forms a stable, partially oxidized mineral called *magnetite* (Fe_3O_4).* Because the iron is oxidized in this process, water itself must be reduced, producing hydrogen, H_2. If CO_2 is present in the water,

*Magnetite, Fe_3O_4, can be thought of as consisting of one part of hematite, Fe_2O_3, and one part of ferrous oxide, FeO. The two iron atoms in hematite are both oxidized, or ferric, iron. By contrast, virtually all of the iron in the original peridotite was reduced, or ferrous, iron.

then it can also be reduced, producing CH_4. We should pay attention to this, as it bears directly on the question raised at the beginning of this chapter, namely, whether methane is a good indicator of life. Clearly, if it is coming from serpentinization reactions, it is not. This issue has also been raised recently with regard to our neighboring planet, Mars, as methane has been claimed to be present in low concentrations, 10–100 ppb, in its atmosphere (see chapter 8). It is thus important to understand whether, and how fast, CH_4 can be generated abiotically.

Although significant abiotic methane sources may exist, most of Earth's methane is clearly biological in origin. It is produced mainly by *methanogenic bacteria*, which are also known as *methanogens*, for short. Methanogens are anaerobic bacteria, meaning that they do not require O_2 to survive. Indeed, they are poisoned by O_2 if it is present in abundance; hence, they cannot live in direct contact with the atmosphere. Instead, they are found in anoxic environments such as the flooded soils beneath rice paddies and the intestines of ruminants, especially cows. Recent, still controversial evidence suggests that some methane is also produced by land plants.[21] But, even if true, this is irrelevant for the Precambrian, as land plants evolved only during the Silurian period, some 440 million years ago.

Would methanogens have been present on the early Earth? The answer is almost certainly "yes." The most direct evidence comes from methane trapped in fluid inclusions in rocks dated at approximately 3.5 Ga.[22] This methane is strongly depleted in ^{13}C relative to ^{12}C, just as is biologically produced methane today.*

Additional evidence in favoring the early evolution of methanogens comes from the modern field of molecular biology. As many readers are no doubt aware, biologists are now able to determine the sequences of nucleotides in DNA and RNA. RNA stands for *ribonucleic acid*; DNA is *deoxyribonucleic acid*. Nucleotides are molecules composed of a sugar (ribose), phosphate, and one of four distinct nitrogen-containing *bases*,

*^{12}C and ^{13}C are the two stable isotopes of carbon. Both isotopes have 6 protons. ^{12}C has 6 neutrons as well, while ^{13}C has 7 neutrons. Abiotic *Fischer-Tropsch synthesis*, in which hydrocarbons are formed from H_2 and CO_2 or CO in the presence of a catalyst, could conceivably produce similar ^{13}C-depleted methane,[23] but the authors of the report argue that this is unlikely to explain their fluid inclusion data.

that are arranged in patterns to form the genetic code. Differences in human DNA are now routinely used to identify victims and perpetrators at crime scenes, for example.

If one wants to look back into biological history, one has to choose the right portion of the (extremely long) DNA or RNA molecule. The sequence that is used most frequently for this purpose, following the pioneering work of Carl Woese at the University of Illinois in the late 1970s,[24] is called *ribosomal RNA* (rRNA). (Technically, it is faster these days to sequence the part of the DNA molecule that codes for rRNA, but the information gained is the same.) The ribosomes are organelles within cells in which proteins are synthesized. All free-living organisms contain ribosomes, and so they can all be compared using this molecule. Because mutations to the protein-manufacturing machinery of a cell are usually fatal to the organism, and hence are not often passed on to the next generation, the rRNA molecule evolves very slowly. Thus, it is useful for looking far back into evolutionary time.

By comparing DNA or RNA sequences from different organisms, it is possible to construct evolutionary trees. Because they can be applied to all organisms, the tree created from ribosomal RNA is sometimes referred to as the *universal tree of life*. A particular example of such a tree, created by Norman Pace at the University of Colorado, is shown in figure 4.6. According to this classification scheme, all organisms fall into one of three different *domains*: the Archaea, the Bacteria, or the Eukarya. The Archaea and the Bacteria are both composed of single-celled prokaryotes that lack cell nuclei. The Eukarya are composed of single- and multicelled eukaryotes that have cell nuclei. Humans, being animals, are members of the Eukarya. They are represented on this diagram by *Homo* (for *Homo sapiens*) and are comfortably placed just between *Zea* (corn) and *Coprinus* (mushrooms). Single-celled *Paramecium* is just one branch back down the tree. This shows that the rRNA tree really does probe far back into evolutionary history.

The methanogens, which produce most of Earth's CH_4, are all found on one branch of the Archaea. Five different methanogens are shown in figure 4.6, all of which have names beginning with "Methano." The fact that they are all on the same branch of the Archaea tells us that

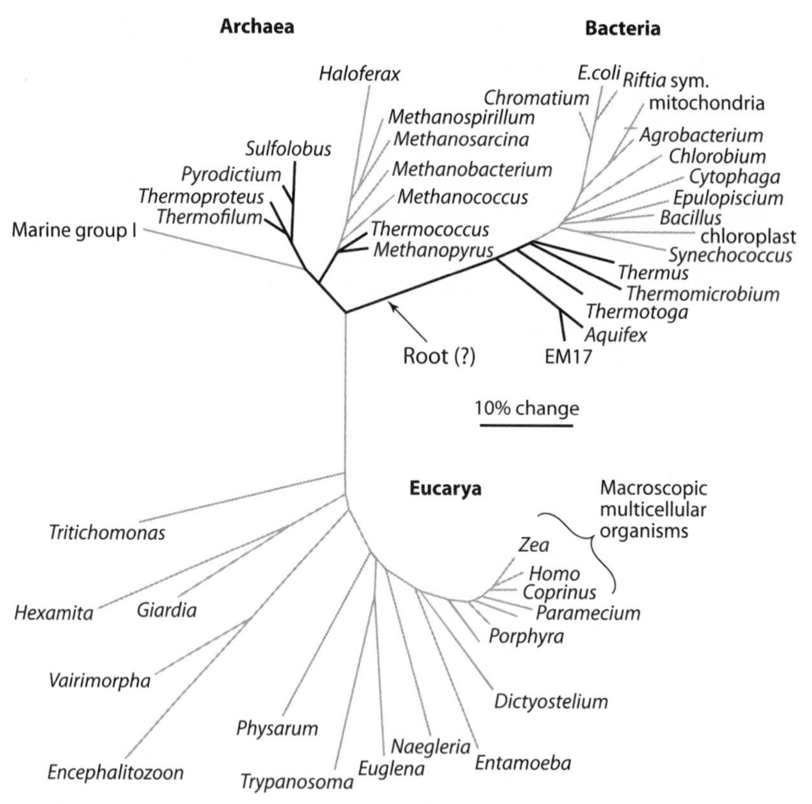

Figure 4.6 The "universal" tree of life based on sequencing of ribosomal RNA.[25]

they were probably not the earliest organisms. However, the relatively short distance between the methanogens and the presumed root of the rRNA tree, thought be near the base of the Bacteria, suggests that they evolved very early—probably during the Archean, or possibly even the Hadean eon.

Figure 4.6 shows something else that relates back to the preceding discussion concerning the origin of oxygenic photosynthesis. Approximately halfway up the Bacterial branch of the tree lies a limb that includes both *Synechococcus* and chloroplasts. *Synechococcus* is a particular species of cyanobacterium. Other cyanobacteria cluster in the same place. Chloroplasts, as readers may recall from high school biology class, are not independent organisms; rather, they are organelles within the cells of plants and algae within which photosynthesis is carried out. Chloroplasts contain their own DNA—a fact that was initially puzzling to biologists, but that today makes perfect sense. The DNA in chloroplasts contains a section that codes for ribosomal RNA and that can therefore be compared with rRNA sequences from free-living organisms. When one does the comparison, one finds that it groups right with cyanobacteria. The only way to explain this is if chloroplasts are directly descended from cyanobacteria. Indeed, just such a hypothesis was proposed in 1905 by the Russian biologist Konstantin Merezhkovsky and was reproposed independently somewhat later by Lynn Margulis, the coauthor of the Gaia hypothesis.[26] Both researchers suggested, based on other lines of argument, that green plants learned to photosynthesize by way of *endosymbiosis*: a living cyanobacterium was somehow engulfed by a eukaryotic organism, thereby conferring on it the ability to generate O_2.

The implication of this by now generally accepted theory is clear: oxygenic photosynthesis was only invented once on Earth. So, like the origin of life itself, we have no way of telling whether it was a likely event or an unlikely one. This leads to one of the key astrobiological questions regarding life on other planets. If such life exists, will it have figured out how to perform oxygenic photosynthesis? This makes a huge difference, both for the gases for which we might search and for the types of life that might evolve. Complex life, as the authors of *Rare Earth* pointed out, can evolve only if O_2 is present.

The Archean Methane Greenhouse

If methane was indeed being produced on the early Earth, it should have been important for climate. CH_4, like CO_2 and H_2O, is a greenhouse gas that absorbs and reemits outgoing thermal-infrared radiation. Unlike H_2O or NH_3 (Carl Sagan's favored gas for warming the early Earth), CH_4 absorption is concentrated in few relatively narrow spectral intervals; hence, it is a good greenhouse gas, but not a great one. Even climatologists are sometimes surprised to hear it described in this manner because, in today's atmosphere, CH_4 produces about 20 times as much greenhouse warming as does CO_2, as measured by its impact on climate over a 100-year timescale.[27] But that is largely because CO_2 is much more abundant than CH_4 in today's atmosphere (380 versus 1.7 ppmv). When the two gases are present at comparable concentrations, CO_2 is actually the better greenhouse gas because it absorbs more strongly over a wider range of infrared wavelengths.

In the low-O_2 Archean atmosphere, and assuming that methanogens were indeed present on the early Earth, CH_4 is predicted to have been quite abundant. Photochemical models suggest that the lifetime of CH_4 in a low-O_2 atmosphere should have been of the order of 10,000 years.[28] By comparison, the CH_4 lifetime in Earth's present atmosphere is only 10–12 years. The biological production rate of CH_4 appears to have been comparable to today, despite the very different nature of the Archean ecosystem[29]; hence, the Archean atmosphere should have contained roughly 1000 times as much CH_4 as is present today. That would increase its concentration from 1.7 ppm to approximately 1700 ppm. This is enough to produce about 15 degrees C (27 degrees F) of greenhouse warming,[30] or about half the amount needed to compensate for reduced solar luminosity in the Archean (see figure 3.3). So this suggests that CH_4 could have played a significant role in keeping the climate warm.

The actual Archean climate problem is even more complex. An atmosphere that was rich in CH_4 should also have contained other hydrocarbon gases, such as ethane (C_2H_6), and these gases could have added to the greenhouse effect.[30] But it may also have contained *hydrocarbon haze*, or photochemical smog, similar to the haze that exists in the atmosphere of Saturn's moon, Titan (see figure 4.7, in color section). Titan's haze is produced by the interaction of its dense N_2-CH_4 atmosphere with solar ultraviolet radiation and with charged particles trapped in Saturn's magnetosphere. This haze cools Titan's surface by absorbing sunlight high up in the atmosphere and reradiating it back to space, a phenomenon that has been termed an *anti-greenhouse effect.*[31]

Earth's atmosphere during the Archean would have been quite different from that of Titan, primarily because it was much warmer. Titan's surface temperature is a frigid 93 K—so cold that both CO_2 and H_2O are effectively frozen out. Earth's early atmosphere, as we have seen, should have contained appreciable quantities of both of these gases, along with CH_4. According to both modeling calculations[28] and laboratory experiments,[32] haze should have formed in Earth's atmosphere only if the ratio of CH_4 to CO_2 was greater than about 0.1. This happens in some, but not all, plausible scenarios. Hence, it is difficult to predict with certainty whether or not such a haze should have been present. If it was present, though, the haze on early Earth must have been thinner than that observed on Titan. A thick haze would have lowered Earth's surface temperature too much, causing the oceans to freeze, and killing off the methanogens that were producing the CH_4. A thin haze should have been stable, however, because any perturbation that increased methane production would have increased haze thickness, thereby cooling the climate and leading to decreased methane production. (Most methanogens tend to reproduce and metabolize faster at higher temperatures.) So the methane greenhouse, like the CO_2 greenhouse discussed in the previous chapter, appears to contain a stabilizing negative feedback mechanism. This one, though, directly involves the biota, and is therefore distinctly "Gaian." Thus, Lovelock may not have been so far off the mark with his Gaia hypothesis. If methane was indeed an important contributor to the greenhouse effect during the Archean, the climate system then may have been under much stronger biological control than it is today.

The Paleoproterozoic Glaciation

We began this chapter by discussing the long-term glacial record, and so it is appropriate to return to that topic now. Recall from figure 4.2 that the Paleoproterozoic glaciation at ~2.4 Ga appears to have occurred right when atmospheric O_2 concentrations first rose. Nowhere is this more evident than in the Huronian rock sequence in southern Canada (figure 4.8). The Huronian rocks contain *diamictites* formed from three distinct glaciations: the Bruce, the Gowganda, and the Ramsey Lake. Diamictites are sedimentary rocks that contain other, smaller rock fragments of various shapes and sizes. They are formed when piles of glacial

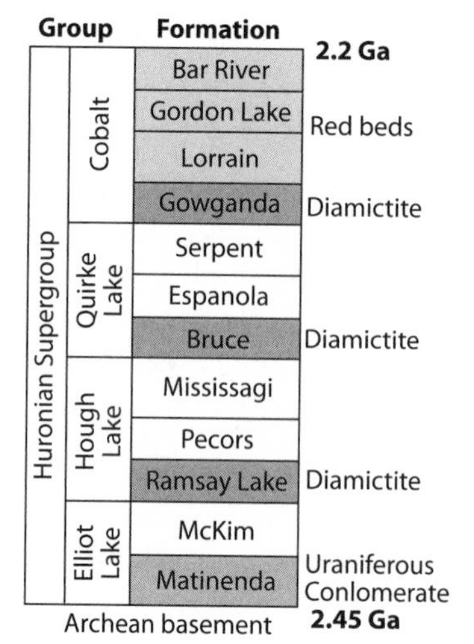

Figure 4.8 Diagram showing the Huronian sequence in southern Canada. This sequence was deposited between 2.2 and 2.45 Ga. Hence, it spans the interval during which atmospheric O_2 levels first rose.[33] (Redrawn by Y. Watanabe, Penn State University.)

till become covered with sand and mud. When their glacial origin is confirmed by other evidence—which is the case for the Huronian—these rocks are called *tillites*. The Huronian sequence has three such tillites, indicating that the glaciers came and went three separate times.

The Canadian geologist Stuart Roscoe, who mapped these rocks initially back in the late 1960s,[34] noticed something very interesting about them. Below the lowermost Ramsay Lake diamictite, in the Matinenda formation, he found detrital uraninite and pyrite. As we learned earlier, these are reduced minerals that did not dissolve during weathering and that can form only when atmospheric O_2 levels are low. Above the uppermost Gowganda diamictite, Roscoe observed the brightly colored redbeds of the Lorraine formation. The oxidized mineral hematite in redbeds is evidence of high atmospheric O_2. Preston Cloud had already proposed his model for O_2 evolution by that time, and so Roscoe was quick to point out that this series of what, to him, were Earth's first recorded glaciations occurred right when atmospheric O_2 levels first rose. To Roscoe, that was simply a coincidence. But if one accepts the methane greenhouse story described above, then it makes perfect sense. When O_2 concentrations went up, CH_4 concentrations went down, so it is not at all surprising that the climate became glacial. Indeed, the ability of this model to predict the Paleoproterozoic glaciation is perhaps the best indication that the Archean methane greenhouse hypothesis is indeed correct.

Before leaving this discussion, let's consider its implications for the more general question of planetary habitability. Nothing that has been said in this chapter undermines the idea, laid out in chapter 3, that the CO_2-climate feedback involved in the carbonate-silicate cycle has played a key role in keeping the Earth habitable. The methane greenhouse model cannot stand on its own—it still requires that atmospheric CO_2 was higher than today by a factor of 100 or more during the Archean.[30] At the same time, the predicted amount of greenhouse warming by CH_4 is large enough to have significantly altered Earth's climate history and to explain, through its disappearance, at least one major set of glaciations. The source for CH_4 is mostly biological, as we have seen, and the source for the O_2 that displaced it is also almost entirely biological. So Gaia is evidently active on these timescales, if one chooses to think of the climate system in that way. Her influence, though, has not always been benign. If biogenic CH_4 helped stabilize climate during the Archean, then biogenic O_2 clearly destabilized it during the Paleoproterozoic. One can hardly imagine a greater planetary catastrophe than the Paleoproterozoic Snowball Earth glaciation. This would have had devastating consequences for all organisms that lived near Earth's surface, including the ones (the cyanobacteria) that caused it. On the other hand, the creation of an O_2-rich atmosphere paved the way for the evolution of complex life, including humans. If that's what Gaia had in mind, then maybe she knew what she was doing.

· Chapter Five ·

Runaway Glaciation and "Snowball Earth"

The previous two chapters have provided a broad overview of Earth's long-term climate evolution, and the news has generally been reassuring. Earth's climate system has at least two stabilizing negative feedback loops, one involving CO_2 and the other CH_4. The CO_2 feedback loop is mostly abiotic, and so it should operate on any planet that has Earth-like endowments of water and carbon, along with sufficient internal heat to drive plate tectonics. The CH_4 feedback loop is inherently biological and operated effectively only during the first half of the Earth's history. The stability it provided was eventually upset by the evolution of cyanobacteria and the subsequent rise in atmospheric O_2.

Other aspects of planetary climates are far less stable and life-friendly, however. Earth's climate system also contains at least two strong positive feedback loops that tend to destabilize it. We need to understand these feedbacks because they probably affect climate stability on other Earth-like planets as well. The first of these is the water vapor feedback loop mentioned in chapter 3: increases in surface temperature increase the amount of water vapor in the atmosphere, further increasing the surface temperature. This feedback loop exacerbates the faint young Sun problem, as we saw in chapter 3, because it implies that a cold early atmosphere would have had even less water vapor, and a smaller greenhouse effect, than the present one. It also roughly doubles the warming that is expected to result in the near future from fossil fuel-produced CO_2. The water vapor feedback has not led directly to any climate instability during Earth's history, as far as we know, but it may well have done so on Venus. The resulting climate catastrophe, which is termed a *runaway greenhouse*, is discussed in the next chapter.

The other strong positive feedback in Earth's climate system is the *ice albedo feedback loop*. The cause of this positive feedback is straightforward: clean snow and ice—thick ice, at least—reflect most of the sunlight that hits them. (That is why they appear white.) By doing so, they increase Earth's albedo, or reflectivity, thereby cooling the climate. This allows even more snow and ice to accumulate, creating a positive feedback loop. As long as the polar caps are confined to high latitudes, the effects of this feedback loop remain under control. If they cover too much of the planet's surface, though, the positive feedback loop can spiral out of control, and the effects can be disastrous. We have already heard in the previous chapter that this may actually have happened several times during Earth's history. Before looking at such extreme cases, however, let's first look briefly at how this feedback loop has affected Earth's more recent climate history. At the same time, we'll examine how variations in Earth's orbit have influenced the advances and retreats of the polar ice caps over the past few million years. We'll need this information later when we examine the effect of the Moon on Earth's climate (chapter 9) and the closely related effect of the lack of a large moon on the climate of Mars (chapter 8).

Milankovitch Cycles and the Recent Ice Ages

As pointed out in the previous chapter, Earth's poles have been ice-covered for about the last 35 million years. During that time, the climate has gradually grown colder and colder, culminating in the glacial-to-interglacial fluctuations of the last 3 million years. We are currently in the midst of a relatively warm interglacial period, the Holocene epoch, which has been going on for about the last 10,000 years.

The comings and goings of these recent ice ages have been studied intensively by geologists for over 200 years. Consequently, a great deal is known about them. A wonderful description of this subject, which covers everything discussed below and much more, is the book by John and Catherine Imbrie.[1] Astronomers and mathematicians have been interested in the ice ages, too. As early as 1875, the Scottish scientist James Croll suggested that the advances and retreats of the ice sheets were linked to variations in the Earth's orbit. Both the shape of Earth's orbit (its *eccentricity*) and the tilt of Earth's spin axis (its *obliquity*) vary with time as a consequence of gravitational interactions with the Sun, Moon, and other planets (see figure 5.1). The dominant periods on which these parameters vary are approximately 100,000 years and 400,000 years for the eccentricity, and 41,000 years for the obliquity. On these timescales, the eccentricity changes from 0 (a circular orbit) to 0.06 (a more elliptical orbit), while the obliquity varies by ± 1 degree from its current value of 23.5°. The spin axis also *precesses* around in a circle every 26,000 years. Currently, it points toward the star Polaris, which is also known as the North Star. 13,000 years ago, it pointed toward the bright star Vega, and it will do so again 13,000 years in the future.

A Serbian mathematician named Milutin Milankovitch is credited with figuring out how these orbital variations affect Earth's climate system. Most of Earth's land area is concentrated in the northern hemisphere, and both North America and Asia extend nearly up to the North Pole. Hence, when the climate gets cold, it is relatively easy for northern hemisphere ice sheets to thicken and expand southward. The most recent peak in such glacial activity occurred about 20,000 years ago when the Laurentide ice sheet covered much of North America and another equally large ice sheet extended across much of Europe. The geography of the southern hemisphere is quite different. The continent of Antarctica straddles the pole and is isolated from other continents by large stretches of ocean (except in the relatively narrow Drake Passage between Antarctica and South America). So, while Antarctica can be, and currently is glaciated, it is difficult for southern hemisphere continental ice sheets to expand toward the equator.

Milankovitch was aware of this hemispheric difference in geography. He also knew that the physics of ice and snow is highly nonlinear, and, as a result, their response to a particular climate forcing can be wholly

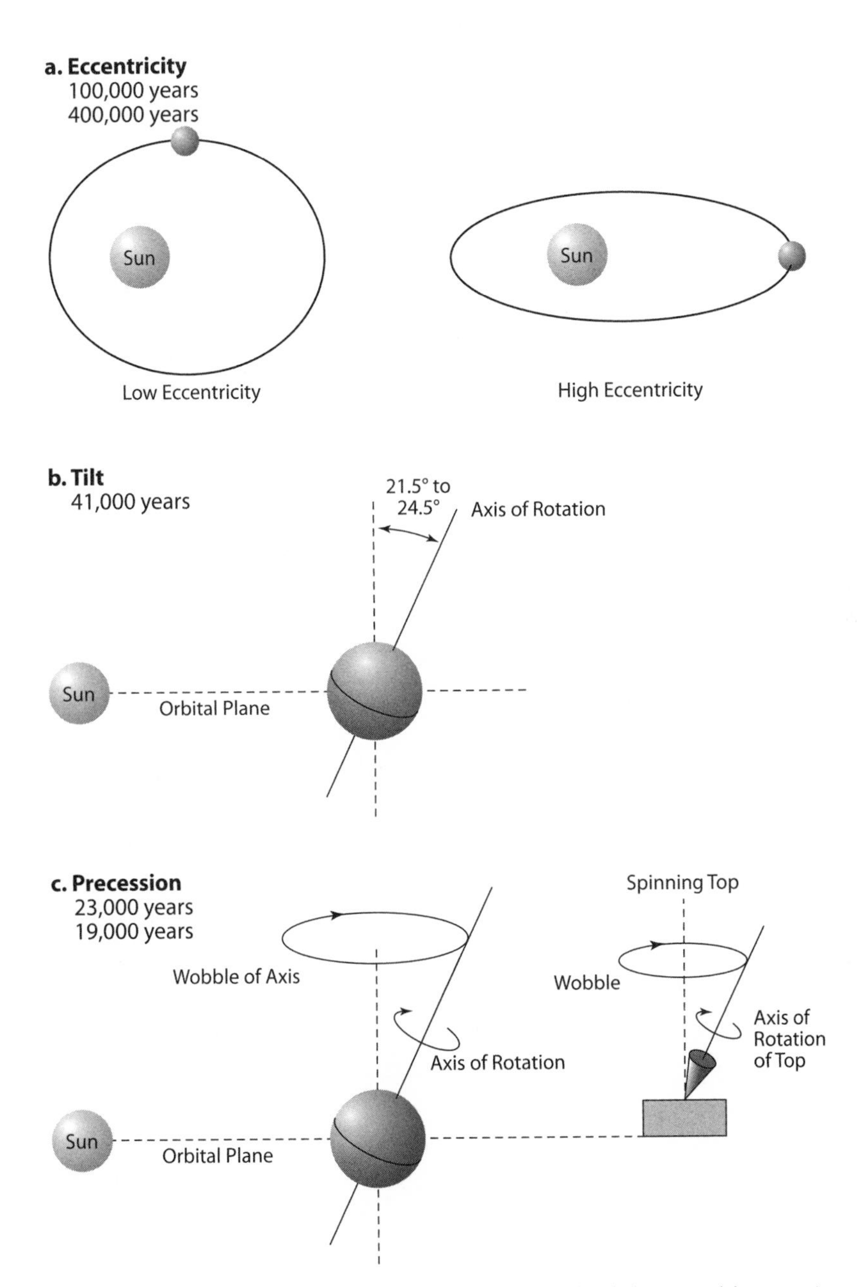

Figure 5.1 Diagram illustrating the variations of Earth's orbital elements: (a) eccentricity; (b) obliquity, or tilt; (c) precession of the spin axis.[2]

out of proportion with the forcing itself. As the surface temperature warms, the ice and snow melt at greatly accelerated rates.* Milankovitch postulated that this was the key to understanding the ice ages. The Earth's orbit is elliptical, rather than circular, and so the distance from the Earth to the Sun varies with time. When the North pole is pointed toward the Sun at the point where Earth itself is closest to the Sun, termed *perihelion*, northern hemisphere summers are particularly hot. This causes the northern hemisphere ice sheets to melt excessively, and hence to retreat. The effect is amplified when the obliquity is higher than normal because the seasonal cycles are then more intense.

These predictions by Croll, Milankovitch, and others were made long before the climate record was well understood. In 1985, oceanographers acquired an important new tool for studying the ice ages: the drilling ship *JOIDES Resolution*. The *Resolution* made cruises around the world, and the researchers aboard drilled cores into carbonate sediments on the seafloor. By measuring the ratio of different oxygen isotopes in these sediments, they were able to determine when the polar ice sheets advanced and when they retreated.** Remarkably, the periodicities recorded in the oxygen isotope data matched up precisely with the periodicities predicted decades earlier by the astronomers, strongly supporting the hypothesis that the glacial–interglacial cycles were driven by changes in the Earth's orbit. This spectacular agreement between theory and observation is one of the shining success stories of modern Earth science.

Orbital forcing alone, however, cannot explain the large (10°C globally) surface temperature swings that characterize the glacial-to-interglacial cycles. Amplification of these orbital forcings by positive feedbacks is required. Some of these feedbacks are understood, whereas others are more speculative. As mentioned already, one feedback mechanism that

*Technically, this is because the vapor pressure of water or ice varies exponentially with temperature, according to the Clausius-Clapeyron equation.

**Oxygen has three stable isotopes: ^{16}O, ^{17}O, and ^{18}O. Recall that isotopes of a given element differ in the number of neutrons in their nuclei. For studying climate, only ^{16}O and ^{18}O are required. Polar ice is depleted in the heavy isotope of oxygen, ^{18}O, relative to the lighter isotope, ^{16}O. When the ice caps expand, excess ^{16}O accumulates in the caps, and so seawater becomes enriched in ^{18}O. This enrichment is passed on to the carbonate sediments that form during that time period. The colder water temperatures enhance the partitioning of ^{18}O into the carbonate sediments, thereby enhancing the climate signal.

is known to have been important is ice albedo feedback. When the polar ice sheets expanded, they reflected additional sunlight, making the climate colder and furthering their own growth. The same feedback operated in reverse when the ice sheets retreated: Earth's albedo dropped, more sunlight was absorbed, the climate grew warmer, and the ice sheets retreated even faster.

Other feedbacks in the glacial-to-interglacial climate system are less well understood. Atmospheric CO_2 is known to have varied by about 30 percent between glacial minimum and glacial maximum, based on measurements of air bubbles trapped in polar ice.[3] CO_2 levels were lower during the glacials and higher during the interglacials; hence, these changes were part of a positive feedback loop. Note that this is just the opposite of how CO_2 affects the climate system on long timescales. On long timescales—millions to billions of years—CO_2 provides a negative, and hence stabilizing, influence on climate by way of the carbonate-silicate cycle. But, on glacial-to-interglacial timescales—tens of thousands of years—CO_2 provides a positive, and hence destabilizing, influence on climate. The climate system is complicated, and its multiple, seemingly contradictory interactions with the carbon cycle are part of that complexity.

The reason why CO_2 levels changed in this way during the glacial-to-interglacial cycles is not known for certain, although many different hypotheses have been suggested. The CO_2 control mechanism may have involved changes in ocean circulation, or biological productivity, or some combination of both.[4] The ice sheets themselves may have provided additional feedbacks on climate. When an ice sheet thickens, it depresses the underlying bedrock, causing both the the ice sheet and the bedrock to sink. Because temperature increases as one moves downward through the lower atmosphere, this warms the ice sheet's surface, which can cause it to melt. The ice sheet can also melt at its base, allowing it to slide, or *surge*. These ice sheet feedbacks may explain why the dominant period of the glacial-to-interglacial cycles observed in the oxygen isotope data switched from 41,000 years to 100,000 years beginning around 900,000 years ago, as sinking of bedrock becomes an important feedback on the 100,000-year timescale. A great deal of effort has gone into understanding these glacial-to-interglacial climate feedbacks, as they may be important for predicting changes in climate and sea level in the near future.

Ice Albedo Feedback and Climatic Instability

The Pleistocene glacial-to-interglacial oscillations are fascinating, and they had a dramatic effect on climate at mid- to high-latitudes, but they did not pose a threat to overall planetary habitability. During glacial maxima, the Earth's tropical regions were cooler than today by perhaps 5 degrees Celsius, but that did not pose any insurmountable problems for life. The end of the last Ice Age, around 10,000 years ago, was marked by a number of extinctions of large land mammals, such as mammoths and saber tooth tigers; however, similar extinctions have occurred episodically throughout the Phanerozoic eon, so there is nothing particularly unusual about this. Climate cycles are merely one of several mechanisms that may contribute to such extinctions. Humans are another, and this may explain why the land mammals disappeared 10,000 years ago.

Climatologists have been aware for a long time, however, that much more severe glaciations are possible. The Swedish climatologist E. Eriksson was the first scientist to publish in English on this subject.[5] The famous Russian climatologist Mikhail Budyko had been working on this topic also,[6] as had William Sellers in England.[7] Budydo's work was motivated by the Cold War arms race between the United States and the Soviet Union and the associated possibility of nuclear winter. Each of these authors came independently to the conclusion that ice albedo feedback can lead to runaway glaciation if the polar ice caps get too large. The reason has already been mentioned: As the polar ice caps expand toward the equator, Earth's albedo increases. This albedo increase becomes very large as the ice sheets advance to lower latitudes, for two reasons. First, there is more surface area at low latitudes than at high latitudes. (Pick up a globe and examine it if you need to convince yourself that this is right. This is also easy to demonstrate mathematically.) And, second, more sunlight hits the Earth at low latitudes, so differences in surface albedo become even more important.

The ice albedo instability can be studied quantitatively by using simplified climate models that keep track of the variable amounts of sunlight that fall between Earth's equator and its poles, along with the size of the polar ice caps. This type of model, termed an *energy balance climate model* (EBM), was pioneered by Budyko and Sellers. Such models are sensitive to ice albedo feedback, because the colder they get, the larger the ice caps grow, and the more sunlight is reflected back to space. If the ice caps get too large—typically, if they extend equatorward of about 30° latitude—they become unstable, and the oceans can freeze right down to the equator.

The results of one such EBM calculation are shown in figure 5.2. Here, the horizontal scale shows either the atmospheric CO_2 concentration (bottom scale) or the solar flux relative to today (top scale), while the vertical scale shows latitude. Both CO_2 and solar flux affect surface temperature in the same way, which is why they can be shown together. The curves represent the southernmost extent of the polar ice sheets. (Such models are typically assumed to be symmetric about the equator, so the southern hemisphere behaves in a similar manner.) The solid curves represent stable solutions of the model; the dashed curves represent unstable solutions. In other words, if the polar cap extends down to a position marked by a solid curve, it will stay there; if it extends down to the position marked by a dashed curve, it will either advance (at low latitudes) or retreat (at high latitudes).

Even though such a model is highly simplified compared to the real climate system, it illustrates a fundamental aspect of its behavior. As one can see from the figure, if the ice line moves equatorward ~30 degrees, the climate becomes unstable, and the polar ice sheets should advance all the way to the equator. Large increases in either solar luminosity or atmospheric CO_2 are needed to escape from that globally glaciated state. If solar luminosity triggers the escape, the recovery is very slow. A roughly 25-percent increase in solar luminosity is needed, and this would require more than 2 billion years of solar evolution, as the solar flux is currently increasing by about 1 percent every hundred million years. CO_2 increases offer a much quicker way to escape from the icehouse. In the climate model shown, a 300-fold increase in atmospheric CO_2 would be needed to deglaciate. At current volcanic outgassing rates,

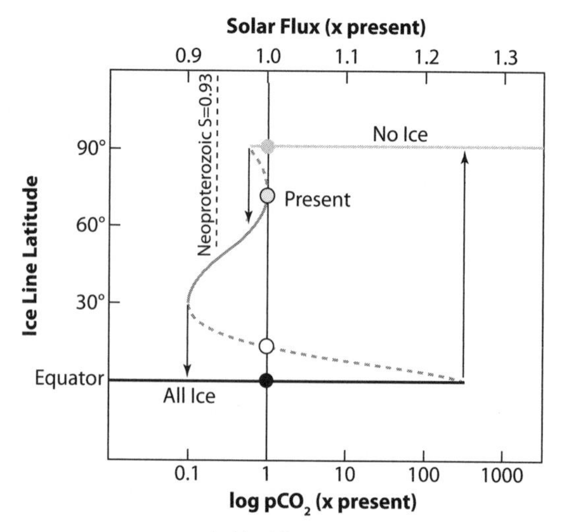

Figure 5.2 Diagram illustrating the ice-albedo instability. Solid curves indicate stable solutions of the energy-balance climate model. Dashed curves indicate unstable solutions. (Figure courtesy of Paul Hoffman, Harvard University.[9] Adapted originally from Caldeira and Kasting 1992.[10])

and with zero assumed loss of CO_2 from organic carbon burial or from silicate weathering, this CO_2 buildup could be achieved in as little as 3 million years.*

Evidence for Low-Latitude Glaciation

Ice albedo instability is not just a theoretical concept; there is also empirical evidence for its great power. As mentioned in chapter 4, if one peers far back in the geologic record, one finds evidence for ice ages that were much more severe than any that have occurred recently. The best examples come from the Neoproterozoic era, just prior to the Cambrian

*The current CO_2 outgassing rate from volcanoes[8] is about 5×10^{12} moles/year, while the current atmospheric inventory of CO_2 is about 5×10^{16} moles, assuming 300 ppmv of CO_2 and a total atmospheric mass of 5×10^{18} kg, or 1.7×10^{20} moles (1 mole is Avogadro's number of CO_2 molecules, 6×10^{23}). Hence, accumulating the amount of CO_2 now present in the atmosphere would require 10^4 years, and accumulating 300 times that amount would require 3×10^6 years.

explosion. During that time, one finds evidence for at least two glaciations, one at about 730 Ma and a second at about 610 Ma, in which glacial deposits were formed on all seven continents.[11,12] Furthermore, some of these continents appear to have been located in the tropics at the time that they were glaciated. The glacial record is particularly well preserved in Australia, which is thought to have been straddling the equator at that time (figure 5.3). These deposits, along with similar ones in Namibia (on the west coast of Africa), tell us that Earth's climate at that time was markedly colder than it has been at any time in the more recent past.

How secure is the evidence for such low-latitude glaciation? To answer this question, we must understand how geologists determine past continental locations. In general, this is a difficult process. The continents continually drift around the globe as a consequence of plate tectonics. A variety of different types of evidence indicate that all of the continents were assembled into a giant supercontinent, *Pangea*, about 200 million years ago. This evidence includes, among other data, the similar shoreline shapes of the east coast of South America and the west coast of Africa—just one of the indications that these two continents were joined in the past.

As one goes back before 200 Ma, it becomes increasingly difficult to trace plate movements directly. However, another approach remains useful—that of *paleomagnetism*. As most people who have used a compass are aware, the Earth has a magnetic field that is shaped more or less like the field of a bar magnet (figure 5.4). Earth's magnetic field is not symmetric: the magnetic north and magnetic south poles are offset from the geographic poles by about 11 degrees of latitude. Over time, however, the magnetic pole wanders around the geographic pole, and this small offset averages out. With this caveat in mind, one can see that at high geographic latitudes the magnetic field lines are more or less perpendicular to the Earth's surface, whereas at low latitudes the field lines are parallel to the Earth's surface. This statement remains true even if the magnetic poles switch *polarity*, as they occasionally do. Paleomagnetists take advantage of this observation to estimate at what latitude continents were located in the distant past. Iron-bearing *igneous rocks*—rocks produced by cooling of magma—become magnetized as they cool. By

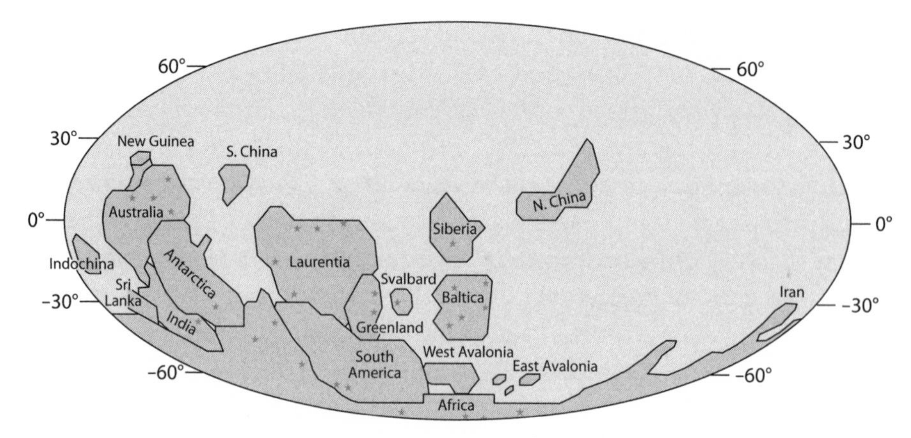

Figure 5.3 Continental reconstruction for 550 Ma.[13] (Continental reconstruction from R. Scotese. Reprinted by permission of *Nature*.)

measuring the direction of the trapped, or *remnant*, magnetic field in such rocks, geologists can determine the latitude at which they formed.

Mechanisms for Explaining Low-Latitude Glaciation

The idea that continental-scale glaciation could occur in the tropics has puzzled geologists for a long time. The Australian geologist George Williams was one of the first to notice this,[15] and so, had he wished to do so, he might have proposed the Snowball Earth hypothesis as early as the mid-1970s. Williams, however, came up with an entirely different explanation for low-latitude glaciation: he proposed that the Earth was tilted on its side at the time. Recall that Earth's current obliquity is 23.5° and that it varies by ± 1 degree every 41,000 years. It can be shown mathematically that for obliquities higher than about 54° the poles become the warmest part of the planet (averaged over the year) and the equator becomes the coldest. This, of course, is just the opposite of the situation today. If Earth's obliquity was $>54°$ in the past, reasoned Williams, then it would be only natural to find glacial deposits at low latitudes.

Williams's hypothesis was attractive for two reasons: (1) it explained the evidence for low-latitude glaciation, and (2) it also explained how

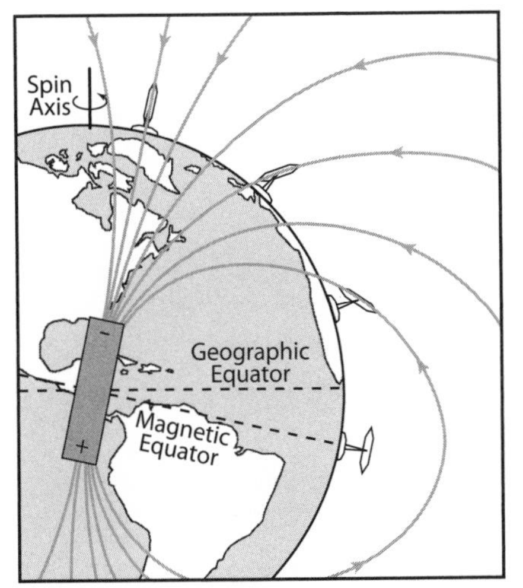

Figure 5.4 Schematic diagram of the Earth's dipole magnetic field.[14] (Courtesy of Adam Maloof.)

marine life could have made it through the glaciations. In Williams's model, the polar oceans remained ice-free, so photosynthetic algae would have had no trouble surviving these episodes. That is a good thing, as the fossils of such algae are seen both before and after the glaciations. However, Williams's model also had a serious problem: How could the Earth be "untilted" down to its present obliquity? Starting the Earth out at a higher obliquity is physically possible, as it could have been knocked that way by a large impact as the Earth was forming. The planet Uranus has an obliquity of 98°, so we know that this type of spin orientation is possible. Getting the Earth back down to low obliquity, though, is more difficult. Another giant impact by a body the size of Mars might do it, but that would have almost certainly destroyed all life on Earth, in addition to melting all surface rocks and thus completely resetting their ages. This clearly did *not* happen, and so such a mechanism can be ruled out. A kinder, gentler mechanism for changing the obliquity has been proposed, however.[16,17] If the polar ice sheets advance and retreat in the right phase with respect to the normal, Milankovitch-type variations of the obliquity, then the obliquity itself can undergo a long-term change. The mechanism has been termed *climate friction*. However, this mechanism has since been studied more carefully by some very competent mathematicians,[18] and it now appears that it does not work, given plausible assumptions about ice sheet thicknesses and about the timing of ice sheet advances and retreats relative to the obliquity oscillations.

Another possible explanation for low-latitude glaciation has been proposed by William Hyde of Texas A&M and his colleagues.[13] These authors constructed a numerical model that predicted that an ice-covered Australian continent could have have existed in the tropics even if the surrounding oceans were ice-free. In their model, glaciers formed up in the mountains where the temperatures were colder and then flowed outward toward the continental margins faster than they could melt. This model, which detractors have termed the "Slushball Earth" model, could also explain how the photosynthetic algae survived the glaciations, as the tropical oceans would never have frozen over. For this reason, and because it appears less extreme than the alternatives, it is favored by a considerable number of modern geologists. It implies, however, that the Neoproterozoic Earth remained poised on the brink of ice-albedo instability (figure 5.2) without ever quite going over the edge. To a climatologist like me, this seems inherently unlikely. There is also little or no evidence for high mountains in Australia during the Neoproterozoic, which may be an even more telling objection. And there is additional geologic evidence, discussed below, that favors a more drastic explanation.

Snowball Earth

If one rejects both the high obliquity hypothesis and the slushball hypothesis for low-latitude Neoproterozoic glaciation, one eventually arrives at a startling third possibility: our planet's entire surface may have been covered with ice at that time, creating a "Snowball Earth." The term "Snowball Earth" was coined by Joseph Kirshvink from Caltech.[19] Kirshvink and his student Dawn Sumner studied *glacial varves* from the Elatina formation in South Australia. Varves are layered sediments that are deposited seasonally in lakes that freeze over during the wintertime. The varves in the Elatina formation, which were originally flat, have been bent, or folded, by tectonic forces since the time that they formed.

By demonstrating that the remnant magnetic field lines had been bent along with the varves, Kirshvink and Sumner were able to prove that the magnetic field was indeed primary, that is, it was established when the varves formed and was not reset at some later time. In describing these results at a scientific meeting at UCLA in 1990, Kirshvink pointed out, half-jokingly, that it appeared as if Earth had been a gigantic snowball at that time, and he showed a slide of a big, ice-covered Earth. The term "Snowball Earth" stuck, although it was not until several years later that other scientists picked up Kirshvink's idea and propelled it into the mainstream of climatological thought.

Although Kirshvink proposed the Snowball Earth hypothesis originally, the geologist most responsible for lending it scientific credibility is Paul Hoffman of Harvard University.[12,20] Hoffman performed extensive field work in Namibia, along the southwest coast of Africa, where the geologic record of the Neoproterozoic glaciations is clearly exposed. The Namibian rocks contain two easily identifiable diamictite layers, along with dropstones and other evidence for glaciation. Dropstones are rocks, often the size of your fist or larger, which are found in otherwise smoothly laminated marine sediments. Rocks can't fly, and there was no one around at that time to throw them far out into the ocean, so researchers deduce that they must have been carried by icebergs which melted and deposited the rocks that they had picked up during their journey (as glaciers) over the continents. The lower and upper diamictite units in Namibia represent the 730- and 610-Ma glaciations, respectively.

The Namibian glacial deposits, as well as corresponding ones found in Australia and elsewhere, are remarkable in one particular respect: they are covered with a thick layer of carbonate rock. The carbonates in Namibia are extremely thick—several hundred meters. The bottommost few meters of the carbonate layer are also very fine-textured. To a sedimentologist like Paul Hoffman, this suggests that they formed rapidly. Such *cap carbonates*, as they are called, are not seen in glacial deposits formed more recently during the Phanerozoic. Indeed, they appear at first glance to be completely bizarre. On the present Earth, carbonate rocks form mainly at low latitudes where the water is warm. This is a consequence of the fact that carbonate minerals, e.g., calcite ($CaCO_3$), are less soluble in warm water. Glacial diamictites, however, are cold-weather deposits that are usually found at high paleolatitudes. But in the Neoproterozoic outcrops in Namibia, Hoffman found carbonates lying directly on top of glacial layers. What exactly was going on?

Hoffman's key insight was to recognize that these cap carbonates were, in reality, the "smoking gun" for Snowball Earth. Helped along by discussions with his geochemist colleague, Daniel Schrag, Hoffman realized that cap carbonates are actually just what one would expect to have formed in the aftermath of a Snowball Earth. As discussed earlier in the chapter, once the ocean surface froze, silicate weathering on the continents should have ground to a halt, and CO_2 released from volcanoes should have accumulated in the atmosphere. After a few million years, or maybe less (see below), the greenhouse effect would have become large enough to melt the ice, and the Earth would have rapidly deglaciated. At this point, the atmosphere should have contained a large amount of CO_2, and the surface should have been dark, like today, because the ice sheets were now gone. Hence, the climate should have become extremely warm—50°C or more in Hoffman's model—and weathering of both silicate and carbonate rocks on the continents would have been intense. The rapid weathering would have continued until the excess atmospheric CO_2 that had built up during the glaciation was put safely back into carbonate sediments. These postglacial sediments are the cap carbonates that we see today. Their existence provides a powerful argument that Snowball Earth really did occur.

Despite the success of the Snowball Earth hypothesis, questions still remain. Perhaps the biggest one is: how did the photosynthetic algae survive? We know that they did because we find the same types of algal microfossils both below and above the glacial sediments. Respect for this observation helped spawn the high-obliquity and slushball Earth hypotheses discussed earlier. And while Hoffman's Snowball Earth model explains most of the geological evidence quite nicely, it is less clear that it can account for the survival of photosynthetic organisms. Hoffman and Schrag[20] argue that pockets of exposed liquid water may have existed near cracks in the sea ice and in volcanic regions such as Iceland where lots of geothermal heat is released. But the ice in their *hard Snowball* model is more than a kilometer thick, even in the tropics, and so it is not obvious that large cracks would have existed. A volcanic island like Iceland might serve as a refugium for photosynthetic organisms, but this depends critically on where the volcanoes were located and whether they were connected to a source of liquid water. Iceland might work in this respect because it is essentially a raised part of the Mid-Atlantic ridge. By contrast, continental geothermal hotspots, like Yellowstone National Park in the United States, would have quickly dried up, because the warm water would have evaporated away, and it would not have been replaced by rainwater.

Another, in my view more likely, hypothesis for explaining how photosynthetic algae survived Snowball Earth is the *thin-ice model* proposed by Christopher McKay.[21] Chris is a former colleague of mine from NASA Ames who was another protégé of Jim Pollack. Chris does fieldwork in Antarctica and has studied the ice-covered lakes that can be found in the "dry valleys" west of McMurdo Sound. Surprisingly, these lakes support a thriving photosynthetic biota, including cyanobacteria, that live beneath as much as 5 m of (very clear) perennial ice. Chris used these observations to develop his own hypothesis for what actually happened during Snowball Earth episodes. In his alternative scenario, the ice in the tropics was relatively thin—only 1–2 m depth. This is thin enough to have allowed a significant amount, roughly 10 percent, of the incident sunlight to penetrate through it—more than enough to keep photosynthesis active beneath the ice. The sunlight, in turn, kept the ice thin because the energy it deposited in the ocean must have been carried by back to the surface by conduction. Conduction of heat would have limited the ice thickness in the hard Snowball model as well, but in that case no sunlight would have penetrated, and so only the geothermal heat would have needed to be transmitted. Hence, the ice in the hard Snowball model is at least a kilometer thick everywhere.*

*The average solar flux hitting the Earth's surface (from chapter 3) is 1365 $W/m^2 \div 4$, or about 341 W/m^2. (We'll neglect the amount reflected by clouds, which may not have been widespread on Snowball Earth.) The flux in the tropics is about 20 percent higher than this, or $\sim$410 W/m^2. Assume that 10% of this energy, or 41 W/m^2, gets through the ice. The geothermal heat flux is about 0.09 W/m^2, so the ratio is $41/0.09 \cong 500$. The ice thickness would have been inversely proportional to the transmitted heat flux, so the ice would have been 500 times thinner in the thin-ice model, i.e., 2 m as compared to 1000 m for the hard Snowball model.

As with any new hypothesis, debate is still active on this topic. Jason Goodman and Raymond Pierrehumbert of the University of Chicago pointed out that thick sea ice, or *sea glaciers*, formed near the poles, should have flowed toward the equator, perhaps preventing the thin ice from remaining stable.[22] This ice from high latitudes would have been more difficult to melt because it would have been covered with highly reflective snow. But dust in the snow, blown in from exposed continental surfaces, might have darkened it and allowed the snow to melt.[24] Some low-latitude ocean basins, like the Mediterranean today, may also have been shielded from sea glacier flow by surrounding land masses.[24] So, while debate continues over the thin-ice model, and over the Snowball Earth hypothesis itself, the good news is that life does seem to be able to survive such events. Carl Sagan would likely have been pleased with this thought. The very real possibility of Snowball Earth episodes is not a reason to be pessimistic about the search for life on other planets.

• Part III •

Limits to Planetary Habitability

• • •

Wherein we explore what went wrong with our neighboring planets, Venus and Mars, and use this information to propose a general theory of habitable planets around different types of stars . . .

· *Chapter Six* ·

Runaway Greenhouses and the Evolution of Venus' Atmosphere

Why is Venus too hot, Mars too cold, while Earth is just right for life? This has been called the "Goldilocks problem" of comparative planetology. (The name was coined by Lynn Margulis, Lovelock's co-conspirator in the Gaia hypothesis.) The obvious answer to this question is that Venus formed too close to the Sun, Mars formed too far away, while Earth formed at just the right distance. And, to be sure, this is a large part of the answer. However, other factors appear to have been at work as well. As we discuss in chapter 8, Mars may actually have been habitable early in Solar System history, despite its large distance from the Sun and despite the fact that the Sun was less bright at that time. And Venus may also have had oceans, like the Earth, although the evidence in this case is more circumstantial. So let's look in more detail at the history of these two planets and try to understand more precisely how they evolved to their present states.

A good place to start, because it is the part of the problem that is better understood, is to try to reconstruct the history of Venus. Venus is often called a "sister planet" to Earth because it is both our closest neighbor (orbital distance = 0.72 AU) and the nearest in mass (M_{Venus} = 0.81 M_{Earth}). But, if one were to visit its surface, it would not seem "sisterly" at all. The mean surface temperature of Venus is a hellish 460°C (860°F), which is hot enough to melt lead. This is well above the critical temperature* for water, 374°C, which means that liquid water could not exist on Venus' surface even if it were present in abundance (which it is not). Hence, any possibility for the existence of life as we know it is ruled out. The atmosphere itself is extremely dense; its surface pressure is 93 bars,** and its composition is mostly CO_2 and N_2, along with trace amounts of SO_2 (sulfur dioxide), H_2O, and CO (carbon monoxide). (See table 6.1.) The SO_2 is photolyzed by solar ultraviolet light high up in Venus' atmosphere, and some of it is converted to H_2SO_4 (sulfuric acid). This sulfuric acid condenses out to form highly reflective cloud particles, causing Venus to appear as a featureless, white ball at visible wavelengths. At UV wavelengths, unknown substances in the clouds absorb radiation, giving the planet the streaked appearance seen in figure 6.1 (see color section).

The History of Water on Venus

How did Venus arrive at its present uninhabitable state? The key to answering this question is to understand its almost complete lack of water. There can be no liquid water whatsoever on the surface, and even Venus' atmosphere contains precious little water vapor. The best estimate, based on measurements from the Pioneer Venus mission in the late 1970s and on modern ground-based spectroscopy,[1] is that Venus has only about 30 ppm of water vapor in its lower atmosphere. If one computes the total mass of this water vapor, it amounts to only about 1 part in 10^5 of the water present at Earth's surface (1.4×10^{21} kg), most of which is in the oceans.

*The critical temperature for any substance is the temperature above which there is no longer any discernible difference between the liquid and vapor phases. For pure water, this is marked by a specific point on a diagram of pressure versus temperature. The corresponding critical pressure for pure H_2O is 22.06 MPa, or 220.6 bar. (See next footnote for a discussion of pressure units.) Seawater, being salty, has a slightly more complicated behavior, but its critical temperature is still in the vicinity of 400°C.

**The standard unit of pressure in the System Internationale version of the metric system is megapascals (MPa). The average sea-level pressure of Earth's atmosphere is 0.1013 MPa. But this is a bit cumbersome, and so planetary scientists (of which I am one) often use a related unit of pressure, the bar. 1 bar = 0.1 MPa, so Earth's surface pressure is just over 1 bar, and Venus' surface pressure is approximately 93 times that of Earth.

TABLE 6.1
Composition of Venus' Atmosphere

Gas	*Percent by volume*
CO_2	96.5
N_2	3.5
SO_2	0.015
Ar	0.007
H_2O	0.003
CO	0.0017

As mentioned in chapter 2, the 1960s and 1970s witnessed a lengthy debate as to whether Venus started out dry initially or whether it started out wet and then lost its water at some later time. John Lewis's equilibrium condensation model for planetary formation predicted a dry origin for Venus because the material from which planets formed was assumed to be derived locally, and because hydrated silicates were not predicted to form at Venus' orbital distance. But, as we discussed earlier, this model was later supplanted by other theories that suggested that all of the terrestrial planets obtained their water from planetesimals that originated farther out in the Solar System, near where the asteroid belt is found today.

An additional piece of observational evidence helped clinch the argument for a wet origin for Venus. The key data came from the Pioneer Venus spacecraft, which dropped a probe into Venus' atmosphere in 1977. The probe contained an instrument called a *mass spectrometer*, which was able to measure the deuterium/hydrogen ratio in Venus' atmosphere. Recall from chapter 2 that deuterium is an isotope of hydrogen that contains both a proton and a neutron in its nucleus. Recall also that the D/H ratio of comets appears to be about twice that of Earth's oceans—an observation that we used to rule out comets as a major source of Earth's water. By comparison, the measured D/H ratio in Venus' sulfuric acid clouds, based on data from the probe mass spectrometer, is about 120 times that on Earth![2] This spectacularly high value is supported by the nearly simultaneous detection of a mass 2 ion in the ionosphere,[3] which was later determined to be D^+. Since then, the D/H ratio in Venus' atmosphere has been measured even more accurately using spectroscopic observations from Earth,[4] which yield an estimate of 150 times the terrestrial value.

These measurements of Venus' D/H ratio were very exciting to planetary scientists, and they went a long way toward resolving the dispute about whether Venus started out wet or dry. Because both Venus and Earth are thought to have received the bulk of their water from the asteroid belt region, their initial D/H ratios should have been approximately equal. But if Venus started out with considerably more water than it has now, its D/H ratio could have increased as the water was lost. The mechanism by which this happened is termed the *runaway greenhouse,*[5] and its details are described below. In a nutshell, though, Venus' surface was so hot that its atmosphere was rich in water vapor, or steam. Some of this water vapor made its way into the upper atmosphere, where it was broken apart, or *photodissociated*, by ultraviolet photons from the Sun. The hydrogen escaped to space, while the oxygen remained behind and reacted with reduced materials (especially ferrous iron) in the planet's crust. Some of the oxygen may have been dragged off to space along with the escaping hydrogen, but we will ignore that complexity here. The key point is that because hydrogen atoms are lighter than deuterium atoms, they should have escaped more quickly, causing the water that remained behind to become more deuterium-rich. Hence, the measurement of a high D/H ratio in Venus' present atmosphere strongly supports the idea that early Venus was water-rich.

Trying to predict exactly how much water Venus had in its distant past based on this one measurement is a difficult proposition. If Venus started out with exactly Earth's D/H ratio, and if only hydrogen atoms escaped, the measured 150 times enhancement in Venus' D/H ratio compared to Earth's would require that Venus started out with at least 150 times more water than it has today. That is still only a small fraction ($<$0.2 percent) of Earth's oceans, so such a model would still imply that early Venus was relatively dry. But if some deuterium was lost along with the escaping hydrogen, as was most likely the case, Venus' initial water endowment could have been much larger—perhaps equivalent to that of Earth.[6] It is difficult to say much more than this, as the results depend sensitively on the relative rates at which H and D atoms escaped. Furthermore, the impact of even one large, icy comet could have reset the D/H ratio,[7] rendering the details of the analysis invalid. So we can say fairly confidently that Venus started out wet, but we cannot say exactly how much water was initially present.

The Classical Runaway Greenhouse Effect

If this story about Venus' water (or lack thereof) is correct, then we can use Venus to derive an estimate for the inner edge of the liquid water habitable zone around the Sun. Recall that we introduced this concept in chapter 1, and we will return to it again in chapter 10. Indeed, from my own viewpoint, the single most interesting thing about Venus is what it tells us about this habitable zone boundary. Evidently, Venus' orbital distance of 0.723 AU is too close to the Sun for a planet to retain its water. If one prefers, one can think of this limit in terms of the flux of radiation from the Sun. The solar flux obeys an *inverse square law*, that is, its intensity decreases by 1 over the square of the distance from the Sun. Hence, the flux at Venus' orbit is higher than that of Earth by $(1\ AU/0.723\ AU)^2 \cong 1.91$. The flux at Earth's orbit is about 1365 W/m^2, so the flux at Venus' orbit is 2607 W/m^2. Based on what we can see has happened to Venus, planets that receive more than this amount of radiation from their parent star are not likely to be habitable.

Theoreticians like to do more, however, than to simply rely on observations. They also like to make numerical models, because doing so allows them to understand the mechanisms by which things occur. The first researchers to do this for the Venus runaway greenhouse were S. I. Rasool and Catherine deBergh of NASA's Goddard Institute for Space Studies in New York, who wrote a classic paper on this topic in 1970.[5] The basic fundamentals of their theory were captured effectively in a slightly simpler model published a few years later by Richard Goody

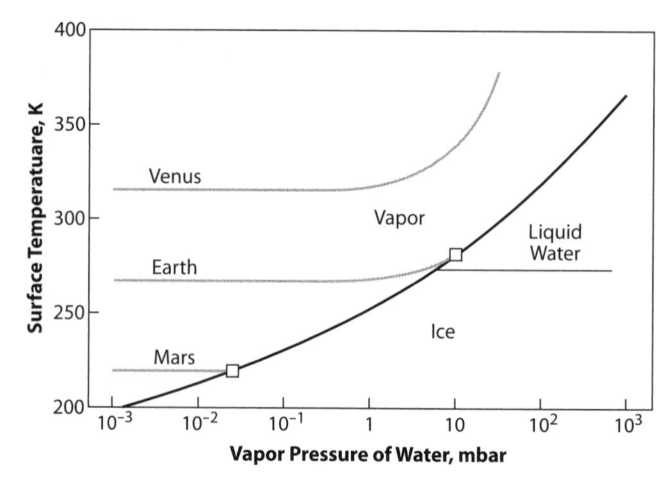

Figure 6.2 Diagram illustrating the "classical" runaway greenhouse effect. In this simulation, the three terrestrial planets are all assumed to have started off as airless bodies and to have outgassed atmospheres of pure water vapor. The horizontal scale shows the surface pressure of the planet's atmosphere (10^6 dyn/cm^2 = 1 bar). The vertical scale indicates temperature in kelvins. The dashed curves show surface temperatures computed with a (very simple) climate model. The dark solid curve shows the saturation vapor pressure of water over liquid or ice. (After Goody and Walker 1972.[9] Reproduced with permission of Pearson Publishing.)

and James Walker[9] (see figure 6.2). Goody and Walker assumed that all three terrestrial planets—Venus, Earth, and Mars—started their lives as airless bodies at their current distances from the Sun: 0.72, 1, and 1.52 AU, respectively. The variation in solar luminosity with time was ignored.* Goody and Walker further assumed that volcanoes on these planets belched out atmospheres of pure water vapor. If enough water vapor was present to create a greenhouse effect, the planet's surface temperature was allowed to rise accordingly, based on a simple model of the greenhouse effect, described further below.

The results of this simple calculation are shown in figure 6.2. The three dashed curves show the surface temperature evolution of the three terrestrial planets as their volcanically outgassed, water vapor atmospheres accumulated. (Goody and Walker ignored Mercury, as I have in

*Sagan and Mullen's paper on this topic was not published until after this work appeared, so Goody and Walker may not have been aware of this complication.

this chapter, because it doesn't have an atmosphere.) The solid curve shows the *saturation vapor pressure* over water (at higher temperatures) or over ice (at low temperatures). The saturation vapor pressure is the maximum amount of water vapor that the atmosphere can hold at a given temperature. Of the three terrestrial planets, Mars was farthest from the Sun; hence, it started out the coldest, roughly 220 K, or −53°C. All three terrestrial planets were assumed to have albedos of 0.17, close to the albedo of Mars today. In Goody and Walker's model, once the surface pressure of Mars' atmosphere reached about 0.02 mbar,* water vapor began to exceed its saturation vapor pressure, and so it condensed out to form ice. This is shown by the intersection of the dashed and solid curves in the figure. And, to a first approximation, this explains what we see today: Mars is basically a frozen desert with a very thin atmosphere.

Earth, being somewhat closer to the Sun, started out a little warmer than Mars, about 270 K, or −3°C. Hence, water vapor was able to accumulate to higher concentrations before it began to condense. In Earth's case, enough water vapor accumulated to produce a slight greenhouse effect. That is why the curve for Earth bends slightly upward as one follows it toward the right. As a result, when water vapor began to condense on Earth, the surface temperature was >0°C, and so it formed liquid water instead of ice. Once again, to a first approximation, this is what we observe on Earth. Earth's large inventory of surface water is almost all contained in its oceans.

The most interesting part of the figure, though, is the predicted surface temperature evolution curve for Venus. Because Venus is even closer to the Sun than is Earth, its surface started out hotter still, about 315 K, or 42°C. This allowed large amounts of water vapor to accumulate in Venus' atmosphere—enough so that the greenhouse effect became large well before saturation was reached. Thus, the surface temperature curve for Venus bends sharply upward at pressures above about 0.01 bar, and so it never intersects the saturation vapor pressure curve. In colloquial terms, Venus' surface temperature "runs away" in this calculation. (This is the origin of the term "runaway greenhouse.") According

*1 mbar = 1 millibar = 10^{-3} bar.

to this calculation, all of the water released from Venus' interior would have remained in the vapor phase, forming a dense steam atmosphere. If one assumes that Venus started out with the same amount of water as did Earth, the surface pressure of this atmosphere would have been about 270 bars. That's about 3 times the surface pressure of Venus' present atmosphere, which is already extremely dense. Once in the atmosphere, the water vapor would have been lost by the mechanism described in the previous section: photodissociation followed by escape of hydrogen to space. The net result would be a dry planet similar to the one we observe today.

An Alternative Runaway Greenhouse Model

The calculation shown in figure 6.2 is ideal for illustrating the basic concept of the runaway greenhouse. However, when examined more closely, it is inaccurate in a number of important details. The numerical treatment of the greenhouse effect was simplistic in these early models, as the absorption coefficient of water vapor was assumed to be constant with wavelength—the so-called "gray atmosphere" approximation. That is not true in reality: as mentioned in chapter 4, water vapor absorbs strongly at some wavelengths and less strongly at others, and this variation needs to be taken into account in order to accurately compute the greenhouse effect. Convection was also neglected in these early runaway greenhouse models. This leads to a significant overestimate of the greenhouse effect, as most of the heat transported upward from the surfaces of both Earth and Venus today is carried by convection, not radiation. These early models also overlooked the fact that solar luminosity was lower in the past and that planets probably begin to form atmospheres as they themselves form, rather than starting their lives as airless bodies. As discussed in chapter 2, current models of planetary formation involve impacts of large volatile-rich planetesimals. When these bodies collided with the growing planets, they should have been largely vaporized, and their volatiles (including water) should have been released directly into the atmosphere.[10] Hence, more recent planetary formation models suggest that both Earth and Venus may have been enveloped in dense steam atmospheres during their formation process.[11-13]

One of the projects that I worked on with Jim Pollack during my years as a research scientist at NASA Ames was to redo the calculation of the runaway greenhouse. Jim had already done one such calculation ten years earlier,[14] and I myself had gone out to Ames partly to learn from him how to do this. We decided to approach the problem from an entirely different perspective. Instead of starting with airless terrestrial planets circling the Sun, we asked the question: What would happen if you started with a fully formed Earth and then gradually pushed the planet closer to the Sun?* This question bypasses the details of how planets like Earth form, but it relates directly to the issue of where Earth-like planets might stably exist. If one were to somehow do this, the solar flux hitting the Earth would increase, following the inverse square law, and this would make the planet's surface hotter. The water vapor feedback should kick in as well, just as it did in the classical runaway greenhouse calculation described above. The destabilizing effect of this positive feedback makes it difficult to find numerical solutions. We got around this problem, however, by using a simple trick. Instead of specifying the solar flux and calculating the planet's surface temperature, as one would normally do, we solved the inverse problem: we specified the surface temperature and calculated the solar flux needed to sustain it. This allowed us to find solutions even when the planet's climate was unstable because of the (runaway) water vapor feedback.

The trickiest part of doing this type of runaway greenhouse calculation is to properly account for the effect of water vapor condensation on the vertical structure of the atmosphere. To do this, we took advantage of a clever analysis[15] that was actually published a year prior to the one by Rasool and deBergh. The author of this paper was Andrew Ingersoll, a renowned planetary scientist now at the California Institute of Technology in Pasadena, California. Ingersoll realized that the latent heat

*In the days of the great German physicists like Einstein, this type of hypothetical question would have been termed a *Gedankenexperiment*, or thought experiment. Einstein solved such problems in his head. Not being as smart as Einstein, we were forced to solve our problem on the computer.

stored by water vapor becomes increasingly important as water vapor becomes more abundant in a planet's atmosphere. *Latent heat* is the energy that must be added to liquid water in order to transform it to a vapor, or to ice to transform it into a liquid. When water vapor condenses to form clouds, the latent heat is returned to the surrounding atmosphere, warming it up. Even today, latent heat release is extremely important in meteorology, particularly in tropical climates. Latent heat release is perhaps most widely known for being the energy source that powers hurricanes.

In our runaway greenhouse model, we allowed the planet's surface to become gradually warmer, and we used Ingersoll's methodology to determine how the atmospheric structure should change. As the surface warms, the abundance of water vapor in the atmosphere increases dramatically because of the increase in saturation vapor pressure, and the amount of latent heat released by condensation increases dramatically as well. This heat warms the upper troposphere, causing the temperature to decrease less rapidly with altitude. (The *troposphere* is the lowest part of the atmosphere in which clouds can form and convection occurs.) As a result, the troposphere expands upward, as shown in figure 6.3a. Currently, the *tropopause* (the top of the troposphere) is located at an altitude of about 17 km in the tropics and about 10 km near the poles. As the atmosphere warms up, though, the tropopause in our model rose to roughly 150 km altitude. This drastically increased the amount of water vapor that could make it up into the stratosphere (figure 6.3b). Or, to put it another way, the efficiency of the tropopause *cold trap* was greatly reduced. A cold trap in an experimental apparatus is a place where flowing air, or some other gas, is cooled in order to cause its moisture to condense out. In Earth's atmosphere, the tropopause acts as a cold trap for water vapor, thereby keeping the stratosphere relatively dry. If the troposphere moves up, though, its drying effect is greatly reduced, even if its temperature remains essentially constant.*

*The water vapor *mixing ratio* at the cold trap is equal to its saturation vapor pressure divided by the ambient atmospheric pressure. If the cold trap temperature remains constant, then the saturation vapor pressure also remains constant. But the atmospheric pressure decreases as the cold trap moves upward, and hence the mixing ratio of water vapor increases.

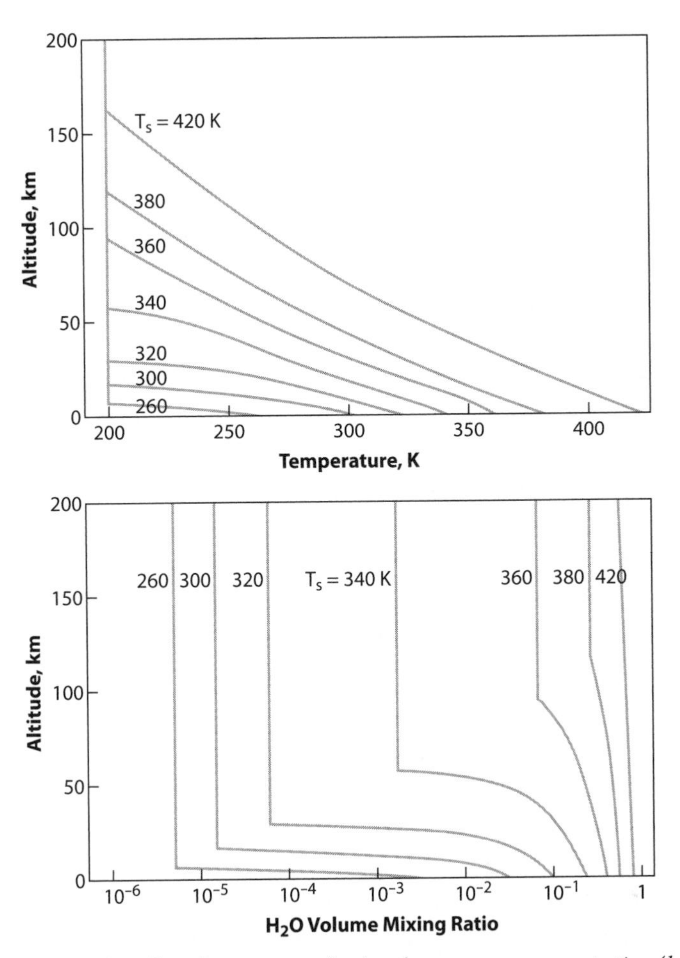

Figure 6.3 Vertical profiles of temperature (*top*) and water vapor concentration (*bottom*) at different surface temperatures. (Redrawn from Kasting 1988.[16] Reproduced by permission of Elsevier Publishing.)

Why should it matter, you might ask, if water vapor makes its way into a planet's upper atmosphere? The answer is that this can render an Earth-like planet uninhabitable, for two reasons. First, the planet's ozone layer can be destroyed by the by-products of water photolysis. Water in the stratosphere is broken apart by solar ultraviolet light to form atomic hydrogen (H) and hydroxyl radicals (OH). Both of these compounds can destroy ozone if they are present in sufficient concentrations. This, of course, would be dangerous to modern Earth life, including humans, because ozone shields the surface from solar near-UV radiation. And, second, a wet stratosphere also allows hydrogen to be lost rapidly to space. This is critical from the standpoint of planetary habitability because if hydrogen is lost, then water would also be lost. On Earth today, the stratosphere contains only a few parts per million of water vapor, and so the hydrogen escape rate is slow. But on a planet with a much wetter stratosphere, the hydrogen escape rate could be quite large,[17] and so an ocean of water could be lost in only a few hundred million years.[18]

The final step in our alternative runaway greenhouse calculation was to calculate the absorbed solar energy flux and the outgoing infrared energy flux for each of our model atmospheres. (Remember that we were essentially starting from the answer—the surface temperature and atmospheric structure—and working backward to find the solar flux that supported this structure.) To do this, we used a climate model similar in principle to the one used earlier by Pollack and Sagan.[14] The results of the calculation are summarized in figure 6.4. In this figure, we have plotted the effective solar flux—the flux relative to that at Earth today—on the horizontal scale, as if we had done a normal forward calculation. The top panel shows the mean surface temperature of the planet. This rises slowly at first and then jumps to extraordinarily high values, approximately 1600 K (1300°C), when the solar flux reaches 1.4 times the value at Earth's orbit. This temperature is well above the critical temperature for water, 647 K; thus, to the right of this point, all of Earth's water would be present in the atmosphere as steam. The solar flux at which this occurs corresponds to a decrease in Earth's orbital distance from 1.0 AU to about 0.85 AU.* As pointed out earlier, the flux at Venus' orbit today is ~1.91 times that of Earth; hence, present Venus is well above the runaway greenhouse limit, according to this calculation. In reality, clouds (which were not included explicitly in the model) might help keep such a planet cool by reflecting incoming sunlight. So one should probably regard this value as the minimum solar flux at

*According to the inverse square law, the solar flux at 0.85 AU would be higher than that at 1 AU by a factor of $1/0.85^2 \cong 1.4$.

which a true runaway greenhouse, i.e., complete ocean evaporation, could occur.

It is not necessary to have a true runaway greenhouse, though, for a planet to lose its water. All that is required is for the planet's stratosphere to become wet. In our model, this happened at a solar flux that is only about 1.1 times higher than the present flux at Earth, as shown by the bottom panel in figure 6.4. The corresponding orbital distance is 0.95 AU (because $1/0.95^2 \cong 1.1$). Hence, the Earth might have become uninhabitable had it formed only 5 percent closer to the Sun. This calculation may be overly pessimistic for reasons mentioned above: clouds, which were neglected in this model, might cool the surface, thereby slowing the loss of water. But it shows that the boundary at which water could potentially be lost is not that far inside Earth's present orbit.

Evolution of Venus' Atmosphere

Let's now go back to where we started at the beginning of this chapter and see if we can understand how Venus' atmosphere may have evolved. Recall from chapter 3 that the Sun was approximately 30 percent less bright at the time when the Solar System formed. So the solar flux hitting Venus at that time should have been about 1.3 times higher than that hitting Earth today (because $0.7 \times 1.91 \approx 1.3$). This is higher than the critical flux for water loss, suggesting that Venus should have started losing its water almost immediately. However, this flux is lower than the runaway greenhouse threshold (1.4 times Earth's flux) at which the oceans evaporate entirely. Hence, Venus could conceivably have begun its life with liquid oceans on its surface. Whether it had oceans or not depends on how much water it received during its formation. If Venus had started with as much water as Earth originally, then it would likely have had an ocean for some time. If its initial water endowment was somewhat less, then its surface may have always been dry. In either case, our model predicts that Venus should have lost its water over time by photodissociation and hydrogen escape.

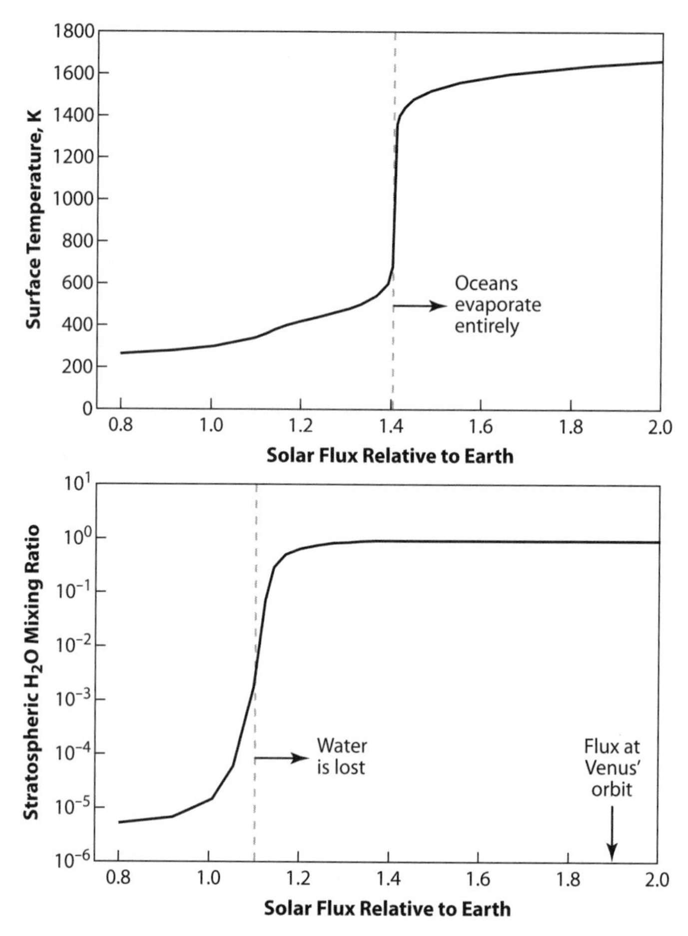

Figure 6.4 Diagram illustrating what would happen if Earth were slowly moved inward toward Venus' orbit. The top panel shows the global mean surface temperature as a function of the effective solar flux (the value relative to that at Earth's orbit today). The bottom panel shows the concentration of H_2O in the stratosphere. Hydrogen escape becomes rapid, and the oceans are lost, when the stratosphere becomes water-dominated. (Redrawn from Kasting 1988.[16])

Once the water was lost, the subsequent evolution of Venus' atmosphere is fairly easy to understand. Without liquid water, weathering of silicate rocks on the planet's surface would have been slow, and CO_2 emitted from volcanoes should have begun to accumulate in Venus' atmosphere. The amount of CO_2 in Venus' atmosphere today, 90 bars, is approximately the same as the total amount of CO_2 tied up in carbonate rocks in Earth's crust. So it appears as if Venus' entire inventory of CO_2 may be present in its atmosphere. The observed high abundance of SO_2, 150 ppm (table 6.1), is also easy to explain. Sulfur gases, being soluble in water, are removed from Earth's atmosphere by rainfall and by direct dissolution in the oceans. Consequently, most of Earth's sulfur is present either as dissolved sulfate in seawater or as sulfate minerals in rocks. Once Venus had dried out, these removal processes would not have operated, and volcanic SO_2 would have accumulated in its atmosphere, just as did CO_2.

To be fair, I should point out that other researchers have proposed alternative explanations for how Venus' atmosphere evolved. Mark Bullock and David Grinspoon suggested that the atmospheric pressure on present Venus is buffered by chemical interactions between the atmosphere and the planet's surface.[19] In their model, atmospheric CO_2 was assumed to be in equilibrium with carbonate minerals in Venus' crust.* In other words, they assumed that CO_2 is constantly being exchanged between the atmosphere and the surface, and the rate of exchange is a function of temperature and pressure. Curiously, though, their model predicted that the present atmosphere of Venus is unstable. That is because carbonates are favored at lower temperatures, whereas gaseous CO_2 is favored at higher temperatures. Thus, the hotter the surface gets, the more CO_2 goes into the atmosphere, and this makes the surface still hotter. Conversely, if the surface cools, atmospheric CO_2 goes into the rocks, making the surface still colder.

This is a classic positive feedback loop, similar to the one involved in the runaway greenhouse. Hence, if this model is correct, Venus' surface pressure is poised precariously at an unstable equilibrium point. This seems unlikely, however, because the surface pressure ought to "run away" in one direction or the other. Furthermore, a simpler explanation is available: chemical reaction rates between gases and dry rocks are exceedingly slow. (That is why the hydrated silicates predicted by Lewis'

*Specifically, they considered the reaction $CaSiO_3 + CO_2 \leftrightarrow CaCO_3 + SiO_2$. Here, $CaSiO_3$ is the mineral wollastonite (a simple silicate mineral), $CaCO_3$ is calcium carbonate (limestone), and SiO_2 is silica, or quartz. This reaction is called the *wollastonite equilibrium*.

equilibrium condensation model should not have actually formed.) Such reactions can be further inhibited by the buildup of weathering products on the surfaces of mineral grains. Once a silicate grain becomes coated with calcium carbonate, no further reaction with atmospheric CO_2 should occur. On Earth, the products of weathering are removed by liquid water, but on Venus this would not happen. So it seems simpler to assume that surface interactions are indeed slow[20] and that Venus' atmosphere has served as a simple collector for CO_2 that was released from its interior.

Control of atmospheric SO_2 by surface-atmosphere interactions is much more plausible, as SO_2 is present in much lower concentrations (~150 ppm), and so less surface rock would be needed to control its concentration. In a subsequent paper,[21] Bullock and Grinspoon suggested that SO_2 in Venus' atmosphere is buffered by the equilibrium reaction $SO_2 + CaCO_3$ (calcite) $\leftrightarrow CaSO_4$ (anhydrite) + CO (carbon monoxide). This reaction could proceed both ways (as implied by the concept of thermodynamic equilibrium) if both calcite and anhydrite were present on Venus' surface. The equilibrium SO_2 concentration predicted from this reaction, though, is only about 1 percent of the observed SO_2 concentration in the atmosphere.[20] Hence, if this reaction occurs, it probably goes in only one direction, from left to right, and continued volcanic outgassing of SO_2 would be needed to maintain its atmospheric concentration.

Alternatively, George Hashimoto and Yutaka Abe proposed that Venus' SO_2 is buffered by reaction with pyrite, FeS_2.[20] The reaction in this case is $3FeS_2$ (pyrite) + 16 $CO_2 \leftrightarrow Fe_3O_4$ (magnetite) + $6SO_2$ + 16CO. As discussed in chapter 4, pyrite is one of the reduced minerals whose presence (in detrital form) is considered an indicator of low O_2 levels on the early Earth. Although we do not know the exact O_2 concentration in Venus' lower atmosphere, it is thought to be well below the ~1 ppm that is present in the upper atmosphere.[22] So this reaction is plausible, and it predicts an equilibrium atmospheric SO_2 concentration close to that which is observed.[20] Pyrite is also one of several minerals that might account for the high radar reflectivity of high-altitude regions of Venus' surface. (The mountaintops appear to be coated with some kind of substance that is highly reflective at radar wavelengths.) So it may be that the SO_2 in Venus' atmosphere is actively controlled by surface interactions, even if CO_2 is not.

Regardless of what exactly is happening on Venus today, the implications of this story are clear: planets like Venus that are located too close to their parent star are likely to lose their water and become uninhabitable. Thus, this part of the Goldilocks paradox has a relatively simple answer. The other part, concerning Mars, is more complicated, as discussed in chapter 8.

· *Chapter Seven* ·

The Future Evolution of Earth

Before we leave the topic of runaway greenhouses, let us briefly return to our discussion of Earth. Unlike Venus, the present Earth is in no danger of experiencing a runaway greenhouse. That's because it is far enough from the Sun that a runaway greenhouse is not possible—at least not under any forcing that is easy to imagine (more on this below). Occasionally, the question will arise as to whether we might trigger a runaway greenhouse by burning large amounts of fossil fuel and causing a big increase in atmospheric CO_2. It is conceivable, for example, that atmospheric CO_2 levels could increase from 380 ppm (today's value) to between 1400 and 2000 ppm within the next few centuries if most of the world's coal reserves are consumed during that time.[1-3] Such a CO_2 increase could trigger a global temperature increase of ~8°C (14°F) or more.[3] This would be catastrophic by human standards, as it might cause the polar ice caps to melt completely over the next several thousand years, raising sea level by as much as 80 meters. But, still, Earth would be nowhere near runaway greenhouse conditions, or even the less stringent conditions required to cause rapid water loss. Humans might be greatly inconvenienced, but life itself would not be in trouble.

High-CO_2 Atmospheres and Temperature Limits for Life

Somewhat surprisingly, even much larger CO_2 increases might not create a runaway greenhouse or cause significant water loss on present

Earth. In another study performed during my years at NASA Ames,[4] we simulated the effect of increasing the CO_2 concentration in Earth's atmosphere to Venus-like levels (100 bars surface pressure). In these calculations, the model oceans never boiled, even though the surface temperature rose to a whopping 230°C, or ~450°F. The stratosphere also stayed reasonably dry, thereby limiting the rate of water loss (although it was on the verge of becoming water-dominated at a CO_2 concentration of a few tenths of a bar). It may seem surprising that the Earth could get this hot without developing a runaway greenhouse, but it makes sense if one thinks about the problem carefully. In such a situation, the atmosphere would act like a giant pressure cooker. As any chef knows, a turkey can be cooked much more quickly in a pot with an airtight lid than in an open vessel. That is because the increase in pressure inside the airtight pot raises the boiling point of water, allowing it to cook the meat at a higher temperature. A similar phenomenon would occur if the CO_2 content of Earth's atmosphere was increased by a large amount. The surface temperature would rise because of the increased greenhouse effect. However, the surface pressure would rise even faster, according to our calculations, and so the oceans would never boil. Boiling occurs only when the vapor pressure of the liquid exceeds the pressure of the overlying atmosphere. When CO_2 is added to Earth's atmosphere, this point is never reached. We might all be cooked, just like the turkey, but Earth itself would retain its water.

The magnitude of the CO_2 greenhouse effect is limited by another factor. This is a technical point, but I mention it here because it turns out to be important as well for understanding the climate of early Mars (see next chapter). CO_2 happens to be very effective at *scattering* incident sunlight. Scattering is the term used when a particle (or molecule) redirects an incident photon without absorbing it. When the particles are much smaller than the wavelength of radiation with which they interact, as is true for air molecules interacting with visible light, the process is termed *Rayleigh scattering*. The name comes from Lord Rayleigh of England, who first described this process mathematically in the late 19th century. Rayleigh scattering is most effective at short wavelengths, which is why Earth's sky is blue. The shorter, bluer wavelengths of sunlight are scattered more effectively than the longer, redder ones.* When the total atmospheric pressure increases, the amount of Rayleigh scattering increases as well, and this increases the planet's albedo. CO_2 is also more effective at Rayleigh scattering than is N_2 or O_2 by a factor of about 2.5. Consequently, the high-CO_2 atmospheres simulated in our study[4] had higher albedos than does the present Earth, and this helped prevent the surface temperature from becoming high enough to "run away."

Life as we know it could not survive the extremely high temperatures that would prevail at the highest CO_2 concentrations. The upper temperature limit for plants and animals is about 50°C.[5, 6] Single-celled eukaryotes (organisms with cell nuclei) can survive up to somewhat higher temperatures,[5,6] ~60°C. Many of the cyanobacteria mentioned in chapter 4 can also survive up to this same temperature, and one particular cyanobacterium, *Synechococcus*, can live at up to 73°C.[7] Such organisms are found today in hot springs environments such as those in Yellowstone National Park. Some *hyperthermophilic* (extreme heat-loving) *bacteria* can survive up to 121°C.[8] Because this temperature is well above the boiling point at 1 bar pressure, these organisms are only found in submarine hydrothermal vents. Whether or not the same temperature limits would apply to life elsewhere is, of course, an open question. But they are useful benchmarks to keep in mind when one discusses the possibility of life on other worlds.

Future Solar Evolution and Lifetime of the Biosphere

While Earth thus appears to be protected against runaway greenhouse effects today, this will not remain true in the distant future. The Sun has brightened by about 30 percent since it formed, some 4.6 billion years ago, and it continues to brighten today at a rate of about 1 percent per hundred million years. If one extrapolates forward in time, the Sun should be about 10 percent brighter one billion years from now (see figure 7.1a). As discussed in the previous chapter, that is the same value of the solar flux that causes rapid water loss in models of early Venus.

*The Rayleigh cross section has a well-known $1/\lambda^4$ dependence, where λ is the wavelength of the incident radiation.

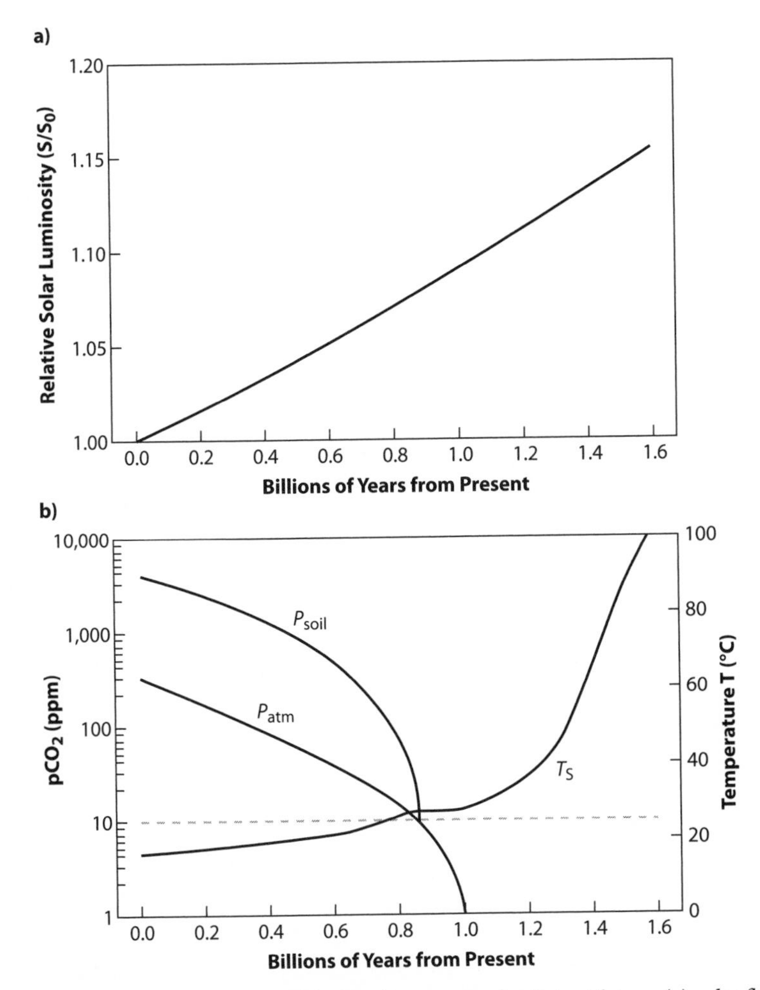

Figure 7.1 Predicted evolution of the Earth system in the distant future: (a) solar flux relative to today; (b) surface temperature, T_s, and CO_2 concentration in the atmosphere, P_{atm}, and in the soil, P_{soil}.[9] (Reproduced by permission of *Nature*.)

Hence, one might expect that Earth will follow Venus' path and lose its water as well, starting at about that time.

The future evolution of Earth's atmosphere is complicated by the fact that atmospheric CO_2 concentrations should change as the Sun brightens. As discussed in chapter 3, silicate weathering should speed up as the climate warms, and so atmospheric CO_2 should decrease with time. This prediction may seem counterintuitive, as CO_2 in the atmosphere is currently *increasing* at a rapid rate as a consequence of fossil fuel burning. The timescale being discussed here, though, is much longer. Fossil fuels will likely be exhausted within a few hundred years unless we start consuming them more slowly or we find large new reserves. The CO_2 produced from burning them will remain in the Earth system for much longer; indeed, the last dregs will linger for up to a million years.[1] But this is still only a blip compared to the timescale for solar evolution. In the long run, the increase in silicate weathering will win out, and atmospheric CO_2 should begin to decrease.

The future evolution of the Earth system has been simulated by two different groups using coupled climate/geochemical cycling models.[9,10] James Lovelock and Michael Whitfield first looked at this problem in 1982.[10] Lovelock, after all, was interested in Gaia, and what could be more Gaian than modification of Earth's climate by humans? In their model, atmospheric CO_2 was predicted to fall below 150 ppm about 100 million years from now. The fossil fuel CO_2 pulse was ignored because its lifetime is much shorter than this. A CO_2 concentration of 150 ppm represents a critical level, called the *CO_2 compensation point*, below which *C_3 plants** cannot live, because they respire faster than they photosynthesize. C_3 plants account for about 95 percent of all plants on Earth, including trees and most agricultural crops. Hence, Lovelock and Whitfield concluded that Earth might be able to support an active biosphere for only another 100 million years. This may sound like a long time, but from the standpoint of planetary evolution it is actually quite short. After all, life itself has probably been around for at least 3 billion years, and plants and animals have existed for more than 500 million years. So a future life span of 100 million years would imply that Earth's biosphere is nearing the end of its existence.

About 15 years ago, my then postdoc, Ken Caldeira,** decided to redo this calculation using our own climate model and our own assumptions

*The name C_3 refers to the length of the carbon chain that is created during the initial step of photosynthesis.

**Ken is now at the Carnegie Institute on the campus of Stanford University and is studying possible *geoengineering* solutions to human-induced climate change.

about how weathering rates vary with temperature. The results of our revised calculation are shown in figure 7.1b. Ken's calculation was more detailed than the earlier one, and it included the so-called *terrestrial biological pump*. As discussed in chapter 3, if one measures CO_2 in the air spaces within soil, one finds that its concentration is typically 20–30 times higher than that in the overlying atmosphere. It is "pumped up" by respiration carried out by the roots of trees and grasses and by decay of organic matter within the soil. In the calculation shown, Ken further assumed that *C_4 plants* would dominate productivity once atmospheric CO_2 concentrations began to decline. C_4 plants, which include many tropical grasses, along with corn and sugar cane, are able to photosynthesize at CO_2 concentrations as low as 10 ppm. They do this by pumping up the CO_2 concentration within their leaves. In Ken's calculation, C_4 plants lasted for about 900 million years before they went extinct. And even C_3 plants lasted for 500 million years—5 times longer than predicted by Lovelock and Whitfield. So perhaps the biosphere has a little more time left after all. Once atmospheric CO_2 declined below 10 ppm in this calculation, its greenhouse effect was essentially exhausted, and so the surface temperature climbed rapidly over the next few hundred million years (figure 7.1b). Unlike the CO_2-induced greenhouse described earlier, water would be lost during this process because the background atmosphere would remain thin. It might take a few hundred million years to lose the entire ocean[11], but the Earth would eventually be left completely dry, as Venus is today. So, if these calculations are correct, the biosphere does indeed have a finite lifetime—1.5 billion years or less—unless humans somehow intervene.

A Geoengineering Solution to Solar Luminosity Increases

Although it is tangential to the main focus of this book, it would be misleading to leave this section without pointing out that there is actually no reason why Earth must necessarily follow the path shown in figure 7.1. If humans are still around at that time, and if technology continues to advance, it may be relatively straightforward to offset the predicted solar luminosity increase by technological means. Roger Angel of the University of Arizona has recently described in some detail how this might be done.[12] The idea itself traces back to an older paper by J. T. Early.[13] Both Early and Angel were actually interested in the more practical problem of counteracting the near-term effects of CO_2-induced global warming—a procedure referred to as *geoengineering*. I personally think that such geoengineering solutions to global warming should be employed only as a last resort, for reasons that I will return to below. The same principles, however, apply to the problem of countering long-term solar luminosity increases. Few people, including environmentalists like me, would harbor qualms about saving ourselves from a runaway greenhouse.

To offset global warming, Early suggested that it might be possible to construct a large solar shield at the L1 Lagrange point in the Earth-Sun system (see figure 7.2, in color section). L1 is one of 5 points of neutral gravitational stability, that is, places where an object has a tendency to fall neither toward the Sun nor toward the Earth. L4 and L5 are *stable equilibrium points*, meaning that an object can remain there indefinitely without the application of any external force. In the mid-1970s, Princeton physicist Gerard K. O'Neill suggested building huge habitats at these locations to try to offset continued populaton growth on Earth. The L5 Society was founded for a brief time to foster this goal. This idea has now been abandoned, as it has since been realized that it is virtually impossible to counter population growth by sending people into space. One could not possibly launch them fast enough. Feeding them is also a problem. L4 and L5 might one day make useful places to set up large space observatories, but sending people there just to offload them from Earth is an idea that hopefully will not be resurrected.

More relevant to the present problem is the unstable Lagrange point L1, which lies directly between the Sun and the Earth. This point represents what physicists call a "saddle point" in the Earth-Sun gravitational field. Just as a saddle curves upward in one direction (front to back) and downward in the other (side to side), the gravitational potential energy, from which the gravitational field is derived, increases in one direction from the saddle point and decreases in the other. So an object placed at L1 will gradually drift away from it if it is given no additional push. For a minimum expenditure of energy, however, it is possible to orbit the L1 point. NASA currently has a spacecraft doing just that: the SOHO spacecraft, which makes observations of the Sun. The IMAX film *Solarmax*, which includes spectacular images of the Sun's active surface, was filmed using a camera mounted aboard SOHO.

Angel has taken Early's original solar shield idea and modified it to try to make it more practical to build and deploy.[12] Rather than a single large shield, he envisions an array of multiple small lenses orbiting near (and around) L1. These would be sheets of semi-transparent material, ~1 μm in thickness, with nonreflective coatings. Because they would be extremely lightweight, the effects of solar radiation pressure are important, and they would need to be flown somewhat inside the L1 point to keep them in place. Angel's lens is designed to offset the climatic effects of one CO_2 doubling; hence, it would need to deflect about 1.8 percent of the Sun's energy that would normally hit the Earth. This could be accomplished with an array of approximately 16 trillion small lenses, each approximately 0.6 m in diameter.* Angel argues that these lenses could be constructed on Earth's surface and launched into space using a type of electromagnetic rail gun. Surprisingly, boosting launch packages to escape velocity from Earth's surface seems feasible, despite atmospheric friction, by using existing heat shield technology. Alternatively, as Early had suggested previously, the lenses could be manufactured on the Moon's surface and launched from there. Given the amount of time available before solar evolution becomes a problem, the prospect of large-scale manufacturing on the Moon does not appear too difficult. Angel estimates that an Earth-launched system could be constructed on a timescale of about 50 years at an average cost of ~$100 billion per year. Although this sounds like (and is) an enormous sum of money, it amounts to only about 0.2 percent of current world gross domestic product.

*The ratio of the total surface area of these lenses compared to the projected surface area of the Earth, πR_E^2, is only about 1 percent. But one gains an additional factor of ~2 from the physics of diffraction. So this array of lenses really would be enough to offset one CO_2 doubling.

The solar shield is probably *not* a good way to offset global warming. Besides being costly and difficult, it would do nothing to reduce the atmospheric CO_2 concentration itself. This could be a problem because high CO_2 levels would make the surface ocean more acidic, and this in turn could cause major damage to marine ecosystems—notably coral reefs, which are made of calcium carbonate and which are therefore easy to dissolve. On the solar evolution timescale, however, the solar shield is a wonderful idea! We would like for CO_2 to remain relatively abundant, so that C_3 plants can persist, and we would also like for the climate to stay relatively cool. If humans manage to survive for even a fraction of the Sun's remaining 5-billion-year main sequence lifetime, large-scale space projects like this one will almost certainly become feasible. So we should probably not spend too much time worrying about the long-term evolution of the Earth's climate. Thinking optimistically, humans might ultimately become the saviors of the biosphere instead of contributing to its premature demise. That is a thought that both Jim Lovelock and Carl Sagan would likely have embraced.

· Chapter Eight ·

The Martian Climate Puzzle

We turn our attention now to Earth's other planetary neighbor, Mars. Mars has long been of interest to both scientists and the general public. That is because, of all the planets in our Solar System, Mars has the greatest potential for harboring life, either now or at some time in the distant past. As a consequence of this continued interest, NASA has sent a whole string of exploratory spacecraft to Mars over the past 30 years. The most recent missions include the two highly successful Mars Exploration Rovers (MER), *Spirit* and *Opportunity*, which reached Mars in 2004, and the Mars *Phoenix* Lander that landed on Mars' northern plains in May of 2008.

Despite all this interest, however, Mars is not really a good analog to Earth. Mars is only about half the diameter of Earth and has only 1/9 its mass. Its reddish-brown surface is devoid of liquid water, although it does have caps of mixed H_2O and CO_2 ice (figure 8.1, in color section) at its poles. Mars' atmosphere, like that of Venus, is composed mostly of CO_2 and N_2 (table 8.1). Its surface pressure, though, is only 6–8 mbar—less than 1 percent of the surface pressure on Earth*—and its mean global surface temperature is a chilly −55°C (−67°F). Surface temperatures at midday near the equator do occasionally rise above the freezing point of water. However, it is considered unlikely that liquid water exists even transiently in these regions, partly because most of Mars' water has migrated to the poles, and partly because water ice should tend to subli-

*Recall that Earth's surface pressure is 1.013 bar, or 1013 mbar.

TABLE 8.1
Composition of Mars' Atmosphere

Gas	*Percent by volume*
CO_2	95.3
N_2	2.7
Ar	1.6
O_2	0.13
CO	0.07
H_20	~0.006 (variable)

mate (go directly from solid to vapor), rather than melt, at the prevailing low surface pressure.[1]

Evidence for Liquid Water near Mars' Surface

Despite Mars' hostile global climate and lack of surface liquid water, planetary scientists continue to hope that life might one day be found there. The reason is that Mars continues to surprise us each time we visit. Some of the biggest surprises have come from high-resolution pictures taken by the Mars Orbiter Camera (MOC) aboard the Mars Global Surveyor spacecraft. Along with the MER rovers, Mars Global Surveyor was one of the most successful spacecraft ever flown to Mars. It orbited the planet, transmitting images and other data, from September 1997 until November 2006. Figure 8.2 shows two pictures of the same crater wall in the Centauri Montes region, taken 6 years apart.[2] The light-colored streak in the panel on the right shows evidence for recent *gully* formation. Here, "gully" is the name given to small-scale, apparently fluvial features found on the walls of some craters, as well as those of larger martian valleys. (Note the 300-m scale bar at the lower lefthand corner of the figure.)

Exactly how such gullies are formed is a matter of debate. Some investigators have proposed that they might be created by landslides, rather

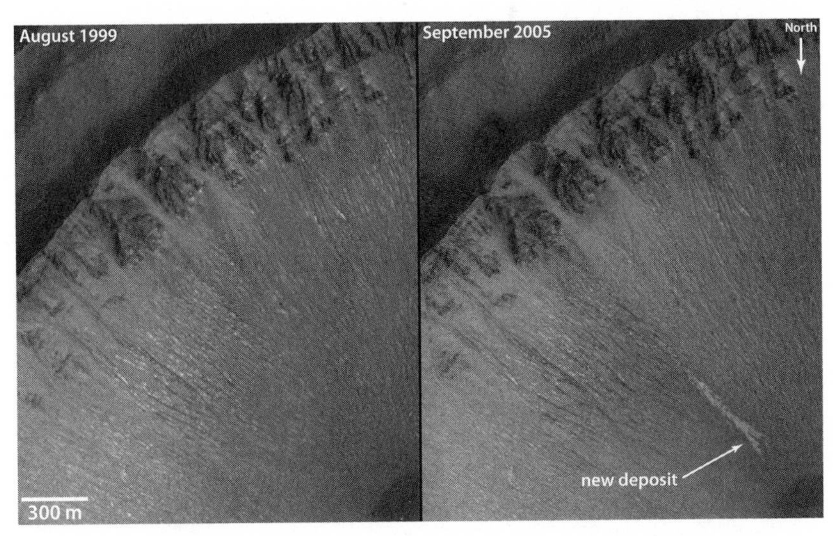

Figure 8.2 Images taken from Mars Global Surveyor indicating fresh gully formation on the wall of an unnamed crater in the Centauri Montes region, located near 38.7 degrees south latitude, 263.3 degrees west longitude. Left: photo taken in 1999. Right: Same location photographed in 2005, showing a fresh gully.[2] (Photo courtesy of NASA.)

than by a flowing liquid. However, detailed analysis of the structure of the flows, specifically, the fact that they tend to go around obstacles, rather than over them, argues otherwise. Others have suggested that they are formed by liquid CO_2, rather than liquid H_2O. But this, too, seems unlikely, as liquid CO_2 is stable only above 5.2 bars pressure, which, of course, is much higher than the ambient surface pressure. By comparison, pure liquid H_2O can exist at temperatures and pressures above its *triple point** (0.01°C and 6.1 mbar pressure).

The argument that the gullies are formed by flow of a liquid is bolstered by the fact that they are found predominantly on equatorward-facing slopes located at >30 degrees north or south latitude. Such slopes receive the full brunt of solar heating during the daytime, as the Sun's rays are more or less perpendicular to the surface. This might allow sub-

*The triple point of a substance is the temperature and pressure at which all three phases (solid, liquid, and vapor) can coexist. Saline (salty) liquid water can exist at significantly lower temperatures, so perhaps the gullies are formed by saltwater.

surface ground ice to melt, although the energetics of this process are unclear. Alternatively, the liquid water could gush up through cracks from regions deeper down in the martian crust where temperatures are warmer. The reason the gullies are found at high latitudes, rather than near the equator, may have to do with the global distribution of martian water. Currently, Mars' obliquity is about 25°, not too different from Earth's value of 23.5°. The polar regions are therefore cold and the tropics are warm, as on Earth. Hence, any water ice near the surface of Mars tends to sublimate away in the tropics and is redeposited as frost or snow near the poles.

Martian climate history is complicated, though, by the fact that Mars' obliquity fluctuates to a much larger extent than does Earth's. As discussed in chapter 5, Earth's obliquity varies by about ±1 degree on a timescale of 41,000 years. By contrast, Mars' obliquity varies by ±10 degrees with two different periodicities of approximately 10^5 and 10^6 yr, respectively[3] (see figure 8.3). At higher obliquities, the summer poles are heated quite strongly, and the equator receives relatively little sunlight. Hence, water ice near the surface probably migrates equatorward at high obliquity and poleward at low obliquity. This may help to explain why water is still present near the surface in some regions, despite billions of years of recycling caused by the fluctuating obliquity and climate.

On even longer timescales, 5 million years or more, Mars' obliquity behaves quite differently from Earth's: it oscillates chaotically. Calculations[5,6] show that on these timescales Mars' obliquity can reach values as low as 0° or as high as 50–60°. As an example of this behavior, figure 8.3 shows that two similar calculations with slightly different parameters can have very different results. In these simulations, the equations governing Mars' obliquity were integrated backward in time from the present, using two different values for the precession rate of Mars' spin axis, both of which are within the range of observations. Evidently, the answer one gets depends sensitively on the precise value of this parameter.

The word *chaos* is used here in its mathematical sense. Technically, it means that small changes in the initial conditions, or parameters, for a problem can lead to large changes in the results after some period of

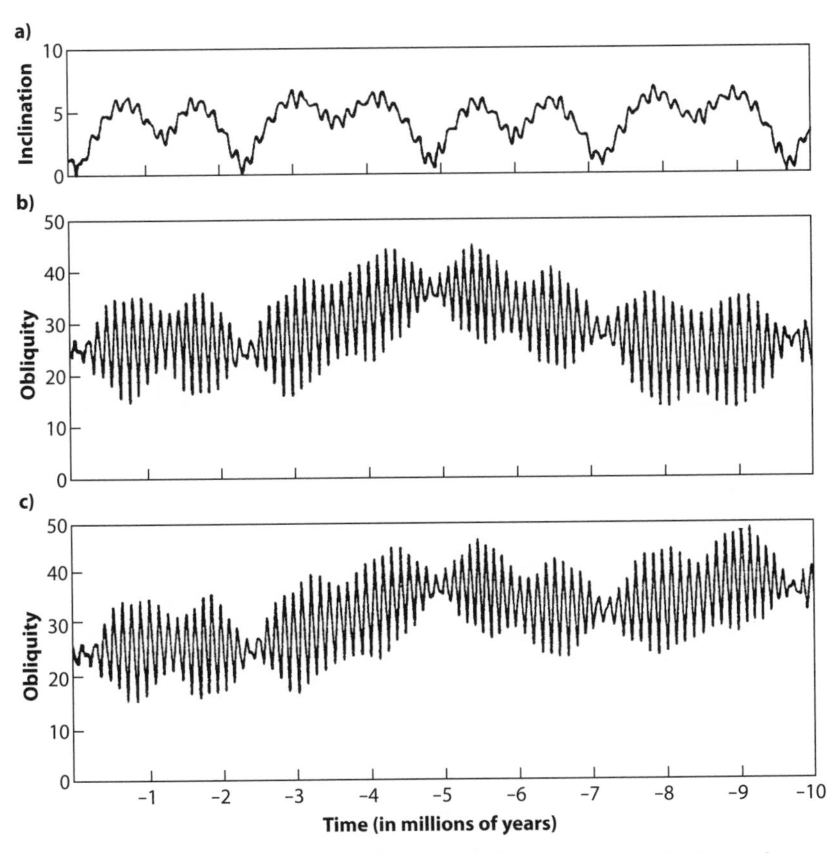

Figure 8.3 Calculations of Mars' orbital inclination (a) and obliquity (b, c) over the past 10 million years. Two slightly different values of the present-day orbital precession constant were assumed in the two obliquity calculations.[4] (Reproduced by permission of Elsevier Publishing.)

time. On a practical level, it means that Mars' obliquity is impossible to predict accurately over timescales longer than about 5 million years. But one can nevertheless demonstrate statistically that Mars must go through periods when its obliquity is very large. During such times, most or all of the ground ice now stored in the polar regions should have migrated to the tropics. The complex surface morphology of Mars may reflect these large shifts in where water ice is stored. We shall return to this topic in the next chapter, as Earth could under some circumstances experience similar large obliquity variations. Indeed, this is a critical element of Ward and Brownlee's "rare Earth" hypothesis.

CH_4 in Mars' Atmosphere?

Another surprise regarding present Mars is the recent spectroscopic evidence for the presence of methane in Mars' atmosphere. The measurements come from three different instruments: two of them mounted on Earth-based telescopes[7,8] and a third that is on board the European Space Agency's *Mars Express* spacecraft.[9] All three studies looked at the reflection of sunlight from the planet at near-infrared wavelengths (near 3.3 μm), where CH_4 exhibits strong absorption. One study[8] found a uniform CH_4 concentration of about 10 ppbv, whereas the other two studies found variable concentrations of up to 200 ppbv. All of these measurements were near the detection limits for the instruments involved, and so none of them can be considered definitive. High spectral resolution measurements from some future Mars orbiter, or mass spectrometer measurements from the upcoming Mars Science Laboratory rover, are probably needed to determine definitively whether CH_4 is actually present.

Why so much interest in these unconfirmed measurements? If methane *is* present in Mars' atmosphere, it might signal the presence of subsurface life. As discussed in chapter 4, the largest source of CH_4 on Earth is methanogenic bacteria. Such organisms could conceivably be present in the martian subsurface as well, living on H_2 produced from the interaction of groundwater with reduced minerals in the martian crust. Alternatively, the CH_4 could be produced entirely abiotically, as has been proposed for the methane emanating from the Mid-Atlantic ridge on Earth (see chapter 4). At this stage, one can only speculate, as the measurements are simply not good enough to discriminate between different hypotheses or even to prove that martian methane exists. If methane is indeed present in Mars' atmosphere, and especially if it varies in concentration from one location to another (which would imply a large, near-surface source), then the idea that it may be produced biologically will gain credence. For now, though, it is probably safest to say that this is an interesting observation that deserves to be followed up.

Evidence That Water Flowed in Mars' Distant Past

The observations of recent gully formation on Mars and the possible identification of CH_4 in Mars' atmosphere are hot topics because they could conceivably lead to a truly spectacular discovery, if life does indeed exist on Mars today. For our purposes here, though, the long-term history of the planet is even more interesting. Regardless of whether Mars is habitable today, it shows evidence of having been much more so in the distant past. The evidence comes partly from older images of Mars taken by *Mariner 9* in 1971 and by the two *Viking* spacecraft in 1976. Each of the *Viking* spacecraft consisted of an orbiter and a lander. The two orbiters were each equipped with a relatively low-resolution camera. The quality of the images pales compared to the high-resolution pictures shown in figure 8.2. However, the fact that the images were low-resolution had its advantages as well, in that it allowed *Viking* to produce global maps of Mars' surface. By contrast, the high-resolution camera on *Mars Global Surveyor* was able to map only a small portion of the planet's surface, despite the fact that it returned more than 240,000 images during its 9-year lifetime.

The *Viking* images showed that parts of Mars' surface—the southern highlands, in particular—are covered with large-scale fluvial features (see figure 8.4). Some of these features (figure 8.4a) are up to 40 km wide. These so-called *outflow channels* are thought to have been created by massive floods, similar to the Spokane flood that created the Channeled Scablands of eastern Washington near the end of the last ice age.[10] The suggested cause in that case was the breakup of a glacial dam that released the remaining portions of glacial Lake Missoula, almost overnight. Such flood features, paradoxically, are thought to indicate relatively cold temperatures, as ice is a necessary part of their formation.

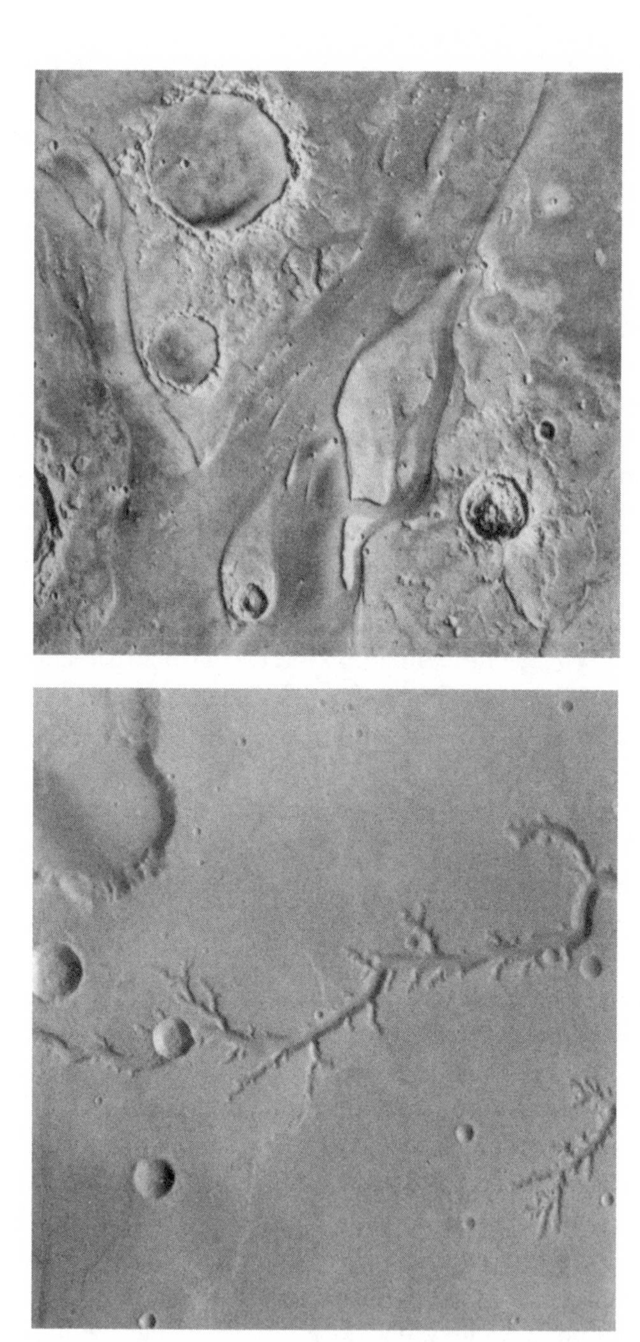

Figure 8.4 (*top*) Ares Vallis (an outflow channel). (*bottom*) Nirgal Vallis (a runoff channel or valley). Both images were taken by *Viking*, so the frames are approximately 200 km on each side. (Courtesy of NASA/JPL-Caltech.)

Other fluvial features, though, look quite different. Nirgal Vallis (figure 8.4b) is a typical example. This valley is several hundred kilometers long and perhaps 2–3 km wide. Clearly, it is not a flood feature like the outflow channel in the left panel. Rather, it was probably formed over a long period of time by water that was flowing in a controlled manner. Consequently, it is termed a *runoff channel*, or *valley*. The short, stubby tributaries suggest that it was formed by a process called *sapping*, in which water flows subsurface before collecting into channels. Similar branching patterns are seen in rivers formed in arid areas on Earth today, such as the American West.

Further evidence that the martian valleys are analogous to terrestrial river valleys has been provided by more detailed images from *Mars Global Surveyor*. Figure 8.5 shows a portion of Nanedi Vallis photographed in high resolution.[11] This valley is similar in its general structure to Nirgal Vallis in the previous figure. At this resolution, though, one can also see a small channel cutting through the middle of the bend in the upper part of the picture. This channel was formed by a river no more than 20–30 m wide. Those who have been to the Arizona's Grand Canyon will immediately recognize the similarities. The Grand Canyon is about a mile (1.6 km) wide and a mile deep. It has been carved by the Colorado River, which is about 20 m wide in most places. Obviously, the canyon that is formed can be much wider than the river that cuts it. This analogy may also provide an indication of how long it takes to form such features. The Grand Canyon has formed over the last 10–20 million years, as the Colorado plateau has been gradually uplifted. If this analogy holds true on Mars, then Nanedi Vallis, and perhaps many of the other valleys as well, probably formed on a timescale of millions to tens of millions of years.

This last inference is important because, if correct, it would rule out some of the suggested formation mechanisms for the valleys. For example, Theresa Segura and coworkers from the University of Colorado and from NASA Ames Research Center suggested in 2002 that the valleys formed early in Mars' history by rainout of steam atmospheres created by giant impacts.[12] The timescale for valley formation in this scenario is

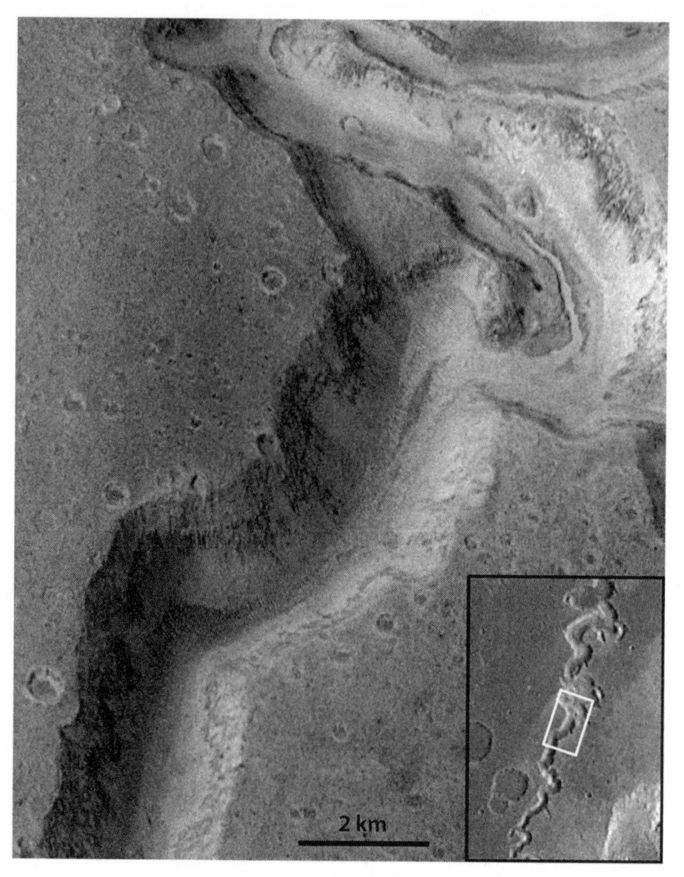

Figure 8.5 Nanedi Vallis—a valley seen up close by Mars Global Surveyor. The frame is approximately 3 km in width.[11] (Photo courtesy of NASA.)

thousands of years, rather than millions of years, which would imply that the analogy to terrestrial formation processes is for some reason invalid. Eventually, we may be able to test both of these hypotheses by sending geologists, or perhaps robots, up and down the canyon walls to collect rock samples, which can then be studied with radiometric age dating techniques. For now, although all hypotheses remain uncertain, it seems reasonable to suppose that martian valley formation required a climate that was warm enough to permit liquid water to flow on Mars' surface for tens to hundreds of millions of years at a time.

When Did the Martian Valleys Form?

If the valleys did form over a protracted period of time, this leads immediately to two further questions. First, *when* was the martian climate this warm? And second, how warm did it have to be? The second question has proven difficult to answer, as discussed below. The first question, though, is fairly straightforward. Although it seems unlikely that giant impacts formed the martian valleys, the evidence for smaller impacts is found all around them. Or, to say this in a different way, the valleys formed primarily in areas where impact craters are abundant. Such craters are concentrated in the southern highlands region of Mars. By comparison, the northern plains are relatively devoid of impact craters and of valleys as well.

The inference from this observation is clear: most of the martian valleys formed early in the planet's history when the cratering rate was much higher than it is today. Fortunately, we have data from our own Moon that tells us approximately when this must have occurred. Most Moon rocks collected by the Apollo astronauts have ages of 3.8 billion years or older. From this, it has been determined that the cratering rate in the inner Solar System was high until about 3.8 Ga and dropped off rapidly after that time.[13] The time period between Solar System formation, at 4.55 Ga, and the age of the Moon rocks, 3.8 Ga, is termed the *heavy bombardment period*. Actually, there is an ongoing debate about this as well. Some observers, including many of the geologists who studied the Moon rocks, believe that an intense pulse of bombardment occurred at about 3.8 Ga.[14] Others have suggested that this was simply the tail end of a long period of bombardment by objects, presumably asteroids, left over after the main accretion period ended.[15]

Proponents of the "pulse" hypothesis got a boost recently when a group of researchers based in Nice, France, suggested a mechanism by which such an event could have been triggered.[16,17] They proposed that Saturn formed somewhat closer to the Sun than it is today and that it migrated outward during the early part of Solar System history. Saturn's semi-major axis is 9.5 AU and its orbital period today is 29.5 years. This is about $2\frac{1}{2}$ times the orbital period of Jupiter, which has a semi-major axis of 5.2 AU and a period of 11.8 years. According to the "Nice model," as it is called, Saturn formed inside of 8 AU and then migrated outward as a consequence of gravitational interactions with smaller planetesimals. In doing so, it passed through a *resonance* in which its orbital period was exactly twice that of Jupiter. At this point, Jupiter and Saturn would have interacted very strongly with each other, and both of their orbits would have become more eccentric. Uranus and Neptune, which also formed closer to the Sun in this model, were scattered out into the outer Solar System, where they perturbed the orbits of icy planetesimals (comets), some of which came in and collided with the terrestrial planets, along with the Moon. Hence, this theory predicts that there should have been a pulse in the bombardment of the inner planets at the time when this resonance occurred, presumably around 3.8 Ga. If this idea is correct, many of the lunar craters and those on the martian highlands formed at about the same time as a result of this singular event in Solar System history.

Regardless of which theory of the heavy bombardment is correct, the implications for the timing of martian valley formation are the same: the valleys must have formed at 3.8 Ga, or earlier. The warm, wet period of Mars' climate history, if there was one, lasted for only a small fraction of the planet's lifetime.

How Warm Was Early Mars?

The second question has befuddled planetary scientists for many years: Just how warm must the climate of early Mars have been to account for the observed fluvial features? Historically, there have been two opposing views on this question: the "cold early Mars" camp and the "warm early Mars" camp. It is amusing to me to attend Mars meetings today and to learn that we are no closer to agreement on this issue than we were 30 years ago. This is probably because we are trying to answer a complex question based on data that come almost entirely from remote sensing, as opposed to in situ observations. I expect that disagreement will persist until robotic landers (or human geologists) do a much more thorough job of exploring Mars' surface.

Theresa Segura and her coauthors,[12] mentioned above, are members of the cold early Mars camp. Their impact-related mechanism for forming the valleys, if correct, would eliminate the need to explain how Mars' climate could have been kept warm. Carl Sagan was also a cold early Mars proponent—or, more correctly, he straddled the issue by writing articles supporting both views. His 1972 paper with George Mullen, mentioned in chapter 2, talked about the necessity of explaining why early Mars was warm. But, in a later paper with David Wallace,[18] he suggested that the valleys could have been formed by ice-covered rivers in a climate not much warmer than that of today. The ice, in their view, would have acted as an insulator that kept the water below from freezing as it flowed for hundreds of kilometers.

One reason that confusion remains about this question is that different types of evidence support different hypotheses. Many of the geochemical data obtained from orbiting satellites, for example, support the cold early Mars hypothesis. Infrared spectra obtained by the Mars Odyssey spacecraft indicate the widespread occurrence of the mineral olivine over the martian surface.[19,20] Olivine is a greenish mineral, abundant in Earth's mantle and in some volcanic rocks, that is rapidly weathered and converted into other minerals (clays) when exposed to liquid water.* The fact that olivine is widely distributed on Mars has been taken by some as an indication that rain never fell over much of the planet's surface. The short, stubby branching patterns of most of the valleys (e.g., figure 8.4b) appear at first glance to support this view. On Earth, most of the land surface is thoroughly dissected by streams and rivers. Other interpretations of both datasets are possible, however. Recent observations by the MER rovers show that most of the olivine at the two landing sites is contained in dust, which blows all over Mars' surface. So its presence in a particular location may not imply that it never rained there. And mineralogical evidence from the *Opportunity* rover strongly suggests that liquid water was present at least episodically in Meridiani Planum.[21] The short, stubby tributaries of the valleys may simply be caused

*The weathering of olivine is not instantaneous, as evidenced by the famous Green Sand Beach on the Big Island of Hawaii. The beach is composed almost entirely of olivine crystals and looks like something right out of the Emerald City in *The Wizard of Oz*.

by the fact that the martian soil is porous and that the rain that used to fall there sank quickly through it before collecting to form streams.

Timescales, again, are of critical importance. If one accepts that millions of years were required to form the valleys, then water must have been continuously resupplied to the surface. This is true even if the flow originated from subsurface aquifers—these would have to have been periodically recharged. On Earth, groundwater is recharged by rainfall. Some models of Mars' hydrologic cycle suggest that groundwater might be recharged by vapor diffusing upward from a global aquifer a kilometer or two beneath the surface.[22] But this seems like a bit of a stretch. More likely, early Mars was warm enough to have appreciable rainfall, or snowfall, which means that the climate could not have been too much colder than the present climate of Earth. (Even if the moisture fell initially as snow, it must have warmed above freezing in order to recharge the aquifers.) Let's assume for the moment that early Mars was indeed relatively warm and see whether or not we can account for it.

Mechanisms for Warming Early Mars

We can begin by drawing an analogy with the early Earth. As discussed in chapters 3 and 4, early Earth was probably kept warm by a combination of higher atmospheric CO_2 and CH_4 concentrations. The CO_2 was produced partly by volcanoes and may have been partly left over from impacts. On Mars, gravity was lower by a factor of almost 3, and so impacts may have been less effective at building up a dense CO_2 atmosphere.[23] But active volcanoes clearly existed, and they likely provided a large source of CO_2. The most striking evidence for volcanism is found in the Tharsis region near Mars' equator. Tharsis is home to Olympus Mons, the largest volcano in the Solar System, as well as several other large, extinct volcanoes.

When we consider the second part of the early Earth story—greenhouse warming by CH_4—the analogy with Mars breaks down. Most of Earth's CH_4 is produced biologically, by methanogens. As we don't know whether life ever originated on Mars, we cannot count on this source having been present. To be sure, abiotic CH_4 sources might have existed, and other greenhouse gases could have been present as well. Let's save this thought for the moment, though, and see how much warming might have been produced simply by CO_2 and H_2O.

Warming early Mars is a challenging problem, both because of the planet's distance from the Sun and because the Sun itself was less bright at the time when Mars is thought to have been warm. Mars' mean orbital distance is 1.52 AU, so the incident solar flux is decreased by a factor of $1/1.52^2$, or 0.43, compared to that on Earth. And, if the valleys formed at 3.8 Ga or earlier, as seems likely, solar luminosity was at most 75 percent of its present value. Hence, the average solar flux at Mars' orbit at the time when the valleys formed was no more than $0.75 \times 0.43 = 0.32$ times the present solar flux at Earth.

In climate calculations performed while I was at NASA Ames,[24] we initially determined that this low solar flux could have been offset by a CO_2-H_2O atmosphere with a surface pressure of about 5 bars. However, we had failed to account for the fact that CO_2 should have condensed in the upper parts of our model troposphere. This does not happen in simulations of present Mars (or in reality), although CO_2 does condense at the winter poles to form ice caps. But at the lower solar flux prevailing in Mars' early history, the upper troposphere should have been filled with CO_2 ice clouds.

When we revised our calculations to include this effect,[25] we got a rather surprising result: we found that it was impossible to warm early Mars with CO_2! The problem is illustrated in figure 8.6. The figure shows calculated mean global surface temperature as a function of surface pressure for four different values of the solar flux. A CO_2-H_2O atmosphere was assumed, with the H_2O concentration set equal to the maximum amount that the atmosphere could hold at saturation. This maximizes the magnitude of the greenhouse effect, and so the calculated surface temperatures are as high as they could possibly be, given this particular model. The results show that for the present solar flux, Mars' surface temperature could be raised to arbitrarily high values by adding CO_2 to its atmosphere. About 2–3 bars of CO_2 would be sufficient to bring the average surface temperature above the freezing point of water,

273 K. Hence, we could potentially *terraform* present Mars, i.e., make it suitable for humans, if we could find enough CO_2 stored in some subsurface reservoir and figure out how to release it.* One still couldn't live freely on Mars under such conditions, as humans cannot breathe at CO_2 partial pressures exceeding about 0.01 bar.[27] But plants might be able to grow there, and humans might walk around without spacesuits if they were willing to carry an oxygen tank.

For early Mars, though, the results of increasing atmospheric CO_2 levels are entirely different. At 3.8 Ga, the latest time when most of the valleys could have formed, the solar flux was still only 75 percent of its present value, or halfway between the two lower curves in figure 8.6. According to these calculations, then, the maximum surface temperature should have been about 225 K, or −48°C. That is about the same value as the mean surface temperature estimated for Snowball Earth (see chapter 5). This, of course, would have made early Mars completely unsuitable for land life. Algae and other single-celled organisms could conceivably have survived beneath the ice, as they appear to have done during Snowball Earth, so we cannot say that life itself would have been precluded. But this scenario definitely falls into the "cold early Mars" camp, and it takes us back to the question: how can we explain the fluvial features? So I, for one, have never been satisfied with this result. It makes me think that we are missing something in our calculation.

If one examines figure 8.6, one sees a curious thing: the curves for the two lower solar flux cases simply end at a surface pressure of between 2 and 4 bars. Why, one might ask, should this be the case? Couldn't one simply add more CO_2 to these atmospheres and thereby make them warmer? The answer is no, for two reasons. One has already been mentioned above: at high CO_2 pressures and low solar fluxes, CO_2 condenses in the upper troposphere of the model to form clouds of CO_2 ice. This is the same "dry ice" that is used to keep ice cream cold. The clouds that it would form would be similar to cirrus clouds on Earth, which are made

*The British dentist/amateur scientist Martyn Fogg has suggested, tongue in cheek, that one could vaporize subsurface carbonate rocks (if they exist) by exploding thousands of hydrogen bombs at regularly spaced intervals deep beneath Mars' surface.[26] This is indeed a conceivable method for terraforming the planet. But then one would need to essentially destroy the planet in order to save it. One hopes that future generations do *not* try this route.

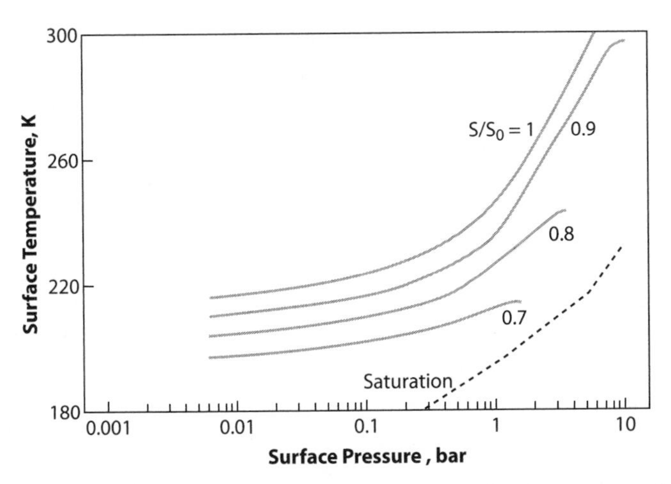

Figure 8.6 Mean global surface temperature as a function of surface pressure for a CO_2-H_2O atmosphere on Mars. The ratio S/S_0 represents the value of the solar flux relative to today. The dashed curve shows the temperatures at which CO_2 would condense at the surface. Before 3.8 Ga, when most of the martian valleys formed, the solar flux was <0.75 times today's value.[25]

of crystals of water ice. It is not the clouds themselves that limit the surface temperature, though, in our model. Surprisingly, CO_2 ice clouds should actually have warmed Mars' surface, had they been present in its early atmosphere[28] (see below). But the process of forming the CO_2 clouds would itself have helped limit greenhouse warming. As discussed in chapter 6, condensation of a gas into a liquid, or a solid, releases latent heat. This applies to CO_2, as well as to H_2O. On early Venus, condensation of water vapor reduced the tropospheric *lapse rate* (the rate at which temperature decreases with altitude), and this caused the tropopause cold trap to move up to high altitudes, allowing water vapor to make its way up into the stratosphere.

In our models of early Mars, condensation of CO_2 also reduces the tropospheric lapse rate in a similar manner. But, instead of changing the concentration of CO_2 in the stratosphere—which could not happen because these atmospheres are assumed to be almost pure CO_2—the important effect of this process is to reduce the magnitude of the greenhouse effect. It is perhaps easiest to explain this by noting that the stratospheric temperature is essentially fixed by the amount of sunlight that it receives, which must be balanced by radiation of infrared energy back to space. The surface temperature is then determined by how the stratosphere is connected to the surface. If the tropospheric lapse rate is less steep, then the surface will be cooler. That is what happens when CO_2 (or H_2O) condenses to form clouds. So this is one reason the curves in figure 8.6 end where they do.

A second, equally important factor in limiting the magnitude of the greenhouse effect on early Mars is the effect of CO_2 on the planet's albedo, or reflectivity. As mentioned in the previous chapter, CO_2 is a very effective Rayleigh scatterer of incident sunlight. Hence, when the atmospheric pressure increases, more sunlight is scattered back into space, and the planetary albedo increases, cooling the climate. Both of these factors make it difficult or impossible to warm early Mars to the freezing point using only gaseous CO_2 and H_2O.

As mentioned above, CO_2 ice clouds may provide a way out of this problem. This prediction is counterintuitive because water clouds on Earth are thought to cool the climate, on average. That is because cumulus and stratus clouds in the lower troposphere (which are made of liquid water) tend to cool the surface by increasing Earth's albedo. Cirrus clouds in the upper troposphere, on the other hand, tend to warm the surface because they let most of the incident sunlight through, but they absorb and re-radiate outgoing infrared radiation, thereby contributing to the greenhouse effect. Low, water clouds contribute to the greenhouse effect, too, but not as much because they are not as cold.*

CO_2 clouds on Mars would behave rather like cirrus clouds on Earth, although technically their effect is different because they tend to scatter infrared radiation, whereas water clouds absorb it. Nevertheless, their "scattering" greenhouse effect can be quite large. Calculations[28] show that if CO_2 ice clouds covered the entire early martian surface, the sur-

*Both warm, low clouds and cold, high clouds radiate infrared energy to space. A cold cloud radiates less energy than does a warm one, though; hence, if cold, high clouds are present, the surface needs to radiate more energy so that the total amount of radiation from the surface plus the clouds is constant. So the surface must therefore be warmer when high clouds are present.

face temperature could have been warmed by as much as 70°C. This would be more than enough to make early Mars warm and Earth-like. But if cloud cover was only 50 percent—a value more typical for Earth—most of this additional greenhouse warming disappears. An atmospheric greenhouse is like a bathtub: if there are even a few holes in it, most of the water (or infrared radiation) escapes rather quickly.

This brings us back to the idea that perhaps early Mars was warmed by other greenhouse gases, in addition to CO_2. CH_4, for example, could have been produced abiotically on Mars by the types of serpentinization reactions mentioned in chapter 4. (Recall that serpentinization reactions occur when warm seawater interacts with basalts during circulation through hydrothermal vents.) Or perhaps life did evolve on early Mars, and methanogens, or their martian equivalents, vented CH_4 into its atmosphere. But neither of these scenarios seems particularly plausible, and, in any case, the additional greenhouse effect caused by CH_4 does not appear adequate to account for the required warming.

Another suggestion is that SO_2 (sulfur dioxide) might have helped keep early Mars warm.[29] SO_2 is a good greenhouse gas if present in concentrations of tens to hundreds of parts per million.[30] Mars is rich in sulfur, and so martian volcanoes probably put out lots of SO_2 and H_2S when the planet was younger. Sulfur gases are soluble in water, however, and so removal by rainfall should have limited their abundance if the climate was truly warm and Earth-like. This is true even if the martian oceans were saturated in SO_2, as has been recently suggested.[29] That is because SO_2 would have been photochemically converted to other sulfur gases and particles, which themselves would have been rained out of the atmosphere. Our own models of early Earth predict only about 1–2 ppb of SO_2 even with an SO_2-saturated ocean.[31] So SO_2 may have provided some warming on early Mars, but it seems unlikely that it could have created a truly Earth-like climate.*

*Since this last paragraph was written, we have performed new calculations, currently under review, which indicate that reflection of sunlight by sulfate aerosol particles, created by SO_2 photolysis, would have more than offset any warming produced by the SO_2 itself. So the SO_2 hypothesis for warming early Mars appears to be incorrect.

Where Are the Carbonates?

One strong point of the SO_2 model is that it may help explain the absence of carbonate rocks on the martian surface. A succession of different Mars orbiter missions have searched spectroscopically for outcrops of such rocks, but they have never been observed, except for traces of carbonate minerals seen in the martian dust.[32] This has been viewed as a major puzzle for the past 30 years or more. Virtually all of the proposed mechanisms for warming early Mars require that its atmosphere was rich in CO_2. And we have already seen that liquid water flowed on many parts of Mars' surface. If both CO_2 and water were present, why did carbonate rocks not form?

In the SO_2 model, carbonates did not form at Mars' surface because dissolved SO_2 (which forms *sulfurous acid*, H_2SO_3) kept the oceans too acidic. Carbonates may still have formed subsurface, however, as groundwater warmed by geothermal heat reacted with silicates in Mars' crust. In a variant of this hypothesis, Fairen et al. propose that *sulfuric acid*, H_2SO_4, kept the oceans acidic and prevented carbonates from forming.[33] These authors suggested that Mars' CO_2 was all lost to space, and so was never incorporated into rocks. Loss of gases, including both CO_2 and N_2, on Mars was facilitated by its lack of a magnetic field, which allowed its upper atmosphere to interact directly with the intense early solar wind. Both of these models are supported by abundant evidence for acidic surface conditions discovered by the MER rover mission.[34,35]

Actually, Mars' surface may have been acidic even without input from sulfur gases. The pH of unpolluted rainwater on Earth today is about 5.6, which is already weakly acidic.* Terrestrial rainwater is acidic because it contains carbonic acid, H_2CO_3, which is in equilibrium with an atmospheric CO_2 concentration of 380 ppm, or 3.8×10^{-4} bars. The pH of rainwater is expected to drop by approximately 0.5 units for every factor of 10 increase in the CO_2 partial pressure. For early Mars, the best estimate of the CO_2 pressure is about 2 bars[28] (although we have already

*pH is defined as the negative logarithm of the H^+ ion concentration. A pH of 7 is considered neutral. Values lower than 7 are *acidic*; values higher than 7 are *basic*.

seen that there is no general agreement on this value). This is more than 5000 times higher than Earth's present CO_2 concentration, and so the predicted rainwater pH should be approximately 1.9 units lower, bringing it down to about 3.7. This is already highly acidic even without any added sulfur gases! So perhaps carbonate rocks did indeed form, but they were dissolved at a later time by exposure to acid rain. The carbonate minerals could then have been redeposited beneath the martian surface as groundwater percolated downward and lost its acidity by reacting with the surrounding rocks. Like the SO_2 hypothesis, this model suggests that carbonates should be found within the martian subsurface. Thus, both models could potentially be tested by drilling deeply into Mars' crust and bringing up samples to analyze. But this would be hard to do robotically. It may have to wait until we can send both humans and heavy equipment to Mars.

Before leaving this chapter, let's take stock of what we have learned. Clearly, Mars' climate and surface evolution history is complex and poorly understood. We have touched on some of the problems and offered some speculative solutions, but most of the important questions remain unresolved. Nevertheless, we have learned two things that may be important. First, Mars could be returned to Earth-like climate conditions today if enough CO_2, roughly 3 bars—3 times the surface pressure of Earth's atmosphere—were pumped into its atmosphere. It is probably impractical, and perhaps unethical, for humans to actually do this, as this would likely require not only finding carbonate minerals below Mars' surface but also blowing them to smithereens with nuclear bombs. So terraforming Mars does not sound like a great idea, despite inspiring science fiction novels that describe how it might be accomplished.[36]

Consider how much easier things would be, though, if Mars was a bigger planet. If Mars was the same size as Earth, it would presumably still have active volcanoes, and they should be pumping CO_2 into its atmosphere just as they do on our planet. Mars would also hold onto its atmosphere more tightly, and so loss of volatiles to space would be less of a problem. Hence, an Earth-sized planet at Mars' orbital distance might well be warm enough to sustain liquid water on its surface, even without human intervention. Or, to put it another way, present Mars is almost certainly within the habitable zone of our Sun. Its chief problem is not that it is too far away from the Sun to remain habitable, but rather that it is too small.

The second take-home lesson from this chapter is that the stabilizing effect of CO_2 on a planet's climate is not without limits. As the solar flux to a planet is reduced, CO_2 becomes more and more likely to condense, thereby limiting the magnitude of the greenhouse effect. Even a 25 percent reduction in solar flux makes it difficult to keep Mars' average surface temperature above freezing. Other greenhouse gases might help explain the apparent warmth of early Mars, but still it is clear that the outer edge of the habitable zone cannot lie too far beyond Mars' orbital distance.

· Chapter Nine ·

Is the Earth Rare?

Thus far, our discussion of planetary habitability has focused largely on climate and on the question of whether liquid water can exist on a planet's surface. These issues are critical, and we will return to them in the following chapter when we discuss habitable zones around other stars. But there are other factors as well that bear on the question of planetary habitability. Peter Ward and Don Brownlee raised a number of them in their book, *Rare Earth*, which was discussed briefly in Chapter 1. As mentioned there, Ward and Brownlee are pessimistic about the chances of finding advanced (animal) life elsewhere in the galaxy because they argue that many of the factors that make life possible on Earth may not be found elsewhere. Here we reexamine some of those same issues, along with several others that they did not discuss.*

Planetary Size / Magnetic Fields

Let's begin with an issue that was raised by Ward and Brownlee, and by others as well: Can a planet remain habitable if it lacks a magnetic field? Earth's magnetic field is thought to be produced by a *magnetic dynamo*

*Much of what follows in this chapter is based on an extended review that I wrote shortly after *Rare Earth* first appeared.[1] Readers who wish to judge matters for themselves are encouraged to read that book as well, so that they can compare our different viewpoints. This would be time well spent, as *Rare Earth* is delightfully written and filled with useful information, whether or not one agrees with the conclusions.

operating in its liquid outer core. The outer core, like the solid inner core, is composed of metallic nickel–iron alloy and can therefore conduct electricity. But it needs to be both rotating and convecting to create a dynamo effect. Convection occurs when a fluid in a gravitational field is heated from below (or cooled from above). In the case of the modern Earth, a significant part of that basal heat flux is provided by crystallization of the solid inner core, which is growing with time as the planet cools. Earth's mantle is also convecting energy upward toward a relatively cool surface, and this also helps to maintain a strong temperature gradient across the outer core. By contrast, Mars does *not* have an intrinsic magnetic field, probably because its outer core is too thin (i.e., it has already solidified to form a solid inner core). Venus does not have an intrinsic magnetic field, either, perhaps because it is spinning too slowly (chapter 6), or perhaps because its interior is still entirely liquid. (Venus has trouble losing internal heat because of its lack of plate tectonics and also because of its large greenhouse effect and correspondingly high surface temperature.) Earth itself may have lacked a magnetic field for the first several hundred million years of its existence if the solid inner core had not yet started to form by that time.[2]

One reason that Earth's magnetic field is considered to be important is because it helps protect the planet from bombardment by *cosmic rays*. Cosmic rays are energetic charged particles (and photons) that originate from space, either from the Sun or from further away in the galaxy. Solar cosmic rays are relatively low-energy particles, mostly protons, that are carried along by the *solar wind*. (The solar wind is a stream of charged particles that flow outward from the Sun's extremely hot, tenuous corona.) Galactic cosmic rays are high-energy particles and gamma rays that originate from outside the Solar System. Protracted exposure to either of these types of radiation can create health risks, including cancer. This is a serious constraint for astronauts out in space—they must limit their total exposure time to this radiation. Astronauts are particularly vulnerable to *solar proton events* (SPEs), which are short bursts of greatly enhanced particle flux from the Sun.

Earth's magnetic field does indeed partially shield the planet from cosmic rays. The reason is that charged particles follow curved paths when traveling perpendicular to magnetic field lines. But the geomagnetic field is effective only in deflecting relatively low-energy particles. That's because the radius of curvature, or *gyroradius*, of a particle's orbit around a magnetic field line is proportional to its energy. The most energetic cosmic rays have energies of many giga-electron volts (GeV), and so they traverse Earth's field along nearly straight paths, which means that they are barely deflected. Solar cosmic rays have lower energies, and hence are more strongly bent by Earth's magnetic field. But the shielding is more effective at low latitudes than at high latitudes because the field lines near the equator are oriented parallel to the Earth's surface (see figure 5.4). At high latitudes the field lines are perpendicular to the surface, and so incoming charged particles simply spiral around them on their way down towards the Earth.

What then shields Earth's surface from such radiation? The answer is the atmosphere itself! Most cosmic rays, including those incident at the poles, are absorbed by the atmosphere at altitudes of 80 km or more. The high-energy ones create cascades of secondary particles, and some of these do indeed make it down to the surface. By the time they get there, however, they have lost most of their punch, and so they account for less than 10 percent of the radiation exposure that an average person receives during the course of a year.[3] One gets a significantly larger radiation dose from *radon* (a gas released by decay of uranium in Earth's crust) and from medical X-rays. And you get an even bigger dose if you catch a train in Grand Central Station in New York, as the granite walls of the station are rich in uranium. You also get a substantially increased dose if you take a lot of plane flights, because you are then spending time above the shield provided by the relatively dense lower atmosphere.

Planetary magnetic fields have another effect that may actually be more important in keeping a planet habitable. Earth's magnetic field prevents the solar wind from interacting directly with its atmosphere and ionosphere, as shown in figure 9.1. (This statement is equivalent to saying that it protects Earth from solar cosmic rays.) Mars, which lacks an intrinsic magnetic field, does not have such protection, and consequently its atmosphere may have been *sputtered* away by the solar wind.[4] "Sputtering" is the term used to describe the process whereby collisions between solar wind particles and ions in a planet's upper atmosphere can lead to loss of atmospheric gases. Based partly on this observation,

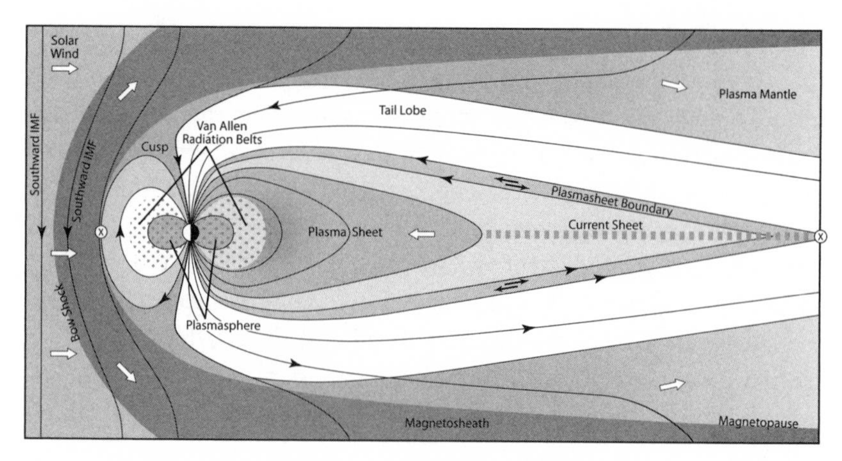

Figure 9.1 Diagram showing how the Earth's magnetic field interacts with the solar wind. The solar wind is deflected at the *bow shock* and then flows around Earth's long *magnetotail.*[5]

Ward and Brownlee suggest that planets that lack magnetic fields are unlikely to be habitable.

This argument, though, overlooks something obvious: Venus, which also has no intrinsic magnetic field, has a very dense atmosphere—100 times thicker than Earth's! Furthermore, Venus is closer to the Sun than is either Earth or Mars, and hence it is exposed to a denser, more energetic solar wind. If solar wind sputtering is so effective, then why didn't Venus lose its atmosphere? The answer probably has to do with planetary size. Venus is roughly the same mass as Earth, and so its upper atmosphere is less extended than that of Mars. That makes it harder for the solar wind to strip away atmospheric gases. Both Venus and Mars do have *induced magnetic fields* (fields generated in their ionospheres by their interactions with the solar wind), and in Venus' case this induced field is evidently sufficient to protect its atmosphere. Mars' problem is the combination of its lack of an intrinsic magnetic field *and* its small size. That double-whammy clearly bodes ill for planetary habitability.

This brings us back to the issue of planetary size itself. In the previous chapter, we pointed out that Mars' small mass (1/9 of an Earth mass) also allowed the planet to cool more quickly. Without enough geothermal heat to sustain volcanism, it was unable to recycle carbonates back into

CO_2. Mars appears to have already cooled off by about 3.8 billion years ago. No one knows exactly how big a planet needs to be to stay volcanically active for 4.5 billion years, as Earth has done, but one can guess that it needs to be 1/3 of an Earth mass, or more. A higher mass should also give a planet a better chance of sustaining a magnetic field because its core should take longer to solidify. And, as we have just seen, a high mass helps a planet hold onto its atmosphere. All of these factors point to an obvious conclusion: when it comes to keeping a planet habitable, size matters! A planet as small as Mars is not likely to remain habitable, except during the earliest part of its history.

How about if one goes in the other direction: Is there an upper limit on the mass of a habitable planet? Steven Dole, whose 1964 book *Habitable Planets for Man* is discussed in the next chapter, estimated that such planets must have a surface gravity $\leq$ 1.5 times that of Earth.[6] This would imply a mass of about 3 Earth masses or less. But this constraint was stipulated so that humans could walk upright. Microbes, and many other life forms, would be subject to no such limitation. Similarly, the climate system imposes no obvious upper limit on planetary size. Large planets would presumably have hotter interiors, and thus more volcanism, than would small ones. All other things being equal, this should cause them to have denser CO_2 atmospheres and warmer climates. But none of this necessarily precludes a planet from being habitable.

A more serious issue is that there may be practical limits to how large a terrestrial planet can grow without turning into a gas giant. In numerical simulations of planetary accretion, objects bigger than about 10–15 Earth masses tend to rapidly accumulate gas from the surrounding solar nebula, provided that such gas is still around. Jupiter and Saturn, for example, are thought have 15- to 20-Earth-mass cores that did exactly that. Uranus and Neptune have masses of 14.5 and 17.2 Earth masses, respectively, roughly comparable to giant planet cores. They accreted more slowly, though, because they were farther out, and hence there was less gas left for them to capture. (This is true even in the Nice model, albeit to a smaller extent.) By analogy, rocky planets formed in other planetary systems could conceivably be larger than 10 Earth masses, if their final formation occurred after the gas in their surrounding nebula was gone. Whether or not this occurs in nature is a question that will ultimately need to be answered by observations.

Ozone and Ultraviolet Radiation

Another habitability issue concerns stellar ultraviolet radiation and screening thereof by ozone in a planet's atmosphere. Earth, with its 21-percent O_2 atmosphere, has a well-developed ozone screen, and so the UV flux at its surface is relatively low. During the 1960s and 1970s humans used a lot of chlorine-containing *freons*, which might have destroyed the ozone layer had this practice been allowed to continue. We wised up quickly, though, after the ozone hole was discovered in Antarctica in 1985. Freons were replaced with other, more ozone-friendly compounds, and we embarked upon a remarkably sensible environmental path. Would that we could say the same about today's problem of global warming!

What, though, about planets with less than Earth's amount of oxygen? How much O_2 would they need to be protected against stellar UV radiation? For Earth, this problem has been studied by numerous atmospheric chemists over the last several decades.[7-11] The more recent models suggest that the ozone screen develops relatively quickly as O_2 concentrations increase (see figures 9.2 and 9.3). This result is not really new, however. Michael Ratner and James Walker showed why this should be the case over 35 years ago using a relatively simple photochemical model of a pure N_2-O_2 atmosphere.[8] They pointed out that, as O_2 is removed from the atmosphere, the short-wavelength solar UV radiation that splits it apart can penetrate more deeply into the atmosphere. Hence, as O_2 decreases, the ozone layer does not immediately disappear—it simply moves downward. Indeed, Ratner and Walker predicted that the ozone *column depth* (the number of ozone molecules in a vertical column of atmosphere) should actually increase as O_2 decreases down to 10^{-3} times the present atmospheric level (PAL). The reason has to do with the 3-body reaction that forms ozone:

$$O + O_2 + M \longrightarrow O_3 + M.$$

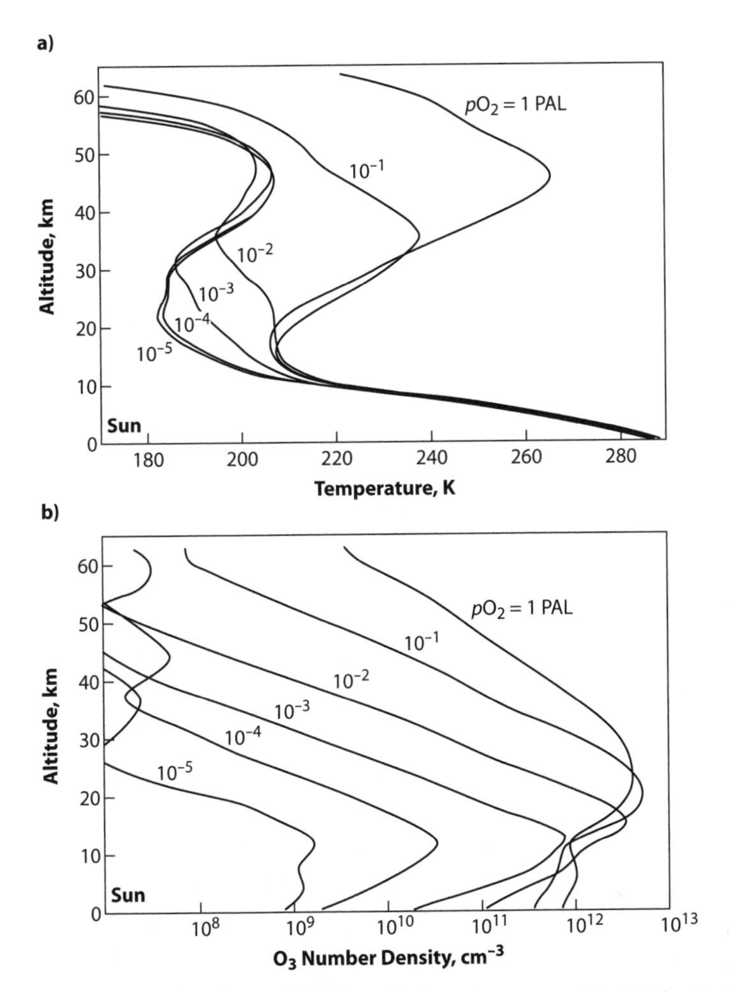

Figure 9.2 Ozone number density (a) and vertical temperature profiles (b) for Earth at different atmospheric O_2 levels. Here, "PAL" means "times the present atmospheric level."[11] (Reproduced by permission of Mary Ann Liebert Publishing.)

Here, M is a third molecule that is required to carry off the excess energy of the collision. As the ozone layer moves downwards, the density of molecules in the air increases, and so the ozone formation reaction proceeds even faster.

In more complex models, this tendency for faster ozone formation at low O_2 levels is offset by other factors, so the effect on ozone column

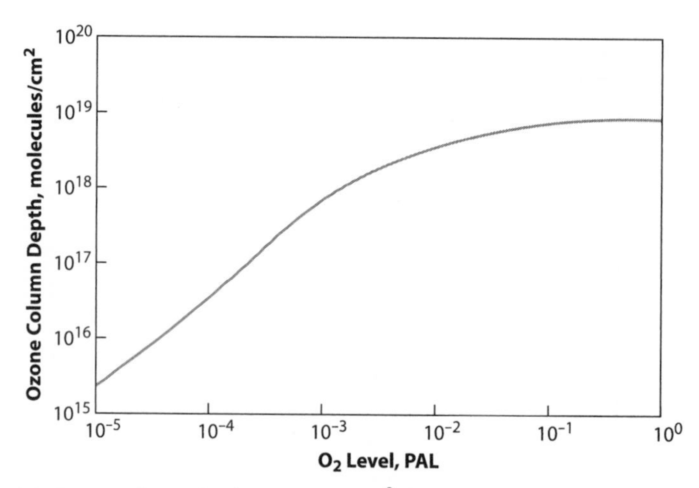

Figure 9.3 Ozone column depth (molecules/cm^2) as a function of atmospheric O_2 level. (Modified from Segura et al. 2003.[11])

depth is less pronounced. In particular, once the atmospheric O_2 concentration drops below about 10^{-2} PAL (times the present atmospheric level), the peak of the ozone layer moves downward into the troposphere (below 10 km). That is where most of Earth's water vapor is located. At this point, H_2O starts to be photolyzed (split apart) rapidly, and the by-products of this reaction then *catalyze* ozone destruction. Catalysis is what happens when a minor chemical species participates in a reaction in such a way as to make it go faster. In the modern atmosphere, chlorine compounds from photolysis of *freons* catalyze ozone destruction. That is why the world has collectively banned most such substances from production.

The bottom line is that the ozone column depth approaches its modern value for atmospheric O_2 concentrations above about 10^{-2} PAL, i.e., 1 percent of present. This suggests that the ozone layer was probably firmly established soon after atmospheric O_2 levels increased around 2.4 Ga (chapter 4). Thus, planets with even modest amounts of O_2 in their atmospheres may be well protected from stellar UV radiation. Turning this problem around, the fact that O_3 is abundant even at relatively low atmospheric O_2 concentrations implies that ozone is a sensitive indicator of O_2—a thought to which we return in chapter 14.

Availability of Nitrogen and the Importance of N_2

To this point, our discussion has almost totally avoided a key element of our own habitable planet: nitrogen. Nitrogen, in the form of N_2, makes up almost 80 percent of Earth's atmosphere. Nitrogen is also essential to life for several reasons. Both the amino acids that form proteins and the nucleic acids that form RNA and DNA are rich in nitrogen atoms. Despite the abundance of N_2 in Earth's atmosphere, it is not that easy for organisms to obtain nitrogen. To make use of N_2, organisms must first split the strong N≡N triple bond, or else depend on other organisms to do this for them. The biologists call this process *nitrogen fixation*. Interestingly, the most important nitrogen fixers in the modern oceans are the cyanobacteria, which we discussed in chapter 4. That's because they are the only aerobic (oxygen utilizing) organisms that can split the N_2 bond. Many anaerobic bacteria can also fix nitrogen, but plants and other eukaryotic organisms cannot. So the biosphere today still depends on the organisms that dominated it billions of years ago in the Archean era. And it has depended all along on the availability of nitrogen.

Nitrogen is an abundant element in stars, and it is abundant in the dusty disks from which planets form as well. Earth's nitrogen was probably obtained in the same way as other volatiles such as carbon and water—as a component of planetesimals originating from the asteroid belt region. In the carbonaceous chondrites discussed in chapter 2, nitrogen is bound up chemically in the same organic compounds that contain the carbon. Extrasolar "Earths" would presumably be supplied with nitrogen in this same manner, provided that the process of planetary formation was similar to what happened in our own Solar System.

Gaseous N_2 itself is useful for other reasons. For one, it slows down the rate of hydrogen escape from a planet by helping to limit the abundance of water in a planet's upper atmosphere. It does so by diluting the concentration of water vapor near the surface. Increasing the amount of N_2 increases the total pressure, and thus reduces the relative concentration of water vapor. As Ingersoll showed many years ago[12] (see chapter 6), water vapor becomes abundant in the upper atmosphere only when its mixing ratio near the surface exceeds approximately 20 percent. Today, the average water vapor mixing ratio is about 1 percent by volume. If N_2 was not present in Earth's atmosphere, the absolute amount of water vapor should still be about the same as at present—because the surface temperature would be largely unchanged—but its mixing ratio would be higher by a factor of 5, bringing the atmosphere closer to losing water, but not yet pushing it over the edge.

N_2 is also important for limiting wildfires. If one removed all (or even half) the N_2 in Earth's atmosphere, but left the O_2 abundance the same, fires would burn extremely rapidly.[13] The reason is that N_2 carries heat away from a flame without feeding it, whereas O_2 feeds the flame by increasing combustion. NASA rediscovered this phenomenon in a painful way when the first 3 Apollo astronauts burned to death in 1967 during an accidental fire on the test pad in Houston. Prior to that accident, NASA had used pure O_2 within the space capsule to keep the pressure down. Afterward, they went back to using air, which is a much safer medium for both men and machines.*

A key issue regarding N_2 is whether or not it would remain in the atmosphere without help from the biota. In his 1991 book, *Gaia, the Practical Science of Planetary Medicine*, Jim Lovelock pointed out that N_2 is removed from Earth's atmosphere by abiotic processes, as well as by biological nitrogen fixation.[14] In particular, lightning converts N_2 and O_2 into NO (nitric oxide). The NO is then oxidized to nitric acid, HNO_3, and rains out of the atmosphere, ending up as dissolved nitrate in the oceans. Without a return flux of N_2 from biological *denitrification*, Lovelock estimated that all of Earth's N_2 would be lost in about 1.6 billion years simply as a result of this (abiotic) process. The same thing would happen even if O_2 were not present, because the N_2 would combine with O atoms from CO_2. If this reasoning is correct, uninhabited planets might not be able to maintain N_2 in their atmospheres. Or, to say it another way, only inhabited planets would be habitable. This, of course, is a very "Gaian" concept—one that fit nicely into Lovelock's world view.

*As an aside, my interest in space and in NASA activities began at an early age, as I grew up in Huntsville, Alabama (home of Marshall Space Flight Center), where I attended Grissom High School. Gus Grissom, Ed White, and Roger Chaffee were the three astronauts who lost their lives in the fire in Houston. Huntsville has a school named after each one of them.

There is another side to Earth's nitrogen cycle that Lovelock neglected, however. Seawater is constantly cycled through the hot, midocean ridge vent systems, as part of the process by which newly created seafloor is cooled. Because the circulating seawater contains dissolved ferrous iron and sulfide, these hydrothermal vent systems are highly reduced. Any dissolved nitrate present within them should therefore be reduced either to N_2 or to ammonia, NH_3. Nitrate is relatively scarce in seawater today because organisms consume it, but it would be much more abundant on an abiotic planet. After exiting from the vents, water containing N_2 and NH_3 would be mixed back up to the surface, and both gases would be released into the atmosphere. There the NH_3 would be photolytically converted back to N_2, as described in chapter 3. If one inserts the appropriate cycling rates and gaseous solubilities, a simple calculation shows that at steady state less than 1 percent of the nitrogen on an abiotic Earth should end up residing in the oceans as nitrate, while the other 99 percent should be present in the atmosphere.[15] If this result is correct, then the ability to maintain an N_2-rich atmosphere may not depend strongly on the existence of life. And that, in turn, would support the idea that even uninhabited planets might be habitable—a notion that was advanced earlier (chapter 3) based on other arguments.

Is Plate Tectonics Common?

A central argument in Ward and Brownlee's analysis is that Earth may be unique in having plate tectonics. They recognized, as I have here, the importance of the carbonate-silicate cycle in regulating atmospheric CO_2 levels. As discussed in chapter 3, plate tectonics plays a key role in this cycle, by dragging carbonate sediments down subduction zones and releasing their trapped carbon as gaseous CO_2. So a planet without plate tectonics might not be able to maintain a stable climate.

To support their argument that plate tectonics might be uncommon, Ward and Brownlee point out that of the 20 largest rocky bodies within the Solar System, including both planets and large moons, Earth is the only one that shows evidence of plate tectonics. They also point out that plate tectonics probably requires liquid water, for two reasons. First, water flowing through the midocean ridge hydrothermal vents cools the newly formed seafloor, and thus helps solidify the oceanic plates. And, second, water lubricates the partially molten *aesthenosphere* that underlies the *lithospheric* plates, allowing them to slide around on top of it. In a sense, water is to plate tectonics as oil is to an internal combustion engine: without it, the engine (or the solid Earth) would seize up and cease to operate smoothly.

Our neighboring planet Venus provides a good example of this. Venus shows no evidence of having ever had plate tectonics, based on radar images of its surface made from the *Magellan* spacecraft in 1994. The contrast with Earth is very clear (see figure 9.4, in color section). Earth has a *bimodal* distribution of elevations, with most of the seafloor being near 4 km below sea level and with continental elevations averaging about 1 km above sea level. By contrast, 60 percent of Venus' surface area lies within 500 m of the mean planetary elevation.[16] Earth has linear mountain chains, such as the Appalachians (not shown), that were created by continental collisions. And, most importantly, Earth has midocean ridges, like the central and southwest Indian ridges shown in figure 9.4b. These are regions where new seafloor is being created. By contrast, Venus (figure 9.4a) has what appear to be low areas (colored in blue) and high areas (colored in brown), but the distinctive tectonic features that mark the Earth are absent. One must be careful in analyzing these images because the map of Venus shown here is based on radar reflectivity, rather than elevation, but a topographic map, while not as pretty, shows much the same patterns.[16]

Further evidence that Venus lacks plate tectonics is provided by the distribution of impact craters on Venus' surface. The *Magellan* synthetic aperture radar that made the image shown in figure 9.4a had a spatial resolution of approximately 100 m, so it was able to look for craters down to a few hundred meters in diameter. It found none smaller than about 3 km in diameter, however. This observation was readily explained: Venus' thick atmosphere shields its surface from any impactor smaller than about 30 m in diameter.[16] More surprising was the fact that the

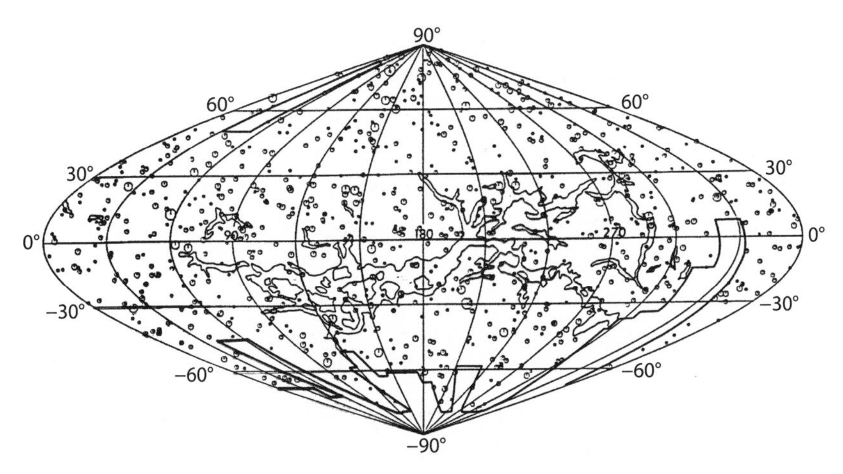

Figure 9.5 Equal-area projection of Venus showing 842 impact craters randomly distributed over the planet's surface.[19] (Reproduced by permission of the American Geophysical Union.)

observed impact craters were distributed randomly over Venus' surface (figure 9.5). Since the end of the heavy bombardment period, around 3.8 Ga, the cratering rate in the inner Solar System is thought to have been more or less constant. Hence, the number density of craters on a given part of a planet's surface can be used to estimate its age. The fact that Venus' craters are randomly distributed implies that all portions of Venus' surface have approximately the same age, between 0.5 and 1 billion years. On Earth, by contrast, the seafloor is much younger than the continents: about 60 million years, on average, as opposed to hundreds of millions to billions of years for the continents. This, of course, is a consequence of plate tectonics: the seafloor is continually being created and destroyed, while the continents float above the fray and survive for much longer time periods.

Why should Venus' surface have a common age? Donald Turcotte of Cornell University proposed some time ago[20] that the reason has to do with Venus' lack of water and lack of plate tectonics. Venus, as we have seen, is nearly as big as Earth, and it probably has a comparable amount of radioactive elements in its mantle and core, so it should produce about the same amount of internal heat. This heat is presumably carried upward by convection in Venus' mantle, as on Earth, but it has to be transmitted through the planet's crust by conduction. Thermal conduction through rock is not very efficient, and so other heat transport mechanisms come into play. On Earth, most of the heat gets out by way of plate tectonics: some 90 percent of this geothermal heat escapes at the midocean ridges where new seafloor is being created. This process is not entirely smooth—the Earth is not as well lubricated as a modern automobile engine—but on geologic timescales it is more or less continuous. The average spreading rate at fast-spreading ridges is of the order of a few centimeters per year.

Venus, by contrast, is dry, and so the plates are poorly lubricated. Consequently, in Turcotte's view, Venus goes through a long and fitful thermal cycle. Most of the time, Venus' surface is as we see it today—more or less static. This means, however, that heat from radioactive decay is building up in Venus' interior. At some point, probably within the next few hundred million years, the interior will become hot enough for large-scale melting to occur. At that time, widespread volcanism will resurface the planet, while at the same time cooling its interior. The interior will then solidify, and the entire cycle will begin again. According to the crater counts, the last cycle of volcanism occurred between half a billion and one billion years ago.

So what does all this imply about the prevalence of plate tectonics? Not very much, I would say. Clearly, a planet must be relatively big to maintain a strong enough internal heat flow to drive plate tectonics. That is why 18 of the 20 largest rocky bodies in the Solar System don't exhibit it—they are all too small. And Venus, which is large enough, doesn't have plate tectonics because it doesn't have liquid water. But that is a separate problem that has to do with its distance from the Sun, as we saw in chapter 6. Consequently, there is simply no logical basis for arguing that plate tectonics is rare. It could just as well be an inevitable development on large, rocky planets that, like Earth, also have liquid water. But it's going to be difficult to prove who is right on this question, isn't it? Determining whether an extrasolar planet has plate tectonics is even harder than studying the composition of its atmosphere.

A Planet's Impact Environment

Ward and Brownlee pointed out that the giant planets in our Solar System, especially Jupiter, affect biological evolution on Earth by altering the rate at which Earth is bombarded by asteroids and comets. Asteroids come originally from the asteroid belt region, as discussed in chapter 2. Comets come from either the Oort Cloud or the Kuiper Belt (see figure 2.3).

In *Rare Earth*, Ward and Brownlee were concerned mostly with comets. That is largely because George Wetherill of the Carnegie Institute in Washington, mentioned in chapter 2, pointed out that Jupiter effectively shields Earth from impacts by comets.[21] Comets in the Oort Cloud are far enough away from the Sun to be influenced by such things as the passage of neighboring stars and by galactic tides. These can produce showers of comets in perturbed orbits that may pass through the main part of the Solar System. Similarly, Kuiper Belt objects are sometimes perturbed by near collisions with each other, or they can be "nibbled" off the inner edge of the belt by gravitational interactions with Neptune. Once in the Solar System itself, these bodies are called comets. A typical comet is several kilometers or more in diameter and is moving very fast (60 km/s or more) relative to planetary orbits; hence, it would create a lot of destruction were it to collide with the Earth. Fortunately, though, most of them never make it into the inner part of the Solar System. The reason, as Wetherill pointed out, is that most comets pass close to Jupiter before getting that far. Once that happens, they either collide with Jupiter, as did comet Shoemaker-Levy 9 in 1994, or they are ejected from the Solar System by Jupiter's strong gravity. It might seem strange that a single planet could protect the entire inner Solar System in this manner, but this happens because most comets make many orbits around the Sun, and hence have many opportunities to encounter Jupiter, before they work their way into the terrestrial planet region.

Wetherill argued in his paper that if Jupiter were not present, the flux of comets into the inner Solar System might be as much as 10^4 times higher than today, and the rate of cometary impacts on Earth might therefore be higher by this same factor. Based on current comet statistics, an impact large enough to cause a severe mass extinction of species is thought to occur roughly once every 100 million (10^8) years.[22] Hence, if Wetherill's analysis is correct, in the absence of Jupiter an impact big enough to cause mass extinctions might occur every 10,000 years. Based on these statistics, Ward and Brownlee argued that advanced life might not be possible on Earth if Jupiter did not exist, because frequent large impacts would likely stifle the evolutionary process. This thought may well be correct. Human civilization has been around now for about 10,000 years. If Earth had experienced a giant comet impact during this time, it is unlikely that you would be sitting there reading this book.

But there is more to the Jupiter story than just this. Earth can be hit by asteroids, as well as by comets. Indeed, the impact that killed off the dinosaurs was almost certainly an asteroid, rather than a comet, based on the large iridium anomaly.[23] This event is referred to as the *K-T impact* because it occurred at the boundary between the Cretaceous (K) and Tertiary (T) periods, some 65 million years ago. A buried impact crater near the town of Chicxulub on Mexico's Yucatan Peninsula is thought to have been created by this impact. The crater is roughly 200 km in diameter, which is just the size that one would expect if the Earth had been hit by a 10-km asteroid—the size needed to explain the iridium anomaly. Because comets are less than half rock, and because their expected impact velocities are much higher than those of Earth-crossing asteroids (~60 versus ~20 km/s), a comet that was the right size to produce the Chicxulub crater would have contained only a small fraction of the observed iridium.

Here is where the impact story begins to get complicated. As we saw in chapter 2, the asteroid belt owes its existence to the presence of Jupiter. (Jupiter formed early and this prevented a planet from forming in the asteroid belt region.) And the way that asteroids get from the asteroid belt into orbits that intersect Earth's orbit is also partly a result of Jupiter's influence. Asteroids suffer collisions with each other, which knock them into orbits that are in *resonance* with Jupiter's orbit. For example, asteroids with semi-major axes near 2.5 AU orbit the Sun 3 times every time Jupiter goes around once.* Once an asteroid is in such a resonance, Jupiter's gravity can then perturb it into an orbit that may eventually cross Earth's path. So Jupiter is responsible for *causing* large impacts on Earth, as well as for protecting us from them. This may actually be a good thing. After all, some impacts are beneficial. We humans would likely not be around if the K-T impact had not knocked off the dinosaurs and thereby allowed mammals to proliferate.

Put this way, this sounds like another beneficial effect of Jupiter. But, on the other hand, some asteroids are large enough to be scary. Ceres is over 1000 km in diameter, and Vesta and Pallas are both around 500 km in size. Because the mass of an object is proportional to its diameter (or radius) cubed, Ceres' mass is roughly 1 million times larger than that of the K-T impactor. The energy of a Ceres-sized impact on Earth would be higher by this same factor. Indeed, it would likely be enough to vaporize Earth's oceans entirely[24] and create a steam atmosphere much like the runaway greenhouse atmosphere on early Venus. Vesta and Pallas are big enough to do this as well. Fortunately, none of these large objects is on an orbit that might allow this to happen. An impact of this magnitude might sterilize the entire Earth, with the possible exception of organisms living in deep subsurface environments.

What, then, is the moral of this story? It appears that having a Jupiter-sized planet present in a planetary system is a mixed blessing. It may encourage biological evolution on neighboring terrestrial planets in some ways, and yet inhibit it in others. On one point, though, astronomers are in near universal agreement: if we wish to understand the habitability of a terrestrial planet orbiting around another star, we need to have information about the giant planets in that system as well. Fortunately, as we shall see later on, finding them is a natural part of looking for other Earths.

*One can demonstrate this from Kepler's third law. If the planet's period, P, is expressed in Earth years, and its semi-major axis, a, is in AU, then the relationship can be written as $P^2 = a^3$. Jupiter's semi-major axis is 5.2 AU, and its orbital period is 11.8 Earth years. An asteroid with a semi-major axis of 2.5 AU has an orbital period of 3.9 Earth years, or 1/3 that of Jupiter.

Stabilization of Earth's Obliquity by the Moon

Finally, let's consider another key point that was raised by Ward and Brownlee in *Rare Earth*: the importance of the Moon in stabilizing Earth's obliquity. Recall from chapter 5 that Earth's obliquity—the tilt of its spin axis—varies between about 22° and 24.5° every 41,000 years. This oscillation is part of the well-studied Milankovitch cycles that have served as the pacemaker for the Pleistocene ice ages. We then saw in chapter 8 that Mars experiences obliquity variations that are much larger than Earth's (±10 degrees) and that vary chaotically by a much greater amount over time scales of tens of millions of years.

Part of the reason why Earth's obliquity variations are small, while Mars' variations are large, has to do with the presence of Earth's large Moon. Mars has two moons itself, Phobos and Deimos, but these are captured asteroids that are too small to have any significant gravitational influence on the planet. The French astronomer Jacques Laskar heads one of the groups who have studied these chaotic obliquity variations in detail.[25] These same authors have studied what would happen to Earth's obliquity if the Moon was not present.[26] The story is told graphically in figure 9.6, which has been adapted from one of their papers.

To preface the discussion, let me relate the story that is told by Ward and Brownlee. If one performs a mathematical experiment in which the Moon is suddenly assumed to disappear, then Earth's obliquity is predicted to behave like Mars' obliquity, only more so. On timescales of tens of millions of years, it would vary chaotically from values as low as 0° up to as much as 85°. (The maximum value is even higher than that of Mars' obliquity, which never exceeds 60°.) At 85° obliquity, Earth would be tilted almost over on its side, much like the planet Uranus, which has an obliquity of 98°. As Earth moved around the Sun in its orbit, its surface temperature would vary dramatically at high latitudes. Each pole would be alternately baked and frozen. Tropical regions would experience four seasons: two summers and two winters. The summers there would occur near equinox, when the Sun was above the equator. The winters would occur at solstice, when the Sun was above one of the poles.

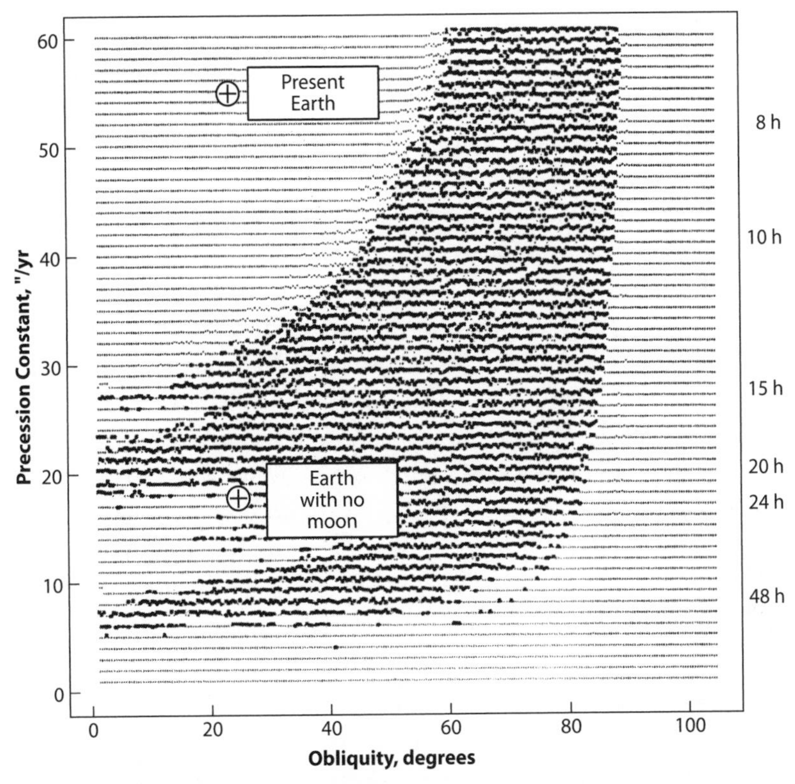

Figure 9.6 Diagram illustrating the effect of the Moon on Earth's obliquity. The vertical scale on the left is a measure of the rate at which Earth's spin axis precesses around in a circle. The right-hand scale shows the Earth's corresponding rotation period if the Moon was not present. The fuzzy region is where chaotic obliquity variations can occur. In the clear region, Earth's obliquity varies regularly. See text for additional explanation. (Modified from Laskar and Robutel 1993.[25])

Ward and Brownlee argue convincingly that such a situation would be hard on land life. Any continent located at high latitudes would almost certainly be uninhabitable, as a consequence of the extreme seasonal temperature variation. Low-latitude continents could conceivably be inhabited, as the temperature extremes there would be smaller. But continents drift around as a consequence of plate tectonics, and so sooner or later each one would pass near one of the poles and would be sterilized of most surface life. So this situation might indeed pose problems for the evolution of land-dwelling animal life—the type of life of interest to the authors.

Ward and Brownlee further argue that the formation of the Moon was a cosmic accident—one not likely to occur during the formation of most terrestrial planets. In saying this, they are entirely correct. The Moon is widely believed to have formed as the result of a very large impact that occurred during the Earth's formation.[27] The evidence includes similar oxygen isotopes in Moon rocks and Earth rocks, the large angular momentum of the Earth-Moon system, the fact that the Moon is deficient in iron, the lack of volatiles (including water) on the Moon, and the inferred high-temperature origin of the lunar crust. Detailed numerical simulations of such a giant impact have been performed by various investigators.[28] These simulations suggest that the planetesimal involved in the Moon-forming event was probably the size of Mars, or even larger. While many large impacts are thought to have occurred during Earth's formation, few of them could have been this big—otherwise Earth itself would be much larger. Furthermore, the chance that the impact would occur at the right angle, or *impact parameter*, is also relatively small. If the impact is too head-on, then the material that is ejected simply falls back upon the Earth. Only a glancing blow at a slow relative velocity leaves a significant amount of material in orbit around the Earth, where it can subsequently accrete to form the Moon.

What does all this imply about the existence of life elsewhere? First, it should be clear that large obliquity variations are primarily a concern for animals that live on land. Microbial life would be much less affected, as microbes can live beneath the planet's surface. And marine life, including fish and dolphins, might also be relatively unaffected. Simulations with climate models have shown that, even at 90° obliquity, Earth's oceans would experience only small temperature changes throughout the year.[29] That's because of the high heat capacity of the oceans, discussed in chapter 1. As most of the biosignature gases that we will discuss later in this book are produced either by microbes or by marine life, the consequences of high obliquity variations for remote life detection would appear to be small.

What, though, of Ward and Brownlee's central argument: Does this really imply that animal life, and intelligent life in particular, is rare? To answer this question, one needs to understand what causes a planet's obliquity to vary in the first place. Briefly, the story is as follows. A planet's obliquity varies for two reasons: (1) the pull of the Sun (and the Moon, for Earth) on its equatorial bulge, and (2) the variation of the inclination of its orbital plane over time. (Recall that a planet's inclination is the tilt of its orbital plane with respect to the average, or *invariant*, plane of the Solar System, as defined on the basis of angular momentum.) Here we are concerned with the first of these two factors. Because both Earth and Mars spin rapidly, they are each a little fatter around the equator than around the poles. This is termed the *equatorial bulge*. The Sun and the Moon both pull on Earth's bulge, and this causes Earth's spin axis to precess around in a circle every 26,000 years. Mars' spin axis precesses as well, but it does so much more slowly, about once every 170,000 years. Mars precesses more slowly than Earth partly because it lacks a large moon and partly because it is farther from the Sun, so that the Sun's pull on the equatorial bulge is reduced as well.

Mars' obliquity varies chaotically because the precession rate of its spin axis is approximately equal to the rates at which the orbits of the other seven planets precess.[25] Without going into detail, the orbit of a planet can be compared to the motion of a dinner plate on a table if one sets it almost upside-down and gives it a little spin, so that it rolls around on its edge. If the precession rate of Mars' spin axis is equal to the wobble rate of another planet's orbit, then a resonance occurs, and the obliquity can undergo large changes.* As we have seen throughout this chapter, resonances of various sorts play all sorts of roles in planetary evolution. And they depend on all of the planets in the system, not just the one being studied. So whether or not a planet in another system will undergo chaotic obliquity variations depends on the nature of that entire planetary system.

We can say more, too, about the problem of Earth's obliquity. The size of Earth's equatorial bulge depends on its rotation period: the faster

*The term "wobble rate" is used here to represent two characteristic motions of a planet's orbit: the rate of precession of perihelion and the precession of the *line of nodes* (the points where the planet's orbit intersects the invariant plane of the Solar System). Resonances with either of these motions are termed *secular resonances* to distinguish them from the *mean-motion resonances* that occur when the orbital periods of two bodies are commensurate.

Earth spins, the larger its equatorial bulge. Because the precession rate depends on the size of the equatorial bulge, this means that Earth's spin rate and its precession rate are proportional to each other. That relationship is shown in figure 9.6 by the two different scales on the vertical axis. The scale on the left is the precession constant, which is related to the precession rate.* The scale on the right represents Earth's rotation period, or daylength, *in the absence of the Moon.* The precession constant and daylength of the present Earth are not related in this manner because Earth, of course, does have a Moon.

If the Earth lacked a Moon, and if it had its present 24-hr day and 23.5° obliquity, then it would plot at the point labeled "Earth with no Moon." Its precession constant would be about 18" (18 arcseconds) per year, and it would lie smack in the middle of the large fuzzy region of the graph.** The fuzzy region is the region where the obliquity is chaotic. If one starts a simulation anywhere within the fuzzy region, and then integrates the mathematical equations for millions of years, the planet's obliquity can vary all the way from the lefthand edge of the fuzzy region to the righthand edge. For Earth, this is 0–85°. It looks from the figure as if there is a gap in the chaotic region over to the left of the point in question; however, that is simply because the equations were only integrated for 18 million years in these calculation. That entire region of the diagram is chaotic.

By contrast, Earth with a Moon plots in the upper lefthand corner of the diagram, where the solutions are regular. That's because its actual precession constant is 55″ per year, or roughly 3 times greater than if the Moon were not present. This precession rate is significantly faster than any of the "wobbles" of the other planetary orbits, and so Earth's obliquity is less strongly affected by them.

Finally, armed with all this additional knowledge, let's revisit the question of what Earth's obliquity might do if the Moon was not present. Now, the answer is not so obvious! In the first place, Earth's spin rate has

*Technically, the precession rate is equal to the precession constant multiplied by the cosine of the planet's obliquity.

**The term "arcseconds" will be defined in chapter 11. For now, let it suffice to say that there are 1,296,000 arcseconds in a complete circle.

been decreasing with time as a consequence of dissipation of tidal energy in the oceans. These tides, of course, are caused by the Sun and the Moon, just as is the spin axis precession. From figure 9.6, one can see that a Moon-less Earth would have a stable obliquity provided that its daylength was shorter than about 12 hours. Earth was probably rotating about this fast sometime back in the early Archean era. So, if one removed the Moon from the Earth-Moon system at that time, Earth's obliquity would have remained stable. But this still does not answer the question. It just leads to another one: how fast would the Earth have been rotating initially if the Moon-forming impact had not occurred? That event probably left Earth spinning with about a 4- to 5-hour rotation period—just short of the spin rate at which the whole planet would fall apart. A Moon-less Earth would presumably have spun slower initially, but it is hard to say how much.

So where does all this lead? Not very far, I'm afraid. It is virtually impossible to predict whether a given planet's obliquity will be stable without knowing its rotation rate, whether or not it has a large moon, and the masses and orbital parameters of all of the other planets in the system in which it resides. The only way that we could possibly obtain such information would be by detailed observational studies with space telescopes well beyond those being considered for the near future. I will speculate about such telescopes in chapter 15, but neither you nor I will likely live long enough to see them. In the meantime, though, one thing seems clear: one shouldn't rule out the possibility of life, or of animal life, just because most Earth-like planets are not likely to have large moons. That's really one of the smaller problems that must be overcome for a planet to remain habitable.

Even though it is difficult to resolve many of the issues discussed in this chapter, we can still offer some general conclusions. Most importantly, orbiting at the right distance from a star is only the first requirement for planetary habitability. Other factors, including the masses and orbits of the other planets within the system, may be equally essential. We should bear this in mind when we start to perform the types of searches described in the last section of this book. But we should also point out that there is no reason for extreme pessimism, or even for the more modest level of pessimism expressed in *Rare Earth*. Nothing that we have learned so far about Earth suggests that our planet should have a unique capacity to support life, be it simple, complex, or intelligent. Carl Sagan voiced this same opinion throughout his career, and we see no reason to overturn it.

· *Chapter 10* ·

Habitable Zones around Stars

In Chapters 6–8, we examined the climate histories of Venus and Mars, neither of which is habitable today, and we also looked at Earth's distant future, during which the planet is unlikely to remain habitable unless humans intervene technologically. Then, in chapter 9, we explored a variety of other factors that bear on the habitability of Earth. Here we return to the issue of long-term climate evolution and use the results from earlier chapters to estimate the boundaries of the habitable zone around the Sun and other stars. Recall from chapter 1 that Harlow Shapley defined the "liquid-water belt" as the region around a star in which a planet can maintain liquid water on its surface. We argued there that restricting our search to liquid water–dependent life is reasonable because of the special chemical properties of the water molecule. We also argued that this water needed to be at a planet's surface for life to be vigorous enough to be remotely detectable. How likely is it that such liquid-water planets exist?

Historical Attempts to Define the Habitable Zone

At about the same time as Shapley was writing his book, a medical researcher named Hubertus Strughold defined what he termed the *ecosphere* around the Sun.[1,2] This region was entirely analogous to Shapley's liquid-water belt, and Strughold also emphasized the importance of surface liquid water. Curiously, Strughold believed that Venus, Earth, and Mars were all within this region. He estimated surface temperatures on Venus to be between −10°C and +100°C, and those on Mars to be between −70°C and +25°C. For Mars, these estimates were not too far off the mark, but for Venus they were, of course much too low. Strughold should not be blamed too much for this error, as this was a common mistake prior to 1962, when *Mariner 2* finally measured Venus' surface temperature using a microwave radiometer. To his credit, Strughold did realize that Venus might not actually be habitable, even if its surface temperature appeared to be in the right range, as he was aware that water vapor was almost entirely absent from Venus' atmosphere.

The astronomer Su-Shu Huang then took up this topic in the late 1950s.[3,4] Huang gave a well-reasoned, if somewhat cursory, discussion of a variety of different issues bearing on the habitability of planets around other stars. He pointed out, for example, that binary or multiple star systems are less likely to harbor habitable planets than single stars because the planetary orbits would be unstable in most cases. And he concluded, correctly, as we shall see, that stars that are similar in mass to our Sun are the most likely to harbor habitable planets. But his most lasting contribution may have been his definition of the term *habitable zone* as another synonym for "ecosphere" and "liquid-water belt." He also coined the term "astro-biology" to describe the search for extraterrestrial life. The first term stuck and is the preferred word today for describing the liquid-water region around a star, while the second term was adopted by NASA in 1996 when it chose to reinvigorate this field by creating a new research area called astrobiology. So perhaps Dr. Huang deserves some of the credit for this development.

Shortly after this, Steven Dole wrote a book entitled *Habitable Planets for Man*,[5] which was mentioned in the previous chapter. Dole was interested in a more focused question: How many nearby stars might harbor planets that could be colonized by humans? His conditions for habitability were thus entirely anthropocentric. They included mean annual temperatures between 32 and 86°F (0–30°C) over 10 percent of a planet's surface, an O_2-rich atmosphere, and a surface gravity less than 1.5 times that of Earth, so that humans could walk upright. Dole's climate models were rather crude: he considered black planets with thin atmospheres (no greenhouse effect) and 45 percent cloud cover. Nevertheless, Dole made a number of points that are still considered to be valid today. In particular, he reiterated Huang's concerns about orbital stability in binary star systems, and he pointed out the problem of tidal locking of planets orbiting dim, red stars, which we discuss later in this chapter. So his book was in many ways well ahead of its time.

None of these early researchers attempted to define the habitable zone using a realistic climate model. The real pioneer in doing so was a researcher named Michael Hart at NASA's Goddard Space Flight Center. Hart wrote two papers on this topic during the late 1970s.[6,7] These papers were influential, not because they were definitive from a climate modeling standpoint—they most certainly were not—but rather because Hart laid out the issues involved in a manner that was clear and that provoked further scientific interest in the topic. I can attest to this personally because I read Hart's papers while I was in graduate school at Michigan, and they directly spurred my own interest. I can also remember being somewhat dismayed by Hart's pessimistic conclusions. I am one of those people who, like Carl Sagan, would like to believe that life is widespread in the universe. Part of my motivation for pursing this subject was to try to figure out if Hart's conclusions might be wrong.

One useful thing that Hart did in his papers was to define a new term, the *continuously habitable zone*, or CHZ, for short. (The conventional habitable zone, then, can be abbreviated as "HZ.") Hart was aware of the fact that the Sun and other main sequence stars brighten as they age. Hence, the HZ must move outward with time, as indicated in figure 10.1. Suppose the Sun's initial HZ at time t_0 covered the range of distances shown schematically in the figure, and that by some later time t_1 it had moved further out, as indicated. Then the CHZ is represented by the overlap between the two regions. Note that the HZ is defined at a single instant in time, whereas the CHZ has meaning only if it is defined over a particular time period, in this case the interval t_1–t_0. For our own Sun, we typically take that period to be the entire 4.6-billion-year history of the Solar System.

In the first of his two papers,[6] Hart attempted to simulate the evolution of Earth's atmosphere and to estimate the width of the HZ and CHZ around our Sun. To do so, he constructed a computer model that

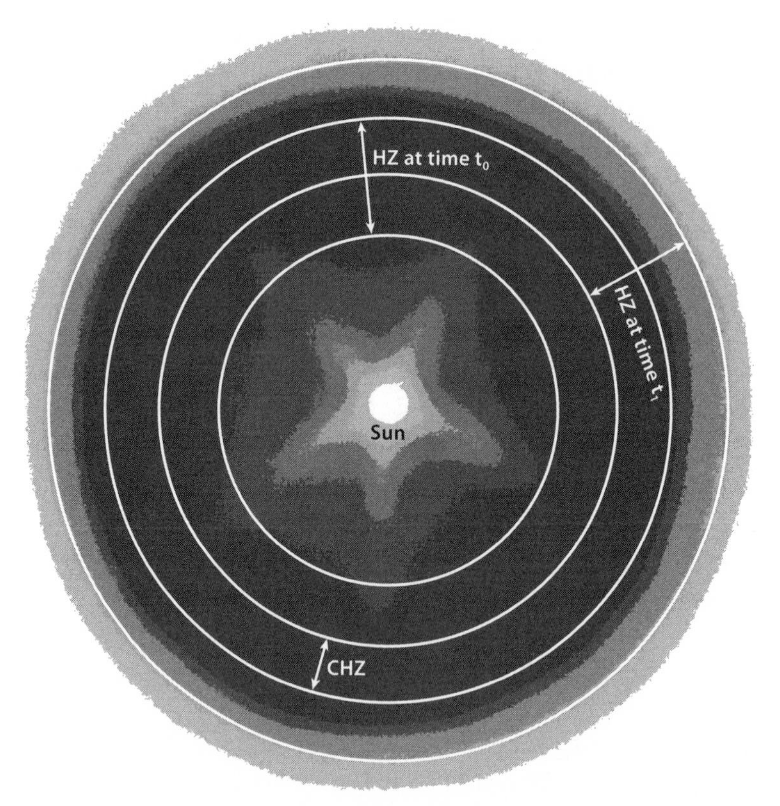

Figure 10.1 Diagram illustrating the concepts of the habitable zone (HZ) and continuously habitable zone (CHZ) around the Sun. (Based on the work of Michael Hart.[7])

incorporated many of the phenomena that we discussed in chapters 3–6. These included outgassing of CO_2 from volcanoes, reactions of CO_2 with surface minerals, the presence of reduced greenhouse gases (CH_4 and NH_3) early in Earth's history, organic carbon burial, the rise of O_2, changes in solar luminosity, and ice albedo feedback. So Hart had a lot of chemistry and physics in his model. Hart's methodology was to start with a planet with specified initial conditions and then to evolve the planet over time by solving a set of time-dependent differential equations. In particular, he started his simulations by assuming an airless Earth at 4.5 Ga, and then he integrated his model equations up to the present in 2.5-million-year increments. Then, he varied his assumptions about the composition of volcanic gases and repeated his computer runs numerous times until he came up with a "best fit" to Earth's atmospheric and climatic evolution. This optimal model successfully accounted for the present surface pressure and composition of the atmosphere, the mass of the oceans, the amount of CO_2 stored in carbonate rocks, and the rise of atmospheric O_2 at ~2.0 Ga. Recall from chapter 4 that Preston Cloud had by this time already outlined the geologic evidence showing when this important event occurred.

Having achieved this satisfying result, Hart then modified his calculations by varying the initial Earth-Sun distance, while leaving all other parameters the same as in his optimal model. When he did so, he found that all simulations that began with the Earth at <0.95 AU from the Sun resulted in a runaway greenhouse. Surprisingly, this catastrophe occurred early in the planet's history, when the solar flux was low, rather than later on. So it is not directly comparable to the calculations discussed in chapter 6, which also indicate that water would be lost inside of 0.95 AU, but for the modern solar constant. Hart's model also developed a true runaway greenhouse, meaning that the oceans evaporated completely, whereas the model described in chapter 6 merely develops a wet stratosphere at that distance. In Hart's model, the runaway greenhouse occurred early because of high concentrations of reduced greenhouse gases in Earth's primitive atmosphere.

Going in the other direction, Hart found that if the Earth had formed at 1.01 AU or beyond, it would have experienced runaway glaciation around 2 billion years ago, when O_2 appeared in the atmosphere, causing the concentrations of reduced gases to decline. Recall from chapter 4 that this is precisely what we now think *did* occur at about this time, based on the evidence for low-latitude glaciation at 2.4 Ga. In Hart's analysis, though, this was considered a fatal problem, as his simulated planet was incapable of recovering from such a situation. Indeed, in one passage he wrote:

> Runaway glaciation occurred in many of my computer runs. In every such case, the computation was continued—often for an additional 2 billion years, or even longer—with CO_2 (from volcanoes) accumulating in

> the atmosphere, and with the greenhouse effect computed. In not a single such case was runaway glaciation reversed.

Exactly why his climate model failed to recover from runaway glaciation is not clear. It was a highly simplified model, though, and its treatment of both radiation and convection left much to be desired. If he had used a more elaborate climate model, he might have gotten different results. It really *does* make a difference in this business how well one's model is constructed!

Putting all these simulations together, Hart estimated that the 4.6-billion-year CHZ around our Sun extends from only 0.95–1.01 AU.[6] In his subsequent paper, Hart extended these calculations to other main sequence stars.[7] His strategy in this second round of calculations was exactly the same: he took his optimal model for Earth and ran the same simulation for planets orbiting stars of different masses and luminosities. (We discuss such stars in some detail later in this chapter.) When he did so, Hart found that the CHZs of stars less massive than the Sun were either extremely narrow or entirely nonexistent. Stars more massive than the Sun had wider CHZs, but these were very short-lived because such stars evolve much faster. The bottom line of Hart's calculation was summarized in the last sentence of his 1979 paper:

> It appears, therefore, that there are probably fewer planets in our galaxy suitable for the evolution of advanced civilizations than has previously been thought.

Indeed, a pessimist reading his papers might well conclude that Earth is the only one.

A More Modern Model for the Habitable Zone around the Sun

Motivated partly by having read Hart's papers in graduate school, I worked on this problem during my years at NASA Ames and for several years thereafter. My coauthors on the paper that eventually came out of this work were Daniel Whitmire from the University of Southwestern Louisiana and Ray Reynolds from NASA Ames.[8] Our calculations are described briefly below.

Estimating the boundaries of the habitable zone for our own Sun is closely related to the problem of understanding long-term climate evolution on Venus and Mars. So much of the discussion in chapters 6 and 8 applies, and we can simply summarize it here. The inner edge of the HZ is determined by when a planet develops a wet stratosphere and loses its water through photodissociation, followed by escape of hydrogen to space. In chapter 6 we described a hypothetical numerical experiment in which we slowly pushed the Earth closer to the Sun. In our model, the planet began to lose water rapidly when the solar flux became more than ~10 percent higher than its present value at Earth. This translates, via the inverse square law, to a distance of 0.95 AU for the inner edge of the modern HZ.

Following this same line of reasoning, finding the outer edge of the HZ is much like trying to understand the climate of early Mars. Although the climate calculations shown in chapter 8 (figure 8.6) were not presented in this way, one can imagine a similar sort of numerical experiment in which one slowly slides the Earth out toward Mars' orbit. Unlike the planet in Hart's model, which froze over irreversibly when it got cold, our model planet has a built-in negative feedback: when the climate gets cold, silicate weathering slows, and volcanic CO_2 builds up in its atmosphere. After all, that is how we think that Earth survived the faint young Sun. In our model, this is sufficient to prevent global glaciation, provided that the solar flux is above a critical value. We can use the climate calculations for Mars from the last chapter to estimate what this value might be. Looking back at figure 8.6, one can see that it is impossible to bring the mean global temperature of Mars above the freezing point of water for solar fluxes of less than ~0.85 times the present value. The calculations shown are for a Mars-sized planet rather than an Earth-sized one, and also for a planet that lacks N_2, but these differences turn out to have little effect on the results.[8] The solar flux at Mars' orbit is about 0.43 times that at Earth's. Hence, the minimum solar flux required to bring the surface temperature above freezing is about $0.85 \times 0.43 \cong 0.37$ times the value at Earth today. According to the inverse square law, this corresponds to a distance of ~1.65 AU for the outer edge of the HZ [because $(1/1.65)^2 \cong 0.37$]. This is well outside of Mars' orbital distance, which is 1.52 AU. So, as pointed out in the chapter 8, Mars might well be habitable today if it were able to recycle its atmospheric CO_2.

Although these values for the inner and outer edges of the HZ were as accurate as we could calculate with our climate model, there are reasons to think that both estimates may be too conservative. As discussed in chapter 6, H_2O clouds may push the inner edge in to 0.8 AU, or even closer, if they cause the planet's albedo to increase significantly as it warms. And the outer edge of the HZ could be as far out as 2.0 AU or more, if CO_2 ice clouds or other greenhouse gases help to warm the planet's surface.[9,10] In one sense, this uncertainty may not matter too much, as even the conservative estimate for the HZ width is fairly broad. If the outer edge is near 1.65 AU and the inner edge is near 0.95 AU, then the width of the HZ should be approximately 0.7 AU. Our own Solar System contains 4 terrestrial planets between 0.4 and 1.5 AU, so the mean spacing between them is only ~0.35 AU. This suggests that, statistically speaking, two of them ought to be in the HZ (which is exactly what we observe, as both Earth and Mars are in it, according to our calculations). If rocky planets exist in other planetary systems, and if they are spaced as they are here, then the chances that at least one of them will be in the habitable zone should be pretty high.

We can also use these climate modeling results to derive an estimate for the width of the 4.6-billion-year continuously habitable zone around the Sun. The Sun is brighter today than it has been at any time in the past. Hence, the inner edge of the CHZ is the same as the inner edge of the modern HZ: 0.95 AU. The outer edge, though, must be closer in because at 4.6 Ga the Sun was only about 70 percent as bright as it is today. Hence, a planet would need to have been closer to the Sun by a factor of $0.7^{1/2} = 0.84$ to receive the same flux that it does at present. If the outer edge of the modern HZ is at 1.65 AU, then at 4.6 Ga it should have been at $1.65 \times 0.84 \cong 1.4$ AU.

Figure 1.3 The giant radio telescope at Arecibo Observatory, Puerto Rico. The dish is 305 m (1000 ft) in diameter. Telescopes such as this one could be used to contact extraterrestrial civilizations, if they exist.[6]

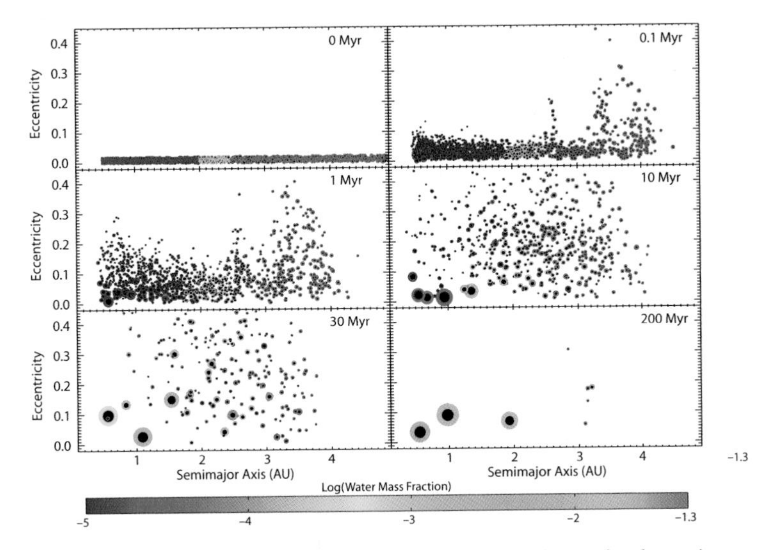

Figure 2.4 Snapshots of a particular planetary accretion simulation for the region inside 5 AU. The horizontal axis is the planet's semi-major axis, i.e., its mean distance from the Sun. The dots represent large planetesimals, termed *planetary embryos*, some of which will grow into planets. The position of the dot on the vertical axis indicates the planetesimal's eccentricity. The size of the dot indicates the mass of the planetesimal or planet, and its color shows the fraction of its total mass that is made up of water. The simulation was terminated after 200 million years. (From S. Raymond et al. 2006.[16])

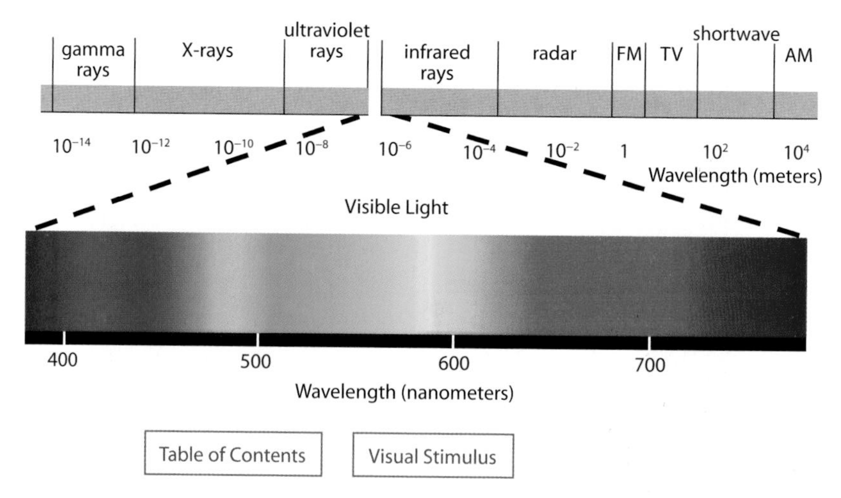

Figure 3.2 The electromagnetic spectrum.

Figure 4.4 A sample of banded-iron formation, or BIF. The width of the fine bands is approximately 1 mm. The iron in this BIF is in the form of hematite. (Courtesy of J. W. Schopf (UCLA).[11])

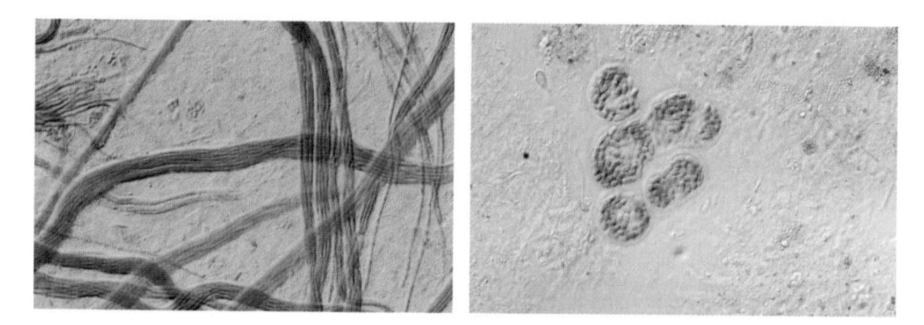

Figure 4.5 Some examples of modern cyanobacteria. Left: *Microcoleus chthonoplastes*; right: Unidentified unicellular cyanobacteria from Guererro Negro, Baja, California. Relatives of these organisms produced the first O_2 in Earth's atmosphere.[15] (Courtesy of NASA.)

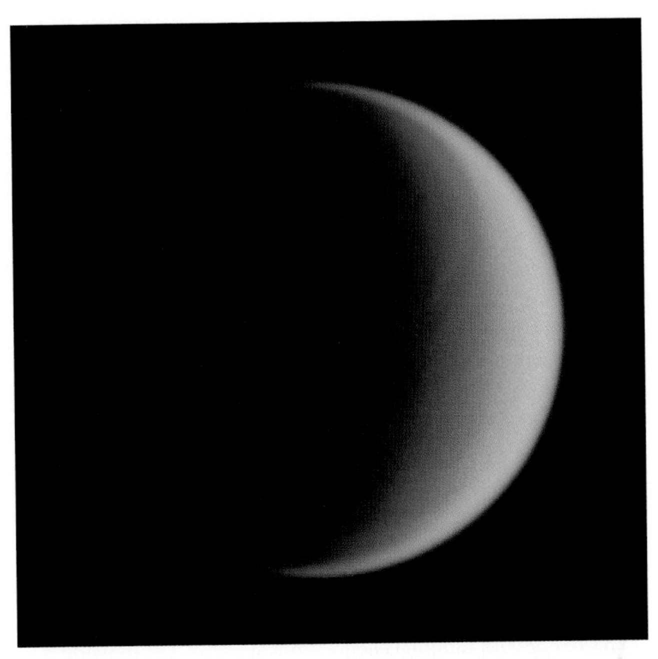

Figure 4.7 Titan's orangish organic haze, as observed from the Voyager 2 spacecraft that flew by Saturn in 1977. (Image courtesy of NASA.)

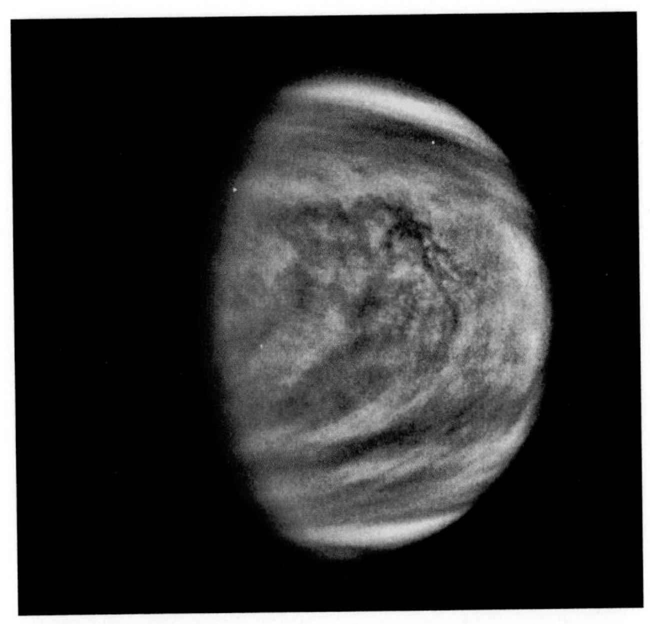

Figure 6.1 Photo of Venus taken at ultraviolet wavelengths by NASA's Galileo spacecraft in 1990. Galileo was actually on its way to Jupiter, but it swung by Venus on the way out to obtain a "gravity assist" on its long journey, and it took this picture as it went by.

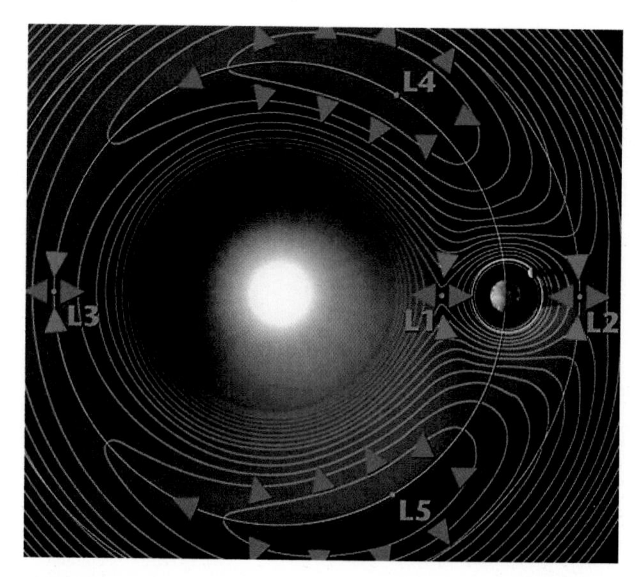

Figure 7.2 Lagrange points in the Earth-Sun system. These are points of neutral gravitational stability where an object could remain stationary, either with no input of energy (L4 and L5) or with minimal energy input (L1–L3). The drawing is not to scale. The distance from L1 to Earth is about 1.5 million km, or about 1 percent of the mean Earth-Sun distance.[14]

Figure 8.1 Photograph of Mars taken from the *Hubble Space Telescope*. (Courtesy of NASA/JPL-Caltech.)

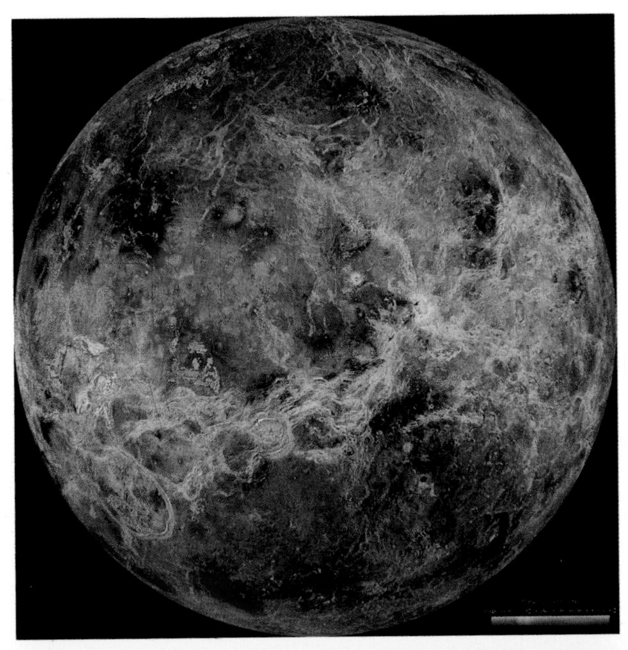

Figure 9.4 Radar map of Venus (top) and topographic map of Earth (bottom). (Venus picture from http://photojournal.jpl.nasa.gov/catalog/PIA00158.[17] Earth picture from http://sos.noaa.gov/gallery/.[18])

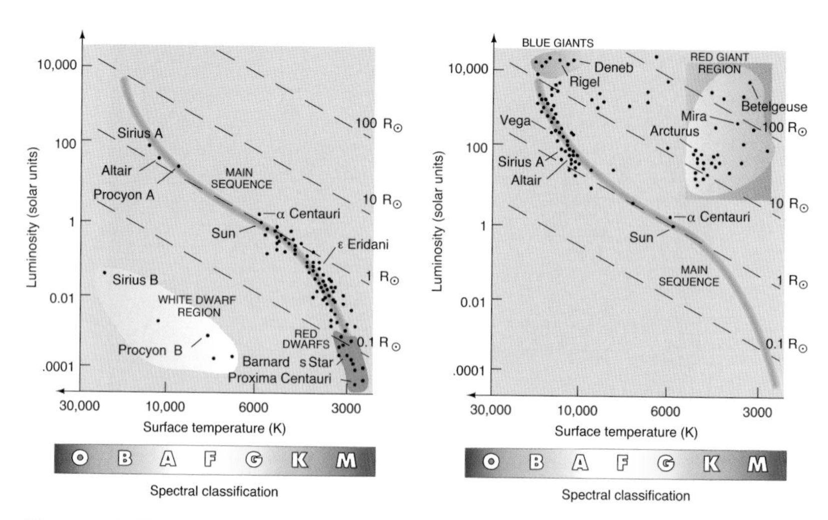

Figure 10.2 Hertzsprung-Russell diagrams showing (a) the nearest stars and (b) the brightest stars. The Sun has a luminosity of 1 on this scale and an effective temperature of 5780 K. Dashed lines show the radius of the star relative to the Sun's radius. (From *Astronomy Today*, 6th ed., Chaisson and McMillan. Reprinted by permission of Pearson Publishing.[11])

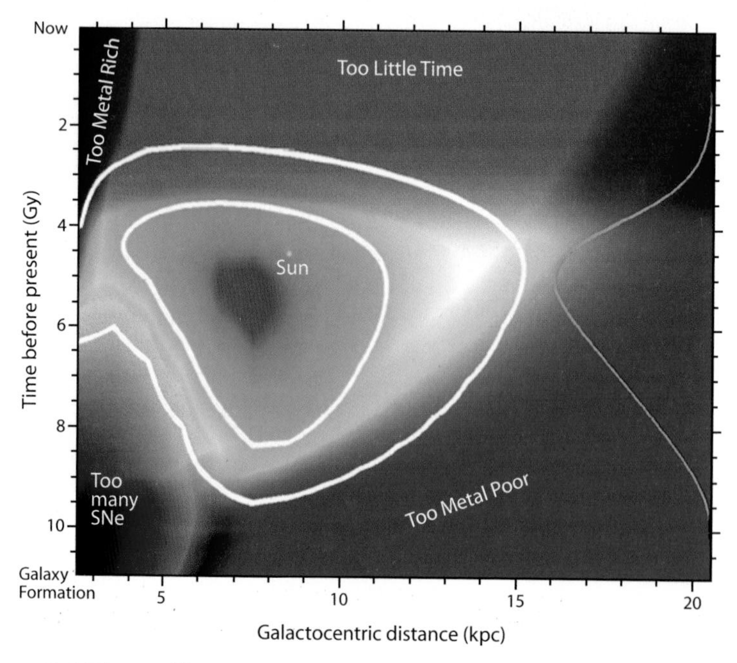

Figure 10.6 Diagram illustrating the galactic habitable zone, or GHZ. Distance units are in kiloparsecs. One parsec ≅ 3.26 light years. Time units are in gigayears, i.e., billions of years. (From C. Lineweaver et al. 2004.[28])

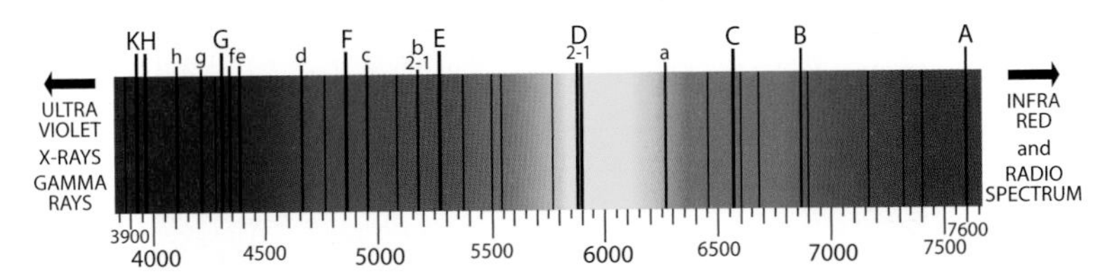

Figure 11.6 The Sun's Fraunhofer spectrum. The horizontal scale represents wavelength in Angstroms. 1 Angstrom = 0.1 nm.[13]

Figure 12.4 Star field to be observed by NASA's *Kepler* telescope, scheduled for launch in October 2008. *Kepler* will monitor the brightnesses of ~100,000 stars looking for transits of Earth-sized planets.[8]

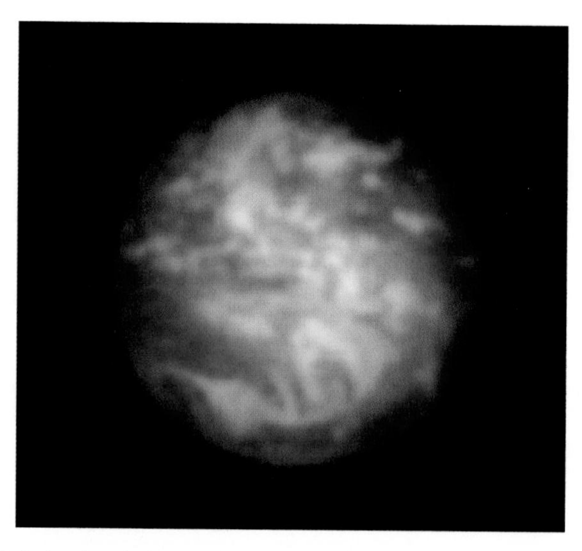

Figure 15.2 A simulated picture of Earth taken by the proposed *Exo-Earth Imager* array from a distance of 3 pc. (From A. Labeyrie 2002.[3])

Figure 15.4 The Allen Telescope Array at Hat Creek Radio Observatory in California.[15]

Hence, our habitable zone model predicts that the 4.6-billion-year. CHZ around the Sun extends from at least 0.95 to 1.4 AU. This is still significantly greater than the mean spacing between terrestrial planets in our Solar System, implying that at least one of them (Earth) should be within the CHZ. This, again, is what we observe. By comparison, Hart's CHZ extended from 0.95 to 1.01 AU, which would imply that only 1 system in 6 might harbor a habitable planet. And the actual numbers would be worse because his model predicted that the Sun was the optimal star for harboring habitable planets. Clearly, the big difference between the two models is the location of the CHZ outer edge. Once again, this illustrates the importance of the carbonate-silicate cycle and the negative feedback it provides on atmospheric CO_2 and climate. Without this feedback, Hart would be right, and habitable planets might indeed be rare.

Hertzsprung-Russell Diagrams and Main Sequence Stars

What about other types of stars—might they harbor habitable planets? In Hart's model, most of them did not, but this was at least partly because his model was fine-tuned to our Sun and our Earth. Before trying to answer this question ourselves, we should first review how different types of stars are classified by astronomers.

Astronomers like to plot stars on a figure called a *Hertzsprung-Russell diagram*, or *H-R diagram*, for short (figure 10.2, in color secton). The figure is so named because it was derived independently by Ejnar Hertzsprung and Henry Norris Russell back in the early 1900s. Each point on an H-R diagram represents an individual star. The horizontal axis is $1/T_{eff}$, where T_{eff} is the effective radiating temperature of the star—effectively, its surface temperature. The vertical axis is the star's luminosity relative to that of the Sun. Occasionally, stellar mass is used on the vertical scale in place of luminosity. More massive stars have hotter interiors, and hence are more luminous, because the nuclear fusion reactions proceed faster. So one can replace luminosity with mass on the vertical scale of the H-R diagram, provided that one understands the relationship between the two variables.*

Also shown on the horizontal axis, at the bottom of the figure, is the star's spectral classification. Originally, this was all that was known about the different stars. They each exhibited different types of absorption features in the visible and ultraviolet, and this allowed them to be sorted into different classes: O, B, A, F, G, K, and M. These spectral classes can be remembered by the classical mnemonic: Oh, Be a Fine Girl, Kiss Me! These days, though, there are more politically correct alternatives, such as Only Boys Accepting Feminism Get Kissed Meaningfully! Or, my favorite: Only Boring Astronomers Find Gratification Knowing Mnemonics! As mnemonics are meant to be repeated silently, it is probably OK to go with whichever of these one can best remember.

In any case, the great insight of both Hertzsprung and Russell was to realize that these different spectral classes represented stars of different surface temperatures and luminosities. Most of the stars on an H-R diagram fall into a band running from the upper left to the lower right. This band is called the *main sequence*, and it includes all stars that are burning hydrogen in their interiors. Recall from chapter 3 that this is how stars spend most of their lifetimes. But giant and supergiant stars that have evolved off the main sequence are also represented on this diagram (upper right-hand corner), as are the white dwarfs (lower left-hand corner) that represent the final stage in most stars' evolution.

Main sequence stars, which are our primary interest here, have a well-defined relationship between their luminosities (or masses) and their effective temperatures. This relationship is reflected by their appearance, because the color of an object depends on its temperature. Brighter (more massive) stars are hotter, and therefore bluer. (This follows directly from Wien's law, which was discussed in chapter 3.) These stars include the exceptionally bright O, B, and A stars, and the somewhat more Sun-like F stars. Dimmer (less massive) stars are cooler, and therefore redder. These include the K stars, along with the extremely dim M stars. Within each spectral category, stars are also ranked numerically

*The relationship between stellar mass and luminosity varies along the main sequence, but on average the luminosity scales approximately as mass to the 4th power.

from 0 to 9, in order of decreasing brightness. Our Sun is a G2 star with an effective temperature of 5780 K.

Another stellar characteristic that is extremely important from the standpoint of planetary habitability is its lifetime. Our Sun, as we have already seen, is expected to remain on the main sequence for about 9–10 billion years. Hence, we are approximately halfway through its main sequence lifetime. Somewhat paradoxically, the more massive stars burn out much faster, despite their greater amount of nuclear fuel. This is because of the steep relationship between a star's mass and its luminosity. A bright, blue O star may exhaust the hydrogen in its core in only 10 million years, after which time it creates a spectacular supernova. Dim, red M stars, by contrast, have predicted main sequence lifetimes of tens of billions of years—longer than the estimated age of the universe (about 14 billion years). More massive stars also evolve in luminosity faster than do their less massive sisters; hence, their habitable zones move outward much more rapidly. The CHZ for a bright blue star thus exists only if one picks a short time interval over which to define it. As virtually everyone who has thought about this question has pointed out, on planets orbiting bright blue stars, life may not have time to originate, much less to evolve. The brightest blue stars also put out most of their energy in the ultraviolet, rather than the visible region of the spectrum, which may pose further problems for planetary habitability. (See further discussion below.) So this far end of the main sequence—the O, B, and A stars—is probably off-limits for life. The F stars, on the other hand, should not be ruled out. Indeed, their intrinsic brightness may make them some of the best candidates for study, as we shall see in later chapters.

Habitable Zones around Other Stars

What can we say about habitable zones around other stars, and what might this imply about the probability of finding Earth-like planets? To approach this question, we followed Michael Hart's lead and performed climate model simulations for planets around other types of stars.[8] However, rather than trying to calculate detailed evolutionary paths of specific planets, as he did, we simply took the Earth, and planets that differed from it in various respects, and placed them at different distances from the star. Obviously, the stellar flux is much higher around a bright blue star, and so the habitable zone must lie at a greater distance. (Astronomers call such stars "early-type" stars because they lie toward the beginning of the main sequence. Similarly, "late-type" stars are the redder K and M stars.) But it is not just the star's luminosity that changes. The spectral distribution of its radiation changes as well, because of the change in the star's surface temperature. This is illustrated in figure 10.3a, which shows the distribution of radiation for a planet orbiting an F2 star, a G2 star (the Sun), and a K2 star. (Sometimes, a "V" is added to the designation of a main sequence star to indicate that it is a small, or dwarf, star, rather than a giant. Hence, "F2V" means F2 dwarf.) As noted earlier, the radiation from an F star is shifted toward the blue, while that from a K star is shifted toward the red. This affects the climate calculations because blue light is more easily reflected from a planet due to increased Rayleigh scattering, whereas red and near-infrared radiation is scattered less and is also partly absorbed by the planet's atmosphere. Hence, the HZ around a blue star is slightly closer in than one would expect based solely on its luminosity, while the HZ for a red star is slightly farther out.

The results of doing such calculations for different types of stars are summarized in figure 10.4. Here, the horizontal axis represents the distance from the star, and the vertical axis is the star's mass, relative to that of our Sun. The habitable zone is shown as a strip running from the lower left-hand part of the diagram to the upper right. Also shown are the nine planets of our Solar System. (This figure was made before Pluto was demoted to the somewhat embarrassing status of "minor planet"). Because the HZ moves outward with time at different rates for stars of different masses, one has to choose a particular time in the star's lifetime in order to make a plot like this. In figure 10.4, the HZ has been plotted at the time when each star first enters the main sequence.

From figure 10.4, one can see that the habitable zone lies farther out for more massive stars and closer in for less massive ones. That is to be expected, as a planet must receive roughly the same amount of starlight

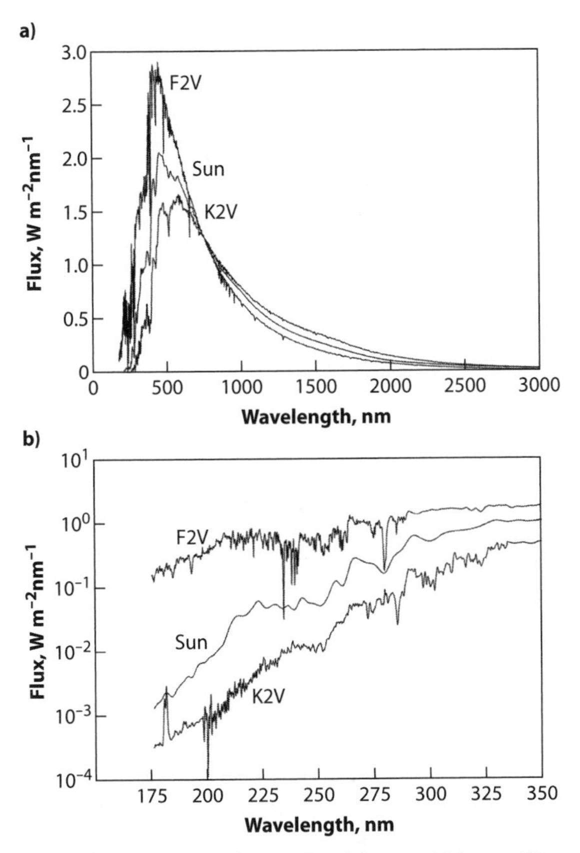

Figure 10.3 Incident stellar flux distribution for a planet orbiting an F2 star, a G2 star (the Sun), and a K2 star. The planet is assumed to receive the same total amount of sunlight as the present Earth. (a) The entire wavelength spectrum; (b) the ultraviolet portion of the spectrum.[12] (Reproduced by permission of Mary Ann Liebert Publishing.)

as does Earth in order to be habitable. But the figure shows some less obvious things as well. For example, it illustrates a point made earlier, which is that the HZ in our own Solar System is relatively wide, as compared to the spacing between the planets. Note that orbital distance is shown here on a log scale. That's because the planets in our own Solar System are spaced logarithmically. This is sometimes called "geometric" spacing, because each planet's orbital distance is larger than that of its inner neighbor by roughly the same amount (a factor of ~1.7, on average).

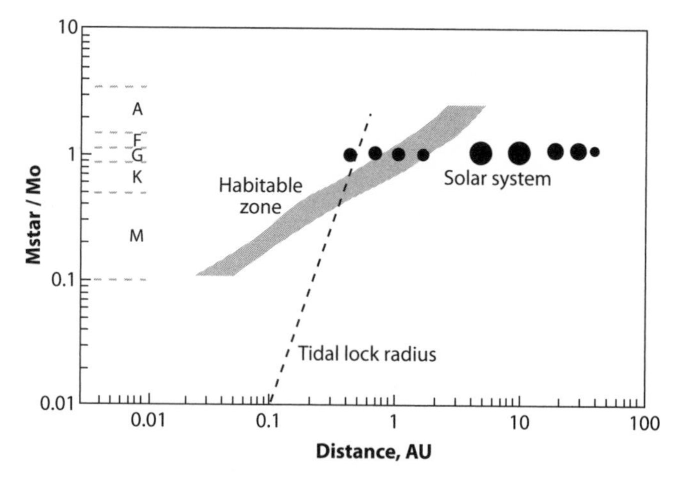

Figure 10.4 Diagram illustrating the extent of the habitable zone around different types of stars. The vertical scale represents the mass of the star relative to the mass of our Sun, and the figure is drawn for the time when the stars have just reached the main sequence. The nine planets of our own Solar System, including Pluto, are also shown. (Adapted from Kasting et al. 1993.[8])

The gap between Mars and Jupiter, where it appears as if a planet should exist, is occupied by the asteroid belt. This observation, of course, is nothing new. The relationship between planetary orbital distances was noticed a long time ago, soon after the invention of the telescope, and was expressed mathematically as "Bode's law," sometimes called the "Titius-Bode law." Bode's law has frequently been dismissed by astronomers, as it was originally proposed as a simple empirical fit to observations. But it is now thought to have a physical basis, at least for the giant planets. One cannot pack planets more closely together than this without making their orbits unstable.[13,14] Whether or not terrestrial planets should be spaced geometrically is a matter of debate. Terrestrial planet spacing may be more influenced by resonances with giant planets than by interactions between each other. This question should ultimately be answered when we are able to observe other inner planetary systems.

The question of planetary spacing becomes extremely important when one is dealing with stars that are significantly different from the Sun. As can be seen from figure 10.4, the HZ is roughly constant in width when plotted against the logarithm of orbital distance. In actual distance units, then, this means that the HZ around an M star is quite narrow compared to the Sun's HZ, whereas that around an F star is quite large. This caused Huang to conclude that the chance of finding habitable planets around M stars was small.[3] But such a conclusion seems premature, or perhaps it is right for the wrong reasons (see below). When expressed in terms of log distance, there is just as much habitable space around an M star as there is around a G star. Whether or not one can populate it with Earth-like planets is another issue.

Problems for Planets Orbiting Early-Type Stars

Planets orbiting stars that are significantly more or less massive than the Sun face a variety of problems that may affect their habitability. For early-type stars, the two main problems have already been mentioned: (1) they have short main sequence lifetimes, and (2) they put out large amounts of UV radiation. The first problem makes the O and B stars, and many of the A stars as well, not very interesting from the standpoint of harboring habitable planets. The F stars, though, are a different matter. The main sequence lifetime of an F0 star is about 2 billion years. This is more than enough time for life to originate and evolve, as we know from our own planet's history (chapter 4). It may not be enough time to develop complex, multicellular life, based on our experience here on Earth, but that is a separate issue. When we do start searching for planets that are habitable, or inhabited, we will probably want to include F stars on our list.

The stellar UV problem is also a serious one for life, but it is not necessarily insurmountable. The problem is illustrated in figure 10.3b. A planet at 1-AU equivalent distance around an F star would receive approximately 4 times as much UV radiation as does Earth at wavelengths less than 315 nm (see table 10.1). This wavelength region includes biologically damaging UV-B and UV-C radiation. Relative fluxes at 250 nm, where absorption by DNA peaks, are even higher—a factor of 10 or more. For a planet like the primitive Earth, that lacks an ozone layer, this

TABLE 10.1
Relative Values of UV Flux, Ozone, and DNA Dose Rate on Earth-Like Planets around Different Stars

Stellar type	*Incident UV flux*	*Ozone column depth*	*Surface UV flux*	*Relative dose rate**
G2 (Sun)	1	1	1	1
K2	0.26	0.79	0.43	0.5
F2	3.7	1.87	0.68	0.38

*Dose rate for DNA damage.
All values from Segura et al. 2003.[12]

could lead to higher rates of mutation for near-surface life. To astronomers like Carl Sagan, this appeared to be a serious problem. However, to many biologists, including Sagan's first wife, Lynn Margulis, it seemed like less of an issue. Margulis argued in her papers[15,16] that organisms could avoid UV damage by forming mats, like the ones that formed the stromatolites mentioned in chapter 4. If this strategy worked on Earth, perhaps it would work on F-star planets as well.

If life was able to originate on an F-star planet, however, and if oxygenic photosynthesis also arose, as it did on Earth, then the problem with stellar UV radiation might take care of itself. The reason is that F stars put out even more radiation at wavelengths <200 nm—roughly 100 times that of the Sun (figure 10.3b). Photons at these very short wavelengths are sufficiently energetic to be able to break apart molecular oxygen:

$$O_2 + \text{photon} \longrightarrow O + O. \quad (\lambda < 200\ \text{nm})$$

The O atoms then recombine with O_2 to form ozone, as discussed in the previous chapter. The result is that F-star planets should develop thick ozone layers, as shown in figure 10.5a. The ozone would then shield the surface from the 200- to 300-nm UV radiation that accounts for most biological damage. The amount of damage can be estimated by combining the UV flux at the surface with the estimated efficiency by which it

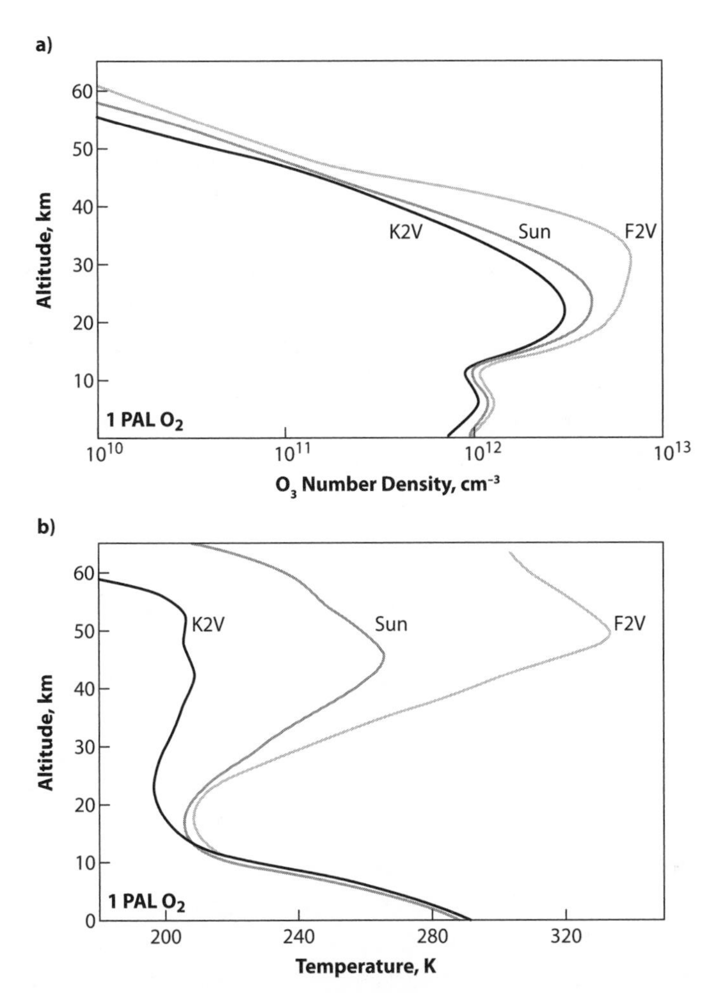

Figure 10.5 Vertical profiles of ozone number density (a) and temperature (b) for an Earth-like planet with a 21% O_2 atmosphere orbiting different types of stars. (Reproduced by permission of Mary Ann Liebert Publishing.)

affects organisms.[17] Surprisingly, the predicted *dose rate* for DNA damage at the surface of an Earth-like planet orbiting an F star is less than 40 percent that for Earth (table 10.1). A K-star planet also has a lower surface dose rate than does Earth. Thus, according to these calculations, stars like the Sun are the *most* dangerous parent stars for planets with atmospheres like that of modern Earth! So Earth may not be the best of all possible worlds. We only think it is because we have nothing with which to compare it!

Also shown in figure 10.5 (panel b) is the way in which temperature varies with altitude in the atmospheres of the F-, G-, and K-star planets. Modern Earth has a warm upper stratosphere because the UV radiation absorbed by ozone heats the air in this region. By comparison, the K-star planet has a cool stratosphere, while the F-star planet has a superhot one. The temperature differences are caused by the different amounts of ozone and the differing stellar UV fluxes. The stratospheric temperature has little effect on climate, as it turns out. However, it has a big effect on the planet's spectrum, that is, on how the planet would appear if observed from afar. This, again, is something to which we will return in later chapters.

Problems for Planets Orbiting Late-Type Stars

Planets orbiting late-type stars face a completely different set of problems. One issue that has been recognized right from the start[5] is shown by the dotted line in figure 10.4. The HZ for an M star and for a late K star as well, lies within the *tidal locking radius* of the star. This means that, after some period of time, the planet would be expected to always show the same face to the star, that is, it should spin once per orbit. To be more precise, the tidal locking radius is a time-dependent concept, like the CHZ. The line shown in figure 10.4 is the distance at which a planet would become tidally locked in 4.5 billion years, if tides dissipated energy at the same rate as on present Earth.

A familiar example of a tidally locked object is Earth's Moon. One always sees the front side of the Moon, no matter when one observes it. That's because the Moon is within the Earth's tidal locking radius. The same thing can happen if a planet is too close to its parent star: it should have one permanently sunlit side, and one side that is permanently dark. The danger, then, is that the planet's atmosphere and oceans could freeze out to form a giant ice cap on the dark side. If that were to happen, none of the planet's surface would remain habitable, not even the *terminator* (the boundary between sunlight and darkness). If this problem is common to most M-star planets, then the potential number of habitable worlds would be greatly reduced, as M stars are the most abundant of all stellar types.

Fortunately, this particular problem can be circumvented in several different ways. One of these is illustrated by the planet Mercury in our own Solar System. As can be seen from figure 10.4, Mercury is within the Sun's tidal locking radius. And yet Mercury is not tidally locked. Instead, it spins 3 times on its axis for every 2 times it orbits the Sun. In technical terms, that is because it has been caught in a *spin-orbit resonance*. The reason this happened, we think, is because Mercury is slightly misshapen as a consequence of violent impacts that occurred during its formation process. These impacts caused Mercury's mass distribution to be slightly nonspherical. One can think of the planet as being shaped somewhat like a football, with one long axis and two shorter ones. Mercury's orbit is highly eccentric ($e \cong 0.21$), and it can be shown that the planet's gravitational potential energy is slightly lower when the long axis is pointed toward the Sun at perihelion. At Mercury's current spin rate, this happens on every orbit, with the opposite side of the planet facing the Sun each time. Mercury probably began its life spinning much more rapidly, and then gradually slowed down in response to tides raised by the Sun.* But it got caught in the 3:2 spin-orbit resonance before it could become tidally locked. Consequently, if Mercury had any significant atmosphere today, it would not form an ice cap on one side. This observation is largely academic, as Mercury has almost no atmosphere, but it might be relevant to hypothetical planets orbiting within the habitable zones of M stars.

Another way out of the tidal locking problem was suggested by Manoj Joshi and his colleagues at NASA Ames Research Center.[18,19] Joshi and coworkers performed climate simulations of tidally locked planets using a 3-dimensional global climate model. They found that if the planet's atmosphere contained at least 30 millibars of CO_2—about 100 times the

*A planet does not need to have an ocean to experience tides. On Mercury, the tides would have been raised within the solid planet itself.

amount in Earth's present atmosphere—it should transport sufficient heat from the dayside to the nightside to prevent the atmosphere from freezing out. High CO_2 concentrations facilitate such atmospheric heat transfer by lengthening the time required to radiate heat off to space. Similarly, one can imagine that if a planet had a deep ocean, like Earth, then ocean currents might also carry heat from the dayside to the nightside. So, from a purely climatic standpoint, M stars should not be excluded as candidates for harboring habitable planets.

Several potential problems for M-star planets have been pointed out, though, by various authors. One of these concerns their ability to retain an atmosphere. M stars exhibit much more flare activity than does our Sun, and they have correspondingly enhanced stellar winds. Furthermore, tidally locked planets around all but the least massive M stars should be rotating fairly slowly, 10–100 days, and hence may be unable to generate strong magnetic fields. As pointed out in the previous chapter, a planet must be rotating to create a magnetic dynamo. Even if it had a similar iron core, a more slowly rotating planet around an M star might not produce a strong magnetic field, and thus its atmosphere could conceivably be stripped away by the intense stellar wind.[20]

Other problems with M-star planets could result from differences in the nebular environment in which they form.[21] As already pointed out, the habitable zone around an M star is very close in. Hence, the amount of material available to be swept up by a growing planet is relatively small, and so the planets that form there may be significantly less massive than Earth. Furthermore, because the orbital times at these distances are short, accretion of planets should occur rapidly, giving the nebula less time to cool. This means that terrestrial planets, as they form, would be further removed from the "snow line"—the distance where icy planetesimals can form. This, in turn, may reduce the amount of water and other volatiles delivered to such planets. The tight orbits within the habitable zone also mean that the planetesimal velocities would have been high, and so impacts would have been even more violent than those that formed the Earth. Such energetic impacts may have swept away, or eroded, more atmosphere than they delivered.[22]

Although none of the above points are show-stoppers, we conclude that while M stars should not be completely excluded from consideration in searching for habitable planets, they appear less likely to harbor them than do F, G, and early K stars. Indeed, the suggestion that one should focus primarily on F, G, and K stars has been echoed consistently by various authors over the last 50 years.[4,5,8] Fortunately, roughly 20 percent of stars in the solar neighborhood have spectral classifications between F0 and K5, so this limitation is not too restrictive. On the other hand, most of the closest stars are M stars, and as we will see later on, these are also likely to be the first stars to be searched for Earth-like planets. So the question of whether habitable planets can exist around M stars could become a hot topic within the next 5–10 years.

Further Extensions of the Habitable Zone Concept

Siegfried Franck and his colleagues from the Potsdam Institute for Climate Impact Research in Germany have proposed an extension of the habitable zone concept in which the properties of the planet are specifically taken into consideration.[23-25] For example, different continental growth rates lead to differences in CO_2 uptake by silicate weathering, and so the greenhouse effect varies in a time-dependent manner in their calculations. These authors find, not surprisingly, that the width of the HZ varies with time for different assumed planetary properties. Similarly, they find that the length of time that a planet can remain habitable depends on its distance from its parent star. For an Earth-like planet orbiting a star like the Sun, they calculate an optimal orbital distance of 1.08 AU. And, indeed, according to the calculations described in this book, Earth is closer to the inner edge of the HZ than to its outer edge. So perhaps we would have been even better off had it formed a little further out. Again, Earth is not necessarily the best of all possible habitable worlds.

To make use of these types of calculations, it is necessary to have lots of information about the planet being considered. Perhaps we will have information on that someday if the planet-finding missions described in the last part of this book eventually succeed. But if we already know that much about the planet in question, then we shouldn't need to predict whether or not it is in the habitable zone. So it is not clear that these extensions of the habitable zone concept are really that useful in practice. For that reason, we prefer to stick with a definition of the habitable zone that is independent of specific planetary properties, other than the basic requirements that we have discussed in the last two chapters.

The Galactic Habitable Zone

Just as there is a region around a star that is optimal for planetary habitability, there may also be a region around the center of our galaxy that is optimal for finding habitable planetary systems. The name given to this mega-concept is the *galactic habitable zone*, or GHZ. This idea was first introduced by Ward and Brownlee in *Rare Earth*[26] and in a related paper with their colleague Guillermo Gonzalez.[27] Charles Lineweaver of Australian National University further elaborated on this idea to include time as well as space.[28] These authors collectively point out that not all stars in the Milky Way are equally likely to harbor habitable planets. Our Sun, they note, is located at a favorable position, a little more than halfway out from the galactic center. (The Milky Way galaxy is estimated to be about 85,000 light years in diameter, and our Sun is 25,000 light years from its center.) Planets orbiting stars that are too close to the center of the galaxy are more likely to have their orbits perturbed by close stellar encounters, and they are also more likely to find themselves in the nearby vicinity of catastrophic events such as supernovae and gamma-ray bursts. Stars too far out toward the rim of the Milky Way spiral galaxy are less "metal"-rich than the Sun and are therefore less likely to be accompanied by rocky planets. (This is another piece of astronomical jargon. A "metal" to an astronomer is any element heavier than hydrogen and helium. Evidently, early astronomers were not overly concerned with chemistry.) Similarly, stars that are born too early in the history of the galaxy are likely to be metal-poor, because not enough hydrogen and helium will have been reprocessed through stars to form the heavier elements.

Figure 10.6 (see color section), from Lineweaver et al.,[28] illustrates one particular concept of the GHZ. In this model, distance from the galactic center is plotted on the horizontal axis, and time before present is plotted on the vertical axis. The Sun is plotted at the position when it formed, some 4.6 billion years ago. This particular figure is designed to be relevant to the search for complex life, and it assumes that ~4 billion years is needed for such life to evolve. (That's a pretty good assumption, based on our experience here on Earth, as the Phanerozoic era began just over 500 million years ago.) Hence, stars that formed more recently than 4 ± 1 Ga are excluded from consideration. The green area near the Sun is the area in which the stars have the proper metal content; in the blue areas, the stars are metal-poor. The red area in the lower lefthand corner—near the center of the galaxy early in its history—is excluded because the frequency of supernovae explosions is too high. The green curve on the right represents the authors' estimate of the age distribution of complex life in the galaxy. As the famous Italian physicist Enrico Fermi pointed out many years ago, most such life (including intelligent life, if it exists) is likely to have been around for longer than we have.

The GHZ concept clearly contains an element of truth, even if its boundaries are rather fuzzy. But it is more relevant to the distant future of galactic exploration than it is to the near-term search for habitable planets. As we discuss in chapter 13, the stars that we are hoping to study within the foreseeable future are all located relatively nearby (within ~50 light years for the planned *TPF* missions). Although Gonzalez's early work[29] suggested otherwise, such stars appear to have roughly the same metal content as the Sun, provided that one restricts the comparison to F-G-K stars.[30] (Gonzalez's original analysis also included M stars, which tend to be older on average, and hence less metal-rich.) These nearby stars are also subject to the same background level of supernovae and gamma-ray bursts as is our Solar System. So, even if the potentially habitable area of the galaxy is indeed limited, this may have little effect on the probability of finding habitable planets in our immediate stellar neighborhood.

Let's pause now and summarize what we have learned in this chapter. Habitable zones around Sun-like (F, G, and early K) stars should be relatively wide because of the natural feedback between atmospheric CO_2 levels and climate—the same feedback loop that kept the Earth habitable early in its history. To benefit from this feedback loop, of course, planets must be volcanically active and they must be endowed with adequate supplies of both water and carbon. Stars earlier than about F0 or later than about K5 are less likely to harbor habitable planets for a variety of reasons. For early-type stars, the major problem is their short main sequence lifetimes. At the other end of the stellar mass scale, planets within the habitable zones of late K and M stars may be small, tidally locked, and deficient in volatiles. Nevertheless, they are worthy of study because they may be the first extrasolar terrestrial planets that can actually be observed. So we know enough at this point to know where to look for habitable planets and to have some preliminary ideas about what we might find. We are now ready to discuss how this search will take place.

• Part IV •

How to Find Another Earth

• • •

Wherein we see what astronomers are doing right now to find planets, including small, rocky ones, around other stars, and we also look forward to see how the search for Earth-like planets is expected to proceed over the next few decades to centuries . . .

· Chapter Eleven ·

Indirect Detection of Planets around Other Stars

In part II of this book we examined the question of why Earth has remained habitable throughout its long history. Then, in part III, we asked why Mars and Venus are different from Earth, and what might this imply about the chances of finding habitable planets around other stars. In part IV, which begins here, we look at what astronomers are doing now to find planets around other stars, as well as what they are hoping to do over the next few decades to centuries. As we shall see, they have already discovered a lot, and the prospects for the future look even brighter. Indeed, the ongoing search for extrasolar planets is one of the hottest new fields, certainly in astronomy, and perhaps in all of science.

Before we discuss what is happening today, though, we should take a brief look at what has happened in the not-so-distant past. Astronomers have been trying to find planets around other stars for over half a century, but it is only in the last 15 years or so that they have been successful. The reason for their recent success is not surprising: technology has improved, and this has made possible a variety of new planet-finding techniques. All of the planet-finding methods discussed in this chapter are classified as *indirect detection methods*, because they all rely on observing the planet's effect on the light from its parent star, as opposed to looking for the light from the planet itself. *Direct detection methods* will be discussed in the chapters that follow.

Barnard's Star

The modern history of searching for extrasolar planets began early in the last century with the discovery of a dim, nearby M star in the constellation *Ophiuchus* (the Serpent Holder). In 1916, the noted astronomer Edward Emerson Barnard determined that not only was this particular star the next closest to our Sun, after the 3-star Alpha Centauri system, but it was also moving toward us at a rapid rate. Indeed, this star—quickly dubbed "Barnard's star"—which is 6 light-years away at present, will pass by the Sun at a distance of only 3.8 light-years some 10,000 years in the future, making it our closest stellar neighbor.[1] Our current closest neighbor, Proxima Centauri, is ~4.3 light-years distant.

Barnard's star is not headed directly toward us, though. Its motion is partly sideways as well, which means that it is partly directed along the plane of the sky. Because it is already close by, and because it is moving quite fast relative to the Sun, Barnard's star has the largest *proper motion** of any star that has been observed, 10.3 arcseconds per year. Barnard himself determined this by comparing photographic plates taken several years apart. When he did so, the star in question had moved relative to the stars around it. This was an early application of a technique called *astrometry*, which refers to the accurate measurement of a star's position relative to other stars. If the stars used as the baseline are distant, then their own proper motions are small, so the apparent motion of the target star is close to its real motion (relative to the Sun, of course).

During the late 1960s, another astronomer, Peter van de Kamp, saw what he believed to be a slight wobble in the position of Barnard's star as it moved along its track. He interpreted these wobbles as an indication that Barnard's star was orbited by two planets with periods of approximately 12 and 20 years, respectively, and with masses somewhat smaller than that of Jupiter.[1] He made this announcement based on a series of photographic plates that he recorded at the Sproul observatory

*The term "proper motion" refers to the actual motion of a star relative to distant background stars. As discussed further below, this should be distinguished from a star's apparent motion relative to more distant stars as the Earth moves around the Sun. The term "arcseconds" is defined in the next section.

at Swarthmore College between 1938 and 1981. However, other astronomers subsequently determined that van de Kamp's data were corrupted by various systematic errors, especially the cleaning and remounting of the telescope lens at Sproul, 25 years after he began his observations. And subsequent observations by investigators using other telescopes, including the *Hubble Space Telescope*, have failed to show any evidence for a large planet orbiting Barnard's star. So the whole endeavor was probably a career-long wild-goose chase.

Van de Kamp's personal crusade to observe Barnard's star did illustrate an important point, however: if one can measure the position of a star accurately enough, then it should be possible, in principle, to determine whether it has planets going around it. This *astrometric method* for finding planets is extremely powerful if one has the right equipment. Van de Kamp's ground-based telescope and primitive photographic plates were simply not up to the task.

The Astrometric Method

The reason why astrometry can be a useful technique for detecting planets is that a star does not remain fixed in space as a planet rotates around it; rather, the planet and the star both rotate around their combined center of mass, which is termed the *barycenter* of the system. When the two objects have very different masses, like the Sun and the Earth, the barycenter is close to the Sun's own center. For the more massive (and more distant) planet Jupiter, the barycenter is just outside the Sun's own radius.[2] Hence, seen from afar, the Sun should tend to wobble by about its own diameter, ~700,000 km, during the course of one Jovian year (11.9 Earth years).

We should take a few moments here to define units and to discuss how stellar distances are measured. This will come in handy later on when we talk about how planets might actually be observed around other stars. Angles can be measured in different units. One that is frequently employed by observational astronomers is the *arcsecond*. Astronomers divide a circle into 360 degrees (°), just as geographers do with longitude. Each degree is then subdivided into 60 arcminutes (‘), and each arcminute is subdivided into 60 arcseconds (“). Consequently, there are 360 × 60 × 60 = 1,296,000 arcseconds in a complete circle. While we're at it, we can also define the milliarcsecond (mas), which is 1/1000 of an arcsecond, and the microarcsecond (μas), which is 1 millionth of an arcsecond. These are both useful units for planet hunters.

Stellar distances are measured by determining a star's *parallax*. This is the apparent motion that the star exhibits relative to more distant stars as the observer, located on Earth, orbits around our Sun (see figure 11.1). A star that moves by an angle of 2 arcseconds as the Earth moves from one side of its orbit to the other (i.e., across its orbital diameter) is defined to be at a distance of 1 *parsec* (pc). The diameter of Earth's orbit is equal to twice its orbital radius, or semi-major axis, which, of course, is 1 AU (=149.6 million km); hence, a star at 1 pc distance appears to move by 1 arcsecond while the Earth moves laterally by 1 AU. Straightforward trigonometry* then shows that 1 pc = 3.0857×10^{13} km, or 3.262 light-years.

It is also necessary to consider the geometry by which the target star is being observed. If a star does have planets circling around it, it is probably safe to assume that these planets are all orbiting in approximately the same plane. (The reason is that these planets should have formed within a protostellar disk, as discussed in chapter 2.) But we do not know ahead of time from what direction we are looking at this system. If we are observing from a direction perpendicular to the orbital plane, then the planets should be moving in the plane of the sky. We say then that the *inclination*, *i*, of the planets' orbits is 0 degrees. Alternatively, we might be looking at the system edge on, that is, we may be observing from within the same plane in which the planets are orbiting. The inclination in this case would be 90 degrees. Typically, neither of these extreme cases will hold true, and the planet's orbital inclination will be at some intermediate value, as shown in figure 11.2. We will see as we go on that this question of orbital geometry is critical if one is trying to detect planets indirectly by observing the motion of the parent star.

*1 pc = 1 AU/sin θ, where θ is the parallax angle of 1 arcsecond. In calculating sin θ, one must be careful to convert the angle back to either degrees or radians, as these are the units most frequently used on scientific calculators.

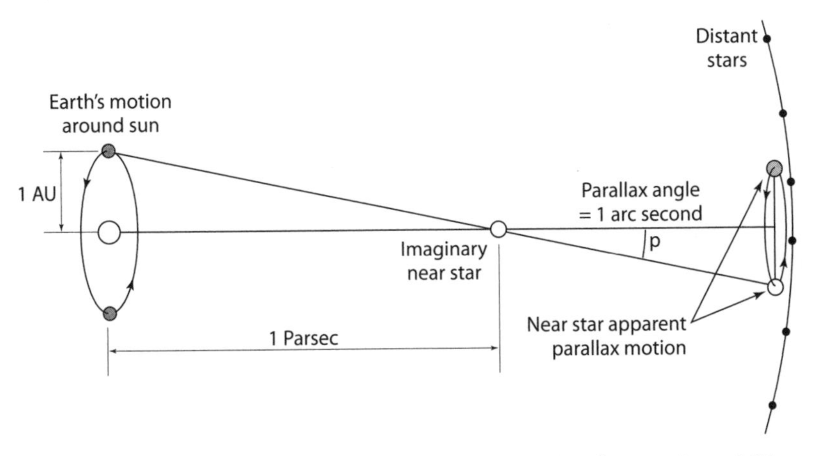

Figure 11.1 Diagram indicating how a star's parallax is measured. A star that exhibits a parallax of 1 arcsecond is 1 parsec (= 3.262 light-years) distant.[3]

With this additional background, let's go back to the question of the Sun's motion and think about what it would look like from a distance. We will assume here that we are observing the Solar System from a position directly above the invariant plane, i.e., perpendicular to the average plane defined by the planet's orbits. The actual motion of the Sun is more complicated than described above because Jupiter is only one of eight planets, each of which tugs on the Sun according to its own mass and orbital distance. (Pluto really does deserve to be ignored here because its mass is so tiny that it has little effect on the Sun's motion.) Figure 11.3 shows the calculated motion of the Sun from AD 1960 to 2025, as it would appear if observed from a distance of 10 pc, or 32.6 light-years. During this time, Jupiter makes about 5¹/₂ revolutions around the Sun. (This can be seen by counting the number of big circles in the diagram.) But the Sun's motion is obviously complex, and a detailed mathematical analysis would be required to figure out what is causing it. In this diagram each tick mark represents 200 microarcseconds. The Sun-Jupiter wobble is about 500 microarcseonds and the Sun-Earth wobble is about 0.3 microarcseconds.

Let's take a moment and consider the significance of this last number. The angle of 0.3 microarcseconds needed to detect Earth from 10 pc distance is truly small, and thus difficult to measure. Indeed, it is virtually

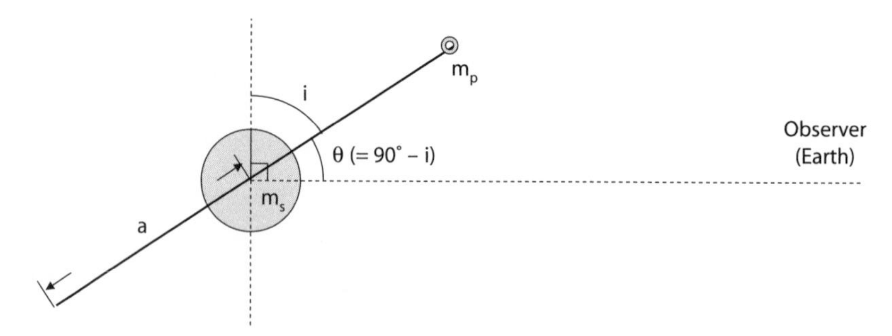

Figure 11.2 Diagram illustrating the geometry of an extrasolar planet observation. Here, i is the inclination of the planet's orbit with respect to the plane of the sky, and θ (= 90° − i) is the angle of the planet's orbit with respect to the observer on Earth. The mass of the planet, m_p, is much less than the mass of the star, m_s.

impossible to measure stellar positions this accurately from the Earth's surface because the atmosphere causes *scintillation*, or what astronomers refer to as difficulties in "seeing." Effectively, a star's position appears to hop around continually as a consequence of turbulence in the air above one's telescope. The turbulence *refracts* (bends) incoming light waves in unpredictable ways. The best way to get around, or above, this problem is to observe from space. Most readers should already be aware of the fabulous pictures taken by NASA's *Hubble Space Telescope*. The spatial resolution of such pictures is much better than can be achieved from ground-based telescopes because the incoming light waves are not distorted by their passage through the atmosphere.*

The Space Age has been under way for several decades now, of course, and astronomers have already made several efforts to measure accurate parallaxes for many stars. The first such mission to do so effectively was the *Hipparcos* mission, launched by the European Space Agency (ESA), which operated from 1989 to 1993. This mission measured parallaxes for about 118,000 stars to a precision of ~1 milliarcsecond. The results now

*There is a clever technique, called *adaptive optics*, that can be used to remove much of the distortion of light waves caused by the atmosphere. In this technique, the perturbations caused by atmospheric turbulence are sensed, and the telescope mirror is continually *deformed* by small adjustments to compensate for them. This technique can eliminate many of the problems associated with "seeing" astronomical objects from the ground. But it is still not as good as observing them directly from space.

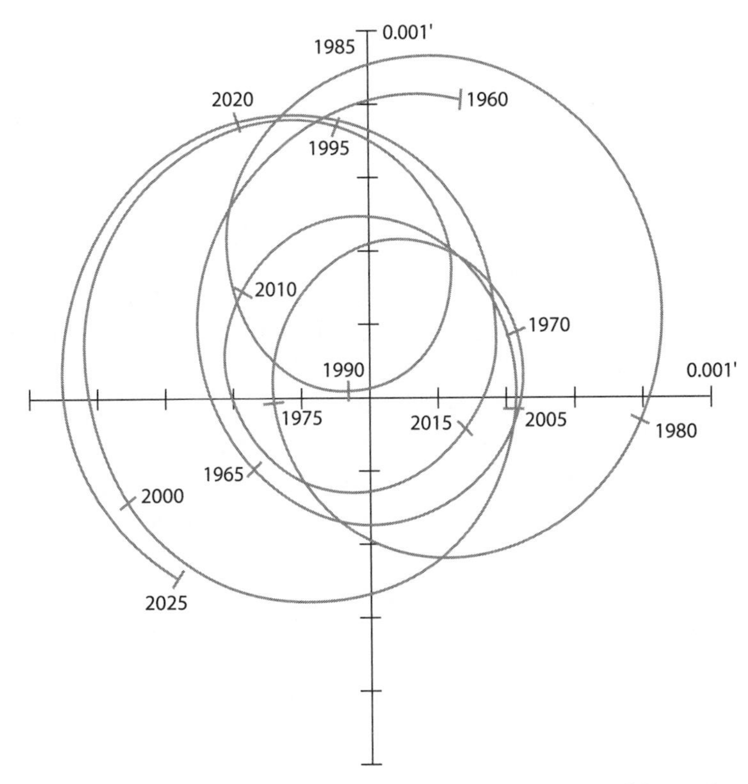

Figure 11.3 Calculated motion of the Sun from 1960 to 2025, as viewed from a distance of 10 pc, or about 32 light-years. The scale is in arcseconds. (From the "Planet Searching" slideshow at http://planetquest.jpl.nasa.gov/Navigator/material/sim_material.cfm.[4])

comprise the *Hipparcos catalogue*, which has been an invaluable tool for astronomers over the last 15 years. The accuracy, though, is still 3000 times less than that needed to find other Earths. ESA will soon fly another space-based astrometry mission, *Gaia*, currently scheduled to launch in December 2011. *Gaia* will measure parallaxes to 20–25 microarcsecond accuracy for almost a billion stars.[5] That, of course, represents a fantastic improvement over *Hipparcos*, and it should allow *Gaia* to find many Jupiter-mass planets. (Recall that Jupiter makes the Sun wobble by 500 microarcseconds, as seen from a distance of 10 pc.) But this will still not be enough to detect an Earth.

An astrometry mission that could detect Earth-sized planets, if it flies, is NASA's proposed *Space Interferometry Mission* (*SIM*). As discussed in more detail in chapter 13, *interferometery* is a technique, commonly employed with radio telescopes, in which the signals from two or more telescopes are combined so as to effectively create a single telescope with a bigger aperture, or diameter. The baseline plan for *SIM* calls for two 50-cm-diameter visible telescopes at opposite ends of a 6-m boom.[6] The quoted accuracy for a single measurement is 1 microarcsecond, so by combining multiple measurements it should be possible to achieve the 0.3-microarcsecond accuracy needed to detect Earth.* This would allow *SIM* to look for Earth-mass planets around stars out to a distance of about 10 pc. Within this range are over 100 single F, G, and early K stars, which, as we saw in chapter 10, are the best candidates for harboring Earth-like planets. There are many more M stars within this range, but these are faint and therefore hard to observe. That is not a major drawback, though, because the F-G-K stars really are the ones of greatest interest.

One political footnote before we leave this section: As of the time of this writing (early 2008), the *SIM* mission is currently back in NASA's funding plan. It was originally scheduled to have flown by 2008, or even earlier, but it has been postponed several times due to lack of funds. As we will see, this same problem plagues other proposed planet-finding missions as well. Planet finding is fun, but it is also expensive. Most big "flagship" missions, including *SIM*, have estimated costs of $1 billion or more. This is not out of the realm of feasibility for a big agency like NASA, which has an annual operating budget of about $17 billion. During the last several decades, NASA has spent the equivalent of $3–4 billion in today's dollars on each of several large planetary missions, including the *Viking* mission to Mars, the *Voyager* mission to the outer planets, and the *Cassini* mission to Saturn. And the total investment in the *Hubble Space Telescope*, including reservicing visits by the Space Shuttle, is now well over $15 billion. So we can afford to do missions like *SIM* if we so choose. But they must compete with other space projects as

*According to the theory of statistics, the error of a measurement decreases as $1/\sqrt{n}$, where n is the number of individual observations. This formula assumes that the errors are random in nature.

well, including manned space exploration, and hence it is not easy to make these missions happen in a timely manner.

Pulsar Planets

Let's go back now to the early days of planet finding and pick up the story where we left off. As we saw, Peter van de Kamp tried very hard to do ground-based astrometry, but in the end nothing tangible came of it. The next big step—arguably the first real step—in the history of planet finding came in 1990 when my colleague Alex Wolszczan of Penn State University announced the discovery of two planets orbiting the *pulsar* PSR B1257+12. Although the interpretation of his data was initially disputed, it was eventually confirmed, and a paper was published some two years later.[7] This was one of those discoveries that come out of left field, as Wolszczan was not looking for planets at all. Rather, he was trying to learn about the structure of *neutron stars*. For that is what pulsars are thought to be: rapidly rotating neutron stars. A neutron star is a collapsed object, only about 10 km in diameter, with a mass approximately equal to that of our Sun. It represents the remaining portion of a star that exploded as a *supernova* at some time in the past. The most famous example is the object at the center of the Crab Nebula in the constellation Taurus. The supernova that formed the pulsar, and the nebula itself, was recorded by Chinese and Arab astronomers[8] in the year 1054.

A neutron star, or pulsar, is typically surrounded by a disk of material left over from the explosion that created it. This material emits electromagnetic radiation as it falls into the neutron star, and that is what we observe from afar. A pulsar also spins rapidly and has a strong magnetic field. The Crab nebula pulsar, for example, spins 30.2 times per second.[8] The material that falls in towards the pulsar is highly ionized (charged), and so it follows the magnetic field lines of the star. What makes the star "pulse" is the fact that the magnetic field is generally not perfectly aligned with the spin axis (see figure 11.4). The matter falling in, and the radiation that is emitted as a result, are "beamed" along the poles of the magnetic field. Hence, an observer viewing the star from a distance sees a

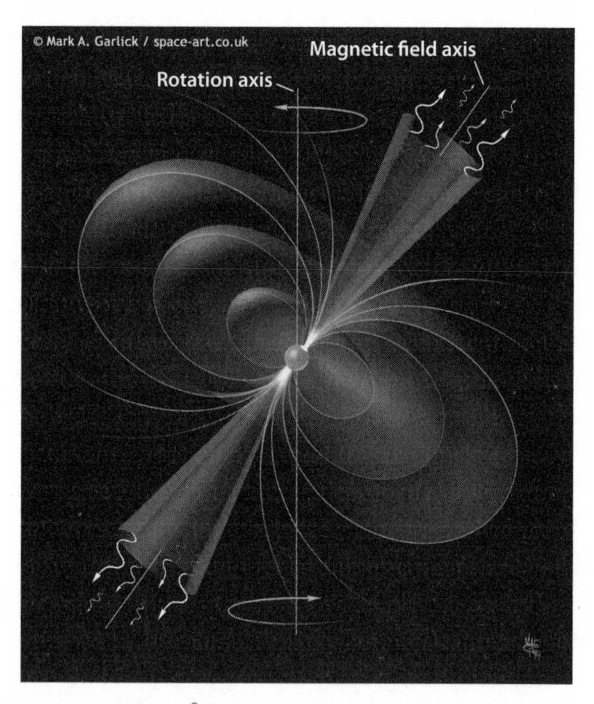

Figure 11.4 Diagram of a pulsar.[9] http://space-art.co.uk/index.html. Note that the magnetic axis is not aligned with the spin axis. (Courtesy of Mark Garlick.)

pulse of radiation every time the magnetic pole swings by. This is sometime termed the "lighthouse effect" because it acts in much the same way as does the rotating searchlight in a lighthouse.

Although pulsars emit radiation at a variety of different wavelengths, they are most easily detected either in the (short wavelength) X-ray region or in the (long wavelength) radio region of the electromagnetic spectrum. Wolszczan is a radio astronomer, and so he was observing this particular pulsar at radio wavelengths. He was also able to measure the timing of the pulses very accurately. PSR B1257+12 rotates roughly 160 times per second, and so it emits a pulse of radiation every 6.2 milliseconds. But Wolszczan found that the pulses were not spaced evenly in time. Sometimes they occurred slightly closer together and sometimes farther apart. He attributed this variation to the *Doppler effect* caused by the neutron star's motion. Because this effect is so important in detecting extrasolar planets, we'll discuss it in more detail below. To make a long story short, however, Wolszczan was able to demonstrate that the star's motion was being caused by the presence of two planets orbiting around it. The smaller one, which had a mass of at least 2.8 Earth masses (M_E), had an orbital period of 98.2 days, and the larger, 3.4-M_E planet had a period of 66.6 days. The semi-major axes of these planets were 0.47 and 0.36 AU, respectively. Since that time, another smaller planet with a mass of $\sim$0.02 M_E has been found in this system, and the (minimum) masses of the two larger planets have been upgraded to 3.8 and 4.1 M_E.[10]

The Doppler Effect

How did Wolszczan's technique work, and why does it yield only a minimum mass for the planet? To answer these questions, we must consider how the Doppler effect works.

An example with which many people are familiar has to do, not with radio waves, but with sound waves. A sound wave consists of a series of alternating compressions and rarefactions of density that propagate through some medium, which could be air, water, or some solid material (figure 11.5). The frequency, or pitch, of a sound wave depends on how closely the compressions are spaced relative to each other. If they are close together, then the pitch is high; if they are far apart, then the pitch is low. As in the case of electromagnetic waves (chapter 3), the properties of a sound wave are related by the formula: *velocity* = *frequency* × *wavelength*. The velocity of a sound wave is much lower than that of an electromagnetic wave, though. As we have seen, electromagnetic waves all travel at the same velocity, the speed of light or c (=3×10^8 m/s). By comparison, a sound wave travels travels through air at only about 340 m/s, or 760 miles per hour.

Now consider what happens when the source of a sound wave is moving. A familiar example is a train that is blowing its whistle. As the train approaches, the whistle has a constant, relatively high pitch. When the train has passed by, the pitch immediately drops to a lower tone. Why?

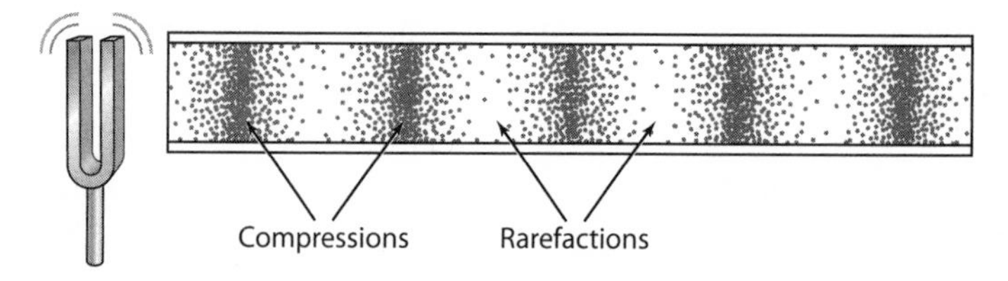

Figure 11.5 Schematic diagram of a sound wave.[11] (Courtesy of Tom Henderson.)

This is the essence of the Doppler effect. During the time that the train is approaching, each successive compression appears to be slightly closer, relative to the stationary observer, than they are to someone riding the train. The wavelength therefore appears shorter, and so the frequency (or pitch) must be higher. When the train is headed away from the observer, the opposite is true: the wavelength appears longer, and the pitch must be lower.

Let's now apply this same logic to the radio waves emitted by a pulsar. The Doppler effect works the same way for electromagnetic radiation as it does for sound waves. When the pulsar is approaching the observer, its radio waves are shifted to higher frequencies. When the pulsar is moving away from the observer, its radio waves are shifted to lower frequencies. As we discussed earlier, with reference to astrometry, the star's motion is a result of the gravitational tug of the planets orbiting around it. But there is an important difference between astrometry and the *Doppler method*, as it is sometime called. In astrometry, one looks for the side-to-side motion of the star in the plane of the sky. In the Doppler method, one looks for back-and-forth motion in the direction of the observer.

Finally, let's return to the question about planet mass. Why does the Doppler method yield only the minimum mass of the planet, and not its true mass? The answer is that in this method one cannot determine the orientation of the planet's orbit (unless the planet passes directly in front of the star, as discussed in the next chapter.) Suppose, for example, that the planet's orbit is entirely in the plane of the sky, that is, we are looking at the system from directly above or below the planet's orbital plane. Then the side-to-side motion of the star is large in both the up-down and left-right directions, so astrometry works well. However, the back-

and-forth motion of the star in the observer's line of sight is precisely zero, regardless of the planet's mass. So the Doppler method would be unable to detect such a planet. If the orbit is now tilted slightly out of the plane of the sky, then the Doppler method can work, in principle. But the star's motion will still be mostly sideways, rather than back-and-forth, and so only a massive planet would be able to move the star fast enough to be able to observe it.*

Let's step back for a moment and compare the Doppler method of planet finding with the astrometric method. In the Doppler method, one obtains only a lower limit on the planet mass, but in the astrometric method one obtains the true mass. Why the difference? In a nutshell, it's because in the astrometric method one measures two components of the star's motion: up-and-down and side-to-side. In the Doppler method, one measures only one component of the motion: back-and-forth. The astrometric method therefore provides more information, and this allows it to provide an accurate mass determination.

We can make one further point without getting overly technical. Just as the astrometric method is most sensitive to large planets like Jupiter, because they move the star by a greater distance, the Doppler method is also sensitive to large planets. But, once again, there is an important difference. The astrometric method is most sensitive to large planets that are relatively *distant* from their parent star. That is because the planet effectively has a longer lever arm with which to tug on the star.** For the Doppler method, however, the situation is reversed: the technique is most sensitive to planets that are *close* to their parent star. The reason is that such planets are orbiting at high velocities, and so they impart a correspondingly high velocity to the star. Planets that are close to their parent stars also have short orbital periods, and this makes it even easier

*For those readers who are familiar with trigonometry, it is easy to express the effects of orbital geometry quantitatively. For the Doppler method, the true mass of the planet is equal to its measured mass divided by sin i, where i is the inclination of the planet's orbit relative to the plane of the sky. When the planet's orbit is close to the plane of the sky, $i \cong 0°$, and sin $i \cong 0$, yielding an extremely high planet true mass. When the planet's orbit is perpendicular to the plane of the sky, $i = 90°$, and sin $i = 1$. In the latter case, the true mass of the planet is equal to its measured mass.

**For a star of mass M and a planet of mass m, the relative distance R that the star moves as the planet orbits it at distance r is given by $R \cong r(m/M)$.

to find massive planets quickly using the Doppler method. We will see below that this has led to spectacular success for this method over the past 12 years.

The Radial Velocity Method

This brings us to what has thus far been by far the most productive, planet-finding technique: the *radial velocity* (RV) *method*. The term "radial velocity" refers to the star's velocity in the observer's line of sight; hence, this is just another name for the Doppler method. But the RV method applies the Doppler effect to visible light, rather than to radio waves. And, furthermore, it can be applied to most main sequence stars, whereas Wolszczan's technique applied only to pulsars. As of the time of this writing, late 2008, the RV method has been used to find over 290 extrasolar planets. These are conveniently kept track of, along with another 20 or so exoplanets discovered by other methods, on a website called *The Extrasolar Planets Encyclopedia*[12] that is maintained by Jean Schneider at the University of Paris. The total number of detected planets currently stands at 318.

Why, you might ask, was this method of planet detection developed only after Wolszczan had used radio waves to detect the pulsar planets? The answer is that measuring the Doppler shift accurately at visible wavelengths is technically more challenging than timing the pulses from a pulsar. Physicists are able to measure time to extraordinary precision by using atomic clocks—clocks calibrated to the internal frequencies of certain atoms, such as cesium. So it was relatively easy for Wolszczan to measure the variation in pulse timing caused by the pulsar's motion. (That's easy for me to say, as I didn't do the work! In practice, I'm sure it was quite tricky.) The hardest part of Wolszczan's experiment was removing all the other Doppler signals from his data. After all, the Earth is spinning on its axis, it is orbiting around the Sun, and the Sun itself is moving relative to other stars. Only by accurately subtracting all of these other effects was Wolszczan able to isolate the Doppler signal that corresponded to the tug of the pulsar planets on their host star.

In contrast to pulsars, main sequence stars do not emit pulses of radiation; rather, they emit radiation of all wavelengths more or less continuously. So, to measure the Doppler shift of a main sequence star, one needs to first break up the light of the star into different wavelengths to form a spectrum. Then, one needs to accurately measure the positions of various spectral features over an extended period of time and determine whether they exhibit a time-dependent Doppler shift.

What makes this possible is that stars like the Sun have numerous *absorption lines* in their spectra. A relatively low-resolution spectrum of the Sun extending all the way through the visible is shown in figure 11.6 (see color section). This is called the *Fraunhofer spectrum,* in honor of the German optician Joseph von Fraunhofer, who first observed these lines back in 1817. The dark vertical lines in the spectrum are caused by absorption of outgoing sunlight by various atoms and ions in the relatively cool outer part of the solar *photosphere.** The pair of lines at ~5900 angstroms (590 nm), labeled "D," is caused by absorption by ionized sodium. Note their position in the yellow region of the spectrum. These same two lines, seen in emission, are responsible for the yellowish light given off by sodium vapor lamps that are used to light streets and parking lots at night.

Just as the radio waves emitted by a pulsar shift their frequency when the pulsar moves back and forth, the absorption lines in a main sequence star's spectrum shift their frequency, or wavelength, for exactly the same reason. If one can measure this Doppler shift sufficiently accurately, one can determine the back-and-forth velocity of the star. From that, one can estimate the mass of any planet orbiting around it. As with the pulsar timing technique, though, only the minimum planet mass can be determined.

The difficult part of doing this lies in measuring the precise positions of the star's absorption lines. One can get an idea for how hard this is by realizing that the fractional shift in the wavelength of a particular absorption line is equal to the back-and-forth velocity of the star divided by the speed of light. In the very best cases, i.e., the easiest planets to find by this method, the velocity of the star is about 500 m/s. Thus, the fractional

*The deeper photosphere is hotter and thus emits more radiation. If this radiation is absorbed and re-emitted in the cooler outer photosphere, an absorption feature results.

change in wavelength is (500 m/s)/(3×10^8 m/s), or about 1.7×10^{-6}. A shift of this small size is very difficult to measure. If one applies this to the sodium D lines seen in figure 11.6, for example, the shift in their positions would be 5900 angstroms $\times$ (1.7×10^{-6}) (0.01 angstrom. Clearly, one would *not* be able to measure this based on a crude spectrum like the one shown in figure 11.5!

Fortunately, some astronomers are clever experimentalists. Two of the cleverest such workers, and the first to demonstrate success with this method, are Michel Mayor of Switzerland and Didier Queloz of France. Working at the Observatoire de Haute-Provence just south of Geneva, in 1995, they announced the discovery of the first extrasolar planet orbiting a main sequence star.[14] The star was called 51 Pegasus, and the planet (following standard notation for binary star systems) was dubbed 51 Pegasus *b*. Its velocity curve, from the original paper, is shown in figure 11.7. The horizontal coordinate, ϕ (phi), represents the *phase* of the planet in its orbit. The points $\phi = 0$ (or 1) are the times when the planet is moving most rapidly away from the Earth; hence, the star is moving toward the observer at a velocity of ~50 m/s. At $\phi = 0.5$, the planet is moving toward the Earth, and hence the star is moving away from the observer at this same rate. One complete orbital period for 51 Pegasus *b* lasts ~4.2 Earth days. Based on the mass of the star, which is close to that of our Sun, one can show that this velocity curve corresponds to a planet with a minimum mass of 0.5 Jupiter masses (M_J), orbiting at a distance of ~0.05 AU.

The discovery of 51 Peg *b*, as it is commonly called, was remarkable for several reasons. In the first place, no one had imagined that such a large planet would be found orbiting so close to its parent star. The planet's large mass suggests strongly that it is a gas giant planet, similar to Jupiter and Saturn in our own Solar System. As discussed in chapter 2, astronomers believe that such planets must form far away from the parent star in regions of the protostellar nebula that are cold enough for water ice to condense. In our own Solar System, this process of giant planet formation happened well beyond Earth's orbit. The closest-in giant planet, Jupiter, has a semi-major axis of 5.2 AU. The orbital radius

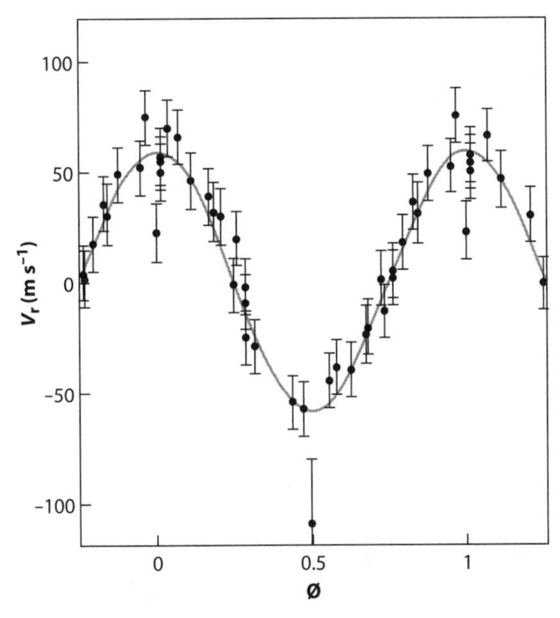

Figure 11.7 Velocity curve for 51 Pegasus *b*. The horizontal axis represents the phase of the planet in its orbit, so one complete orbit is represented by the range $0 < \phi < 1$.[14] (Reproduced by permission of *Nature*.)

of 51 Peg *b* was 100 times smaller, putting it well inside the orbit of Mercury. (Mercury's semi-major axis is about 0.4 AU.)

Although other astronomers were initially very surprised, and somewhat skeptical, at this announcement, the discovery was quickly confirmed by the U.S. planet-finding team of Geoff Marcy and Paul Butler. Marcy, then at San Francisco State University, had been looking for planets for several years using a similar RV technique. Indeed, at the time, Marcy and Butler's *spectrograph*—the instrument used to measure the position of the stellar absorption lines—had a greater sensitivity than did the one used by Mayor and Queloz. But Marcy and Butler had never analyzed their data to look for orbital periods as short as 4 days. When they turned their spectrograph on 51 Pegasus, they immediately found a strong RV signal, with a phase that was identical to that measured by the Swiss team.[15] So the discovery of 51 Peg *b* was spectacularly confirmed. Marcy and his team subsequently discovered several more short-period planets in their data that they could have identified earlier, had they been looking for them. To this day, this team continues to be among the most successful of modern-day planet finders. The tables have turned, though, in terms of measurement precision: at the time of this writing, Mayor's group now has the most sensitive instrument! Science, like many other aspects of life, tends to be very competitive.

We now know that 51 Peg *b* is just one member of a class of planets known as "hot Jupiters" (see figure 11.8). Once its existence was confirmed, it did not take long for the theoreticians to come up with an explanation for how such planets form.[16,17] Indeed, as has happened before in other areas of physics, the basic idea had already been developed several years earlier. Planets that form quickly, while the gas and dust of the protostellar nebula are still present, can *migrate* inward from the position in which they initially formed. They do so by interacting gravitationally with the surrounding gas and dust.[18,19] Even after the nebula has dissipated, planets can migrate via gravitational interactions with remaining planetesimals[20] or with other large planets that are not yet in stable orbits.[17] Technically, astronomers who study disk evolution speak of two types of migration: type 1 (which applies to small planets) and type 2 (which applies to larger ones).[21] In type 2 migration, the planet is large enough to clear a gap in the nebula as it migrates inward. Hence, this mechanism could be potentially observable once our observations of protostellar disks become detailed enough to identify such gaps.

Not all of the discovered planets were hot Jupiters. Figure 11.7 shows that some of the earliest planets to be found have semi-major axes of 2 AU or more. These planets still probably migrated inward from where they were formed initially. As time has gone by, however, planets have been discovered at larger and larger distances and with smaller and smaller masses. The smallest known planet at the present time, based on RV detection by the Swiss team, is a 0.016 Jupiter-mass planet ($= 5.1\ M_E$) orbiting the star Gliese 581.[23] This planet attracted considerable media attention when it was first announced because it is small enough to be a rocky planet like Earth, and because it was initially thought to lie within the habitable zone of its parent star. Recall from the previous chapter that planets with masses greater than $\sim 10\ M_E$ are considered likely to be

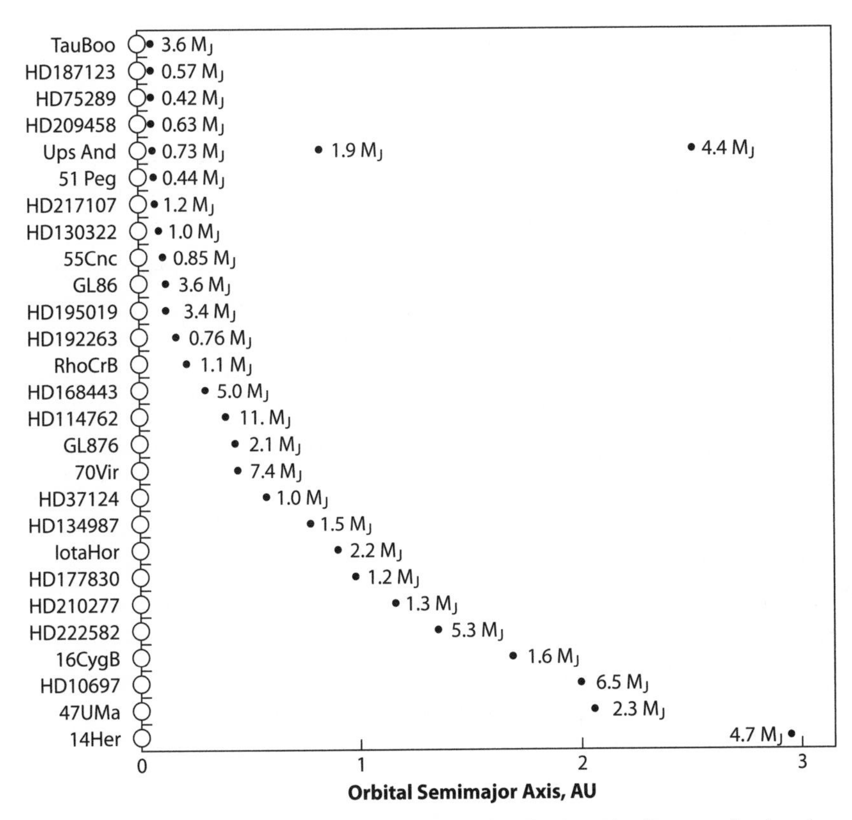

Figure 11.8 Diagram showing the first 29 extrasolar planets to be discovered using the radial velocity method. M_J is the mass of Jupiter. The numbers represent minimum planet masses. (Figure courtesy of Geoff Marcy.[22])

gas giants, because they should be able to efficiently capture gas from the stellar nebula. Gliese 581c, as it is called, is probably below that limit, unless its orbit is within 30 degrees of the plane of the sky, which is unlikely.* The parent star, Gliese 581, is a dim red M star, and so this planet must orbit very close to it in order to be habitable. The luminosity of the star is ~0.013 times that of our Sun, and the planet's orbital distance is about 0.073 AU. Unfortunately, if one crunches the numbers, the flux of starlight hitting the planet is about 2.5 times that hitting the Earth, and

*Recall that the true mass of the planet is equal to its measured mass divided by sin i, where i is the orbital inclination. The sine of 30° is 0.5, so if $i < 30°$, the true mass of Gleise 581c is $> 10.2\ M_E$.

about 30 percent higher than that hitting Venus. So this planet is probably inside the inner edge of the habitable zone. It may be a super-Venus, but it is probably not a super-Earth.[24,25]

Despite the fact that we have not yet located an Earth-like planet by this particular search method, it would be a mistake to end this section on a negative note. The radial velocity method has been incredibly successful already, and its potential for finding many more exoplanets seems high. The different planet-finding teams—Mayor's group in Switzerland, Marcy's group in the United States, and others who have since joined them—are becoming better and better at making these measurements. The discovery of Gliese 581c, just mentioned, is groundbreaking in the sense that it shows that many other similar planets are likely to be detected in the near future. Planets as small as or smaller than Gliese 581c are most easily detected around M stars, and so the RV technique is not likely to yield a true twin of the Earth. But it may well turn up a number of planets not much bigger than Earth, orbiting within the habitable zones of their parent M-stars. This is an exciting possibility, which, if it occurs, seems certain to fuel interest in the more ambitious planet-finding missions discussed in the following chapters.

Gravitational Microlensing

One additional indirect planet detection method deserves mention before we move on. (Actually, there are two more such methods, but we shall save transits for the next chapter.) This last technique is called *gravitational microlensing*. This one is an extremely clever method. Indeed, the basic physics of the process was elucidated by one of the cleverest scientists of all time—Albert Einstein.

As one consequence of his theory of general relatively, Einstein showed that light passing by a star will be bent slightly by the star's gravity. Or, if you prefer, the star's mass bends space–time in its vicinity, and the light follows a straight path in the resulting curved coordinate system. The angle at which the light is bent depends on the star's mass and can be calculated from Einstein's theory.

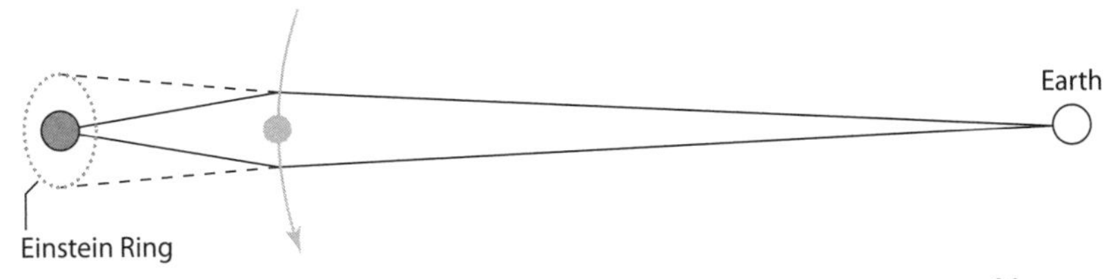

Figure 11.9 Diagram illustrating the principle of gravitational lensing.[26]

Stars, as we have seen, are moving relative to each other, and occasionally one star will pass directly in front of a more distant one, as seen from Earth. When this happens, the closer star can act as a *gravitational lens*. The principle is illustrated in figure 11.9. When the geometry of the stellar encounter is correct, the light from the more distant star will be focused by the nearer star (the "lensing" star), making it appear as a ring, which is called, appropriately enough, the *Einstein ring*. Importantly, the light from the more distant star is also magnified by as much as a factor of 20. (The reason is that the lensing star helps to capture some of the source star's light that was emitted in slightly different directions.) As the lensing star passes in front of the source star—a process that typically takes a few minutes to a few hours—the apparent luminosity of the source star first increases and then decreases, as shown in figure 11.10. The magnification is this high only when the two stars are almost exactly aligned. For a typical encounter, the peak magnification is lower, but it is still large enough to identify itself as a *microlensing event*.*

The frequency of microlensing events is not high. One has to monitor hundreds of thousands, or even millions, of stars to have a decent chance of detecting one. This can now be done, however, using modern *CCD detectors*.** The trick is to look at a patch of sky that includes lots of stars. If one stares through the Milky Way into its thickest part, the *galactic bulge*, one can see enough stars simultaneously to make the technique work.

*The term "microlensing" is used here because gravitational "lensing" refers to the focusing of light from a distant galaxy when another galaxy passes in front of it.

**"CCD" stands for *charge-coupled device*. It is basically a semi-conductor chip that counts individual photons and converts them into electrical signals. CCDs have replaced photographic plates in modern telescopes.

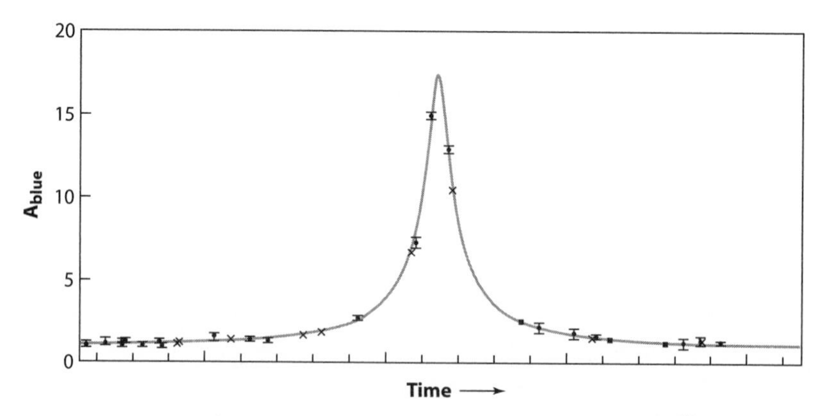

Figure 11.10 Light curve for a candidate microlensing event.[27]

It is not enough, however, to simply detect the microlensing events themselves. After all, doing so simply proves once again that Einstein was right—something that we have known for a long time (although physicists still delight in testing his theory to ever greater levels of precision). The real trick is to use microlensing events to detect planets. The way it works is the following: Suppose that the *lensing star* (the one in the middle) has one or more planets orbiting around it. If the orbital radius of the planet is near the Einstein ring radius, then it can alter the amount of light being received on Earth from the source star. This results in a small bump in the light curve. An example, detected by the OGLE and MOA microlensing networks[28] in 2004, is shown in figure 11.11. The presence of the planet is indicated by the two spikes on the rising (left-hand) side of the light curve. This particular planet, named OGLE235-MOA53 b, is thought to be a 2.6 Jupiter-mass planet orbiting at about 5 AU from its parent star. The star itself is a K star, with about 0.6 times the mass of the Sun. Thus, this planet appears to be a fairly close analogue of our own planet Jupiter.

The gravitational microlensing technique is still in its infancy at the time of this writing. Only 8 planets have been discovered so far, compared to over 290 by the radial velocity method. But the microlensing community is starting to get organized. Up until now, individual microlensing events, when discovered, were reported by email to the rest of

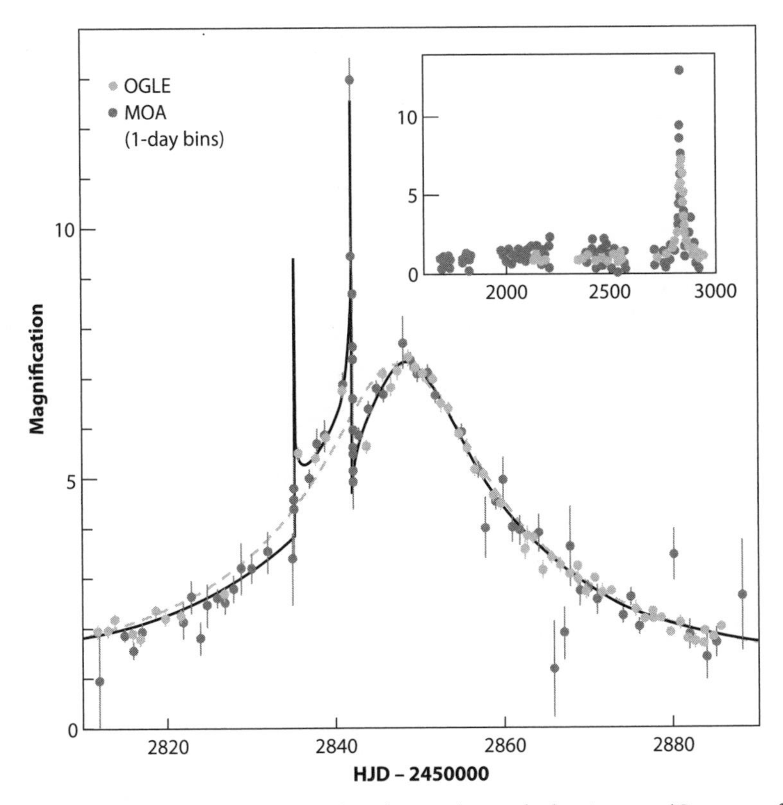

Figure 11.11 Light curve for a star with a planet orbiting the lensing star. (Courtesy of David Bennett, University of Notre Dame.[29])

the community. Many potential lensing events were only partially observed, or missed entirely. Once this process becomes more automated, which should happen within the next few years, the number of detected planets is expected to increase dramatically.

A disadvantage of the microlensing technique is that the stars that are being examined (the lensing stars) are many thousands of light-years distant. Hence, the chance of doing follow-up observations using another technique is practically nil, at least for the foreseeable future. A second disadvantage is that the Einstein ring radius for most lensing stars is at several AU distance—well outside the habitable zone. So, while this technique is capable of detecting Earth-mass planets, it is unlikely to find any close analogs to Earth. It is nonetheless a powerful tool that can potentially be used to estimate the frequency with which planets, including small ones, exist around other stars. This would have surprised Einstein, as he himself considered stellar microlensing to be just a passing fancy. In a paper[30] written in 1936, he wrote:

> Of course, there is no hope of observing this phenomenon directly. First, we shall scarcely ever approach closely enough to such a central line. Second, the angle β will defy the resolving power of our instruments.

This quotation should take its place in the list of blanket statements by famous people that turn out to be dead wrong.

In the next chapter, we discuss the final, and in some ways the most exciting indirect planet detection technique—the *transit method.*

· *Chapter Twelve* ·

Finding and Characterizing Planets by Using Transits

In the previous chapter, we discussed three fundamentally different methods of finding planets around other stars: the astrometric method, the Doppler method, and the gravitational microlensing method. (Both the radial velocity method for main sequence stars and the pulse-timing method for pulsars are variants of the Doppler method.) All of these techniques are classified as "indirect" methods, because they rely on looking at the effect of the planet on the light emitted from the star, rather than looking for the light from the planet itself. The transit method, discussed below, is also indirect in the sense that it relies on the dimming of a star's light when a planet passes in front of it, as observed from Earth. We shall see, though, that by taking advantage of a few tricks one can actually begin to observe the planet's own light by this technique. Hence, the transit method forms a bridge between the indirect detection methods of the previous chapter and the direct detection methods discussed in the chapters to follow.

Transits of Mercury and Venus

The transit method of observing planets has actually been used for a very long time. It was first applied, though, to planets within our own

Solar System. Mercury and Venus are both closer to the Sun than is Earth, and so both planets occasionally transit the Sun, i.e., they pass directly in front of it.

Transits of Venus are exceedingly rare. They tend to occur in pairs that are separated by about 8 years, and the pairs themselves are roughly a century apart. Hence, there have only been 6 transits of Venus since the invention of the telescope in 1608. The last transit of Venus occurred in 2004, and the next one will occur in 2012[1] (see figure 12.1). The next pair after this one will occur in 2117 and 2125. Although they might appear to be a curiosity, Venus transits have actually been put to practical use by scientists. The famous astronomer Edmund Halley realized a long time ago that transits of Venus could be used to establish the absolute scale of the Solar System. Surprisingly, the absolute orbital distances of the planets were difficult to determine, even with telescopes, because the mass of the Sun was not known. However, by timing the passage of Venus across the Sun's face simultaneously from two or more widely spaced locations on the Earth's surface, and by knowing how far apart those points were, it was possible to estimate the distance of Venus from the Sun, and hence the size of the Solar System. Indeed, scientific expeditions were mounted in 1761 and 1769 to observe Venus' transit and to take advantage of this opportunity.

Transits of Mercury occur much more frequently—about 13 or 14 times each century. They, too, can be used to measure the size of the Solar System, but the measurement is more difficult because Mercury is closer to the Sun, and hence the time required for Mercury to transit (i.e., to pass across the face of the Sun) is smaller. So they have not been discussed nearly as much as have transits of Venus.

Transits of Extrasolar "Hot Jupiters"

The real interest in transits these days, of course, is not in looking at planets within our Solar System, but in looking at planets outside of it. The technique was pioneered in 1999 by a graduate student at Harvard University named David Charbonneau. Charbonneau is now on the fac-

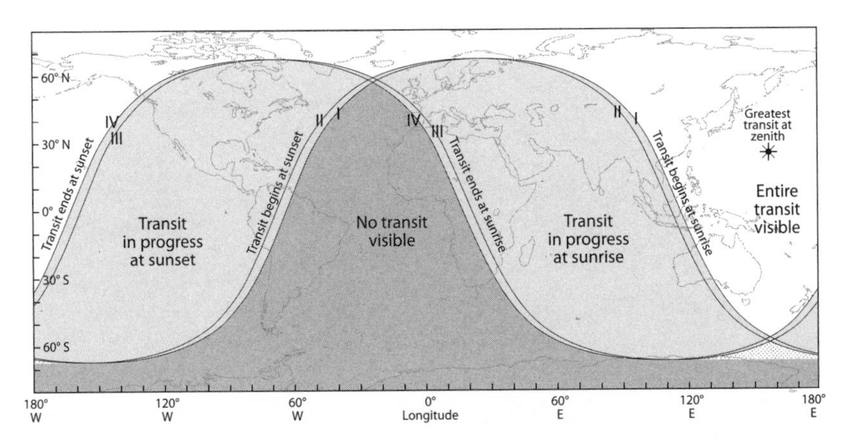

Figure 12.1 Locations on Earth where it will be possible to observe the 2012 transit of Venus.[1] (Courtesy of NASA/Goddard.)

ulty there, demonstrating that it is still possible to get ahead quickly in this world if you have good ideas.

Charbonneau knew, as did other astronomers, that there were a number of "hot Jupiters" orbiting nearby stars. At that time, approximately 10 such planets had been discovered, all by the radial velocity method described in the previous chapter. Charbonneau also knew that the chance that one of them might pass in front of its parent star was pretty good. One can demonstrate mathematically* that if you observe a one-planet system from an arbitrary angle and from a large distance, the odds of seeing a transit are equal to the radius of the parent star, R_s, divided by the mean orbital distance of the planet, a. The geometry is indicated in figure 12.2. Here, i is the inclination of the planet's orbit relative to the plane of the sky. For the planet to transit, the angle $\theta = 90°—i$ must be less than θ_o, where θ_o is defined in the figure. It is also easier to find big planets like Jupiter, than small ones like Earth, by this method for an obvious reason: they block out more of the light from the star.

Given this information, consider first what the chances would be of detecting Jupiter in our own Solar System. The radius of the Sun is about

*Technically, one integrates the inclination angle, i, weighted by sin i, over the range of angles for which a transit occurs and divides by the integral of sin i over all possible angles, 0 to 90°. Equivalently, one can integrate $\theta = 90°—i$ from 0 to θ_o, weighting by cos θ, and divide by the integral of cos θ from 0 to 90°. The latter quotient yields $\sin \theta_o = R_s/a$ as the answer.

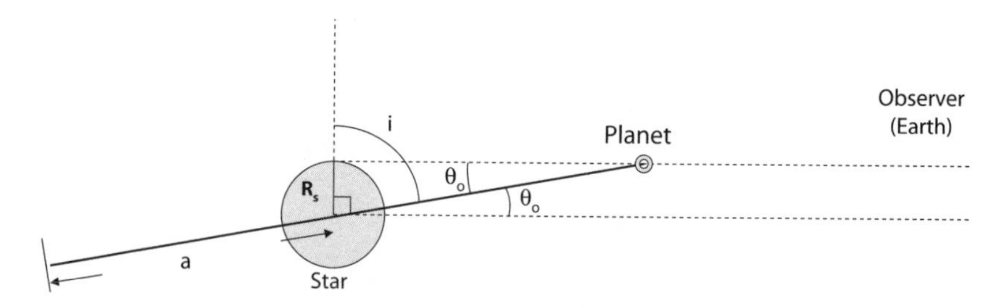

Figure 12.2 Diagram illustrating the geometry of a transiting planet relative to an observer on Earth.

7×10^5 km. Jupiter orbits the Sun at 5.2 AU, and 1 AU $\cong 1.5 \times 10^8$ km. Thus, the chance of detecting a Jupiter transit from outside our Solar System is $7 \times 10^5/(5.2 \times 1.5 \times 10^8) \cong 1 \times 10^{-3}$, or roughly 0.1 percent. This probability is obviously rather small. It means that to have a good chance of finding even one Jupiter by this method, one would need to look at $1/(1 \times 10^{-3}) \cong 1000$ stars.

Consider what happens, though, if one uses this technique to search for "hot Jupiters." Such planets orbit their stars at a typical distance of 0.05 AU. Hence, for a star the same size as our Sun, the chance of observing a transit is $7 \times 10^5/(0.05 \times 1.5 \times 10^8) \cong 0.1$, or 10 percent. Or, to put it a different way, the chance of finding a hot Jupiter is 100 times higher than the chance of finding a normal Jupiter, because it is 100 times closer to its parent star. Hence, to have a good chance of observing a hot Jupiter by this method, one needs only to look at about 10 stars (knowing beforehand, of course, that they have a hot Jupiter in some kind of orbit). This, of course, was just how many hot Jupiters were already known from radial velocity searches when Charbonneau began his project. So Charbonneau put two and two together and reasoned that if he looked at each of these 10 stars, he'd have a good chance of observing a transit. This is the kind of thinking that gets you onto the faculty at Harvard. It's relatively simple, in retrospect, but at the time no one else had thought to do it.

Charbonneau needed one more piece of information. He had to know how deep the transit would be, that is, how much of the star's light would be blocked out by the transiting planet. That calculation is also straightforward. Jupiter, it turns out, has a radius of about 7×10^4 km, which is very close to 1/10 that of the Sun. The area of each object projected on the sky is proportional to its radius squared. (This is because the area of a circle is πr^2, where r is the radius of the circle.) Hence, a planet the size of Jupiter transiting in front of the Sun should block out a fraction of its light given by $(0.1)^2$, which equals 0.01, or 1 percent. And a change of 1 percent in a star's brightness is relatively easy to measure. One simply needs a good CCD detector, similar to those used for the radial velocity measurements discussed in the last chapter.

So Charbonneau borrowed a $50,000 CCD, attached it to a small "backyard" telescope, and made his measurements. And, sure enough, he soon struck gold. Figure 12.3a shows his results for the star HD 209458. The "HD" stands for the Henry Draper catalogue—a compendium of stars that was compiled during the early 1920s. The planet that Charbonneau observed is a 0.7-Jupiter-mass planet with a 3.5-day orbital period. In this case, we know the planet's mass quite accurately, because the plane of the planet's orbit has to be nearly edge-on to our line of sight; otherwise, it would not transit. Despite the fact that this planet is somewhat less massive than Jupiter, the light from the star drops by about 1.6 percent when the planet passes in front of it, indicating that the planet is somewhat larger than Jupiter. The reason for this is not hard to understand. This planet is close to its parent star, and so its upper atmosphere is hot and extended, whereas Jupiter's upper atmosphere is cold and relatively compact.

Once it was established that HD 209458 had a transiting planet, it did not take long to confirm Charbonneau's measurement. As the timing of the transits was known, it was a relatively simple matter to program the *Hubble Space Telescope* (*HST*) to look at the star at the appropriate time.[3] A friend of mine from my postdoc days at the National Center for Atmospheric Research, Ron Gilliland, was the *HST* contact on this project. The result was the graph shown in figure 12.3b. This is the same light curve as measured originally but it has far greater precision, for two reasons. First, *HST* is a bigger telescope than the one that Charbonneau

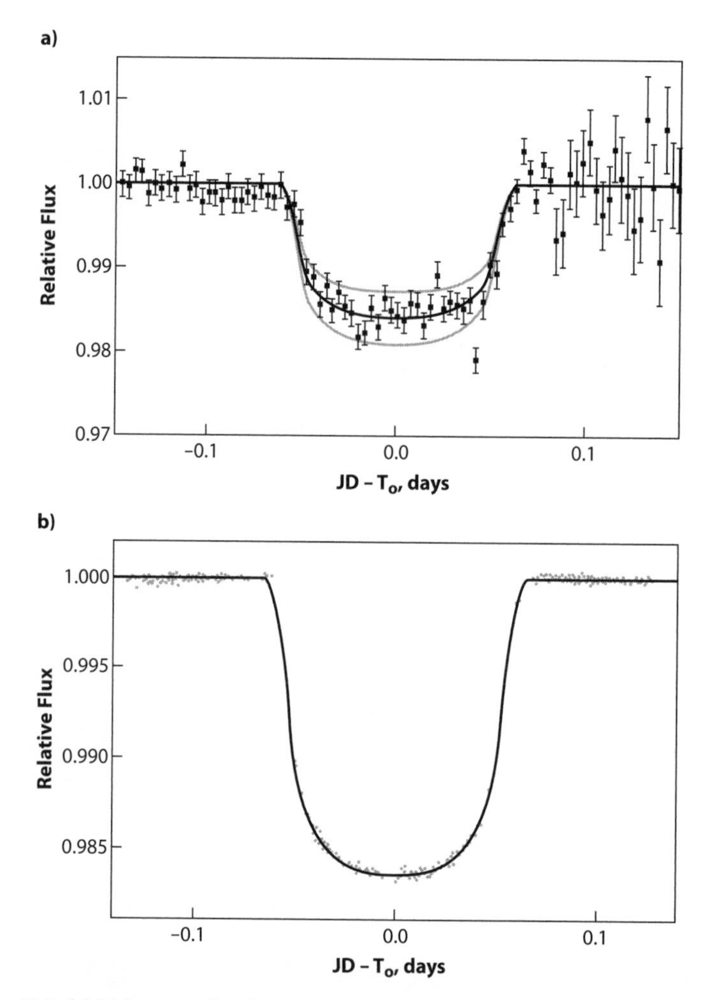

Figure 12.3 (a) Light curve for the star HD 209458, as measured by David Charbonneau and coworkers.[2] (b) The same light curve measured a few months later using the *Hubble Space Telescope*.[3] (Reproduced by permission of the AAS.)

had used in his initial measurement, and so it can collect more photons. More importantly, however, making the measurement from space removes the variability imposed by Earth's turbulent atmosphere. As discussed further below, transits can be observed much more effectively from space, allowing this method to potentially be extended to planets much smaller than the one observed by Charbonneau.

Space-Based Transit Searches: CoRoT and Kepler

We'll return to the HD 209458 system below, as there is more to the story. Before we do so, though, let's think about what this implies about the prospects for searching for extrasolar Earths.

Detecting Earth-like planets by the transit method is more difficult than finding Jupiters for two reasons. First, Earth is much smaller than Jupiter. Earth's radius is 6371 km ($\sim$4000 miles), which is roughly 10 percent that of Jupiter. Its effective area is thus smaller than that of Jupiter by a factor of 10^2, or 100. Hence, for an Earth-like planet transiting a star like our Sun, the star's brightness is expected to drop by only about 1 part in 10^4, or 0.01 percent. Measuring such a tiny variation in a star's luminosity is virtually impossible to do from the ground because the fluctuations in light intensity caused by Earth's atmosphere are much larger than this. (Think about it—stars twinkle when you gaze at them!) So the only way to observe transits of Earth-like planets around Sun-like stars is from space. One still needs a good CCD photon detector, but such devices exist, and so measuring stellar brightnesses to this accuracy is theoretically possible.

The second problem with finding transiting Earths has to do with probabilities. Recall that the odds of seeing a transit are equal to the radius of the planet's orbit divided by the radius of the star. For Earth around the Sun, this value is $(7 \times 10^5 \text{ km})/(1.5 \times 10^8 \text{ km}) \cong 5 \times 10^{-3}$, or 0.5 percent. That is not as bad as looking for Jupiters at 5 AU, but it is still highly unfavorable. To have a good chance of observing a transiting Earth, one would need to look at $\sim 1/(5 \times 10^{-3}) = 200$ stars. And that number is based on the assumption that each one of the stars has an Earth-like planet going around it! Unlike the case of the hot-Jupiter planets that Dave Charbonneau was observing, we do not yet know for sure that Earth-like planets even exist. So it is clearly *not* safe to make such an assumption! Consequently, we need to observe a much larger number of stars to be reasonably certain of finding a planet like Earth. Or, to take the pessimist's point of view, we want to look at a large enough sample of stars to be reasonably sure that Earth-like planets *do not* exist if we don't see them.

Both NASA and ESA (the European Space Agency) have recognized the potential of looking for transits from space, and both agencies have designed space missions to do so. ESA took the first step in this endeavor by launching a small (27-cm aperture) telescope called *CoRoT* in December 2006. *CoRoT* is too small to search efficiently for other Earths, and it has other limitations as well. Specifically, it is in orbit around the Earth, and it must point away from the Sun. Consequently, this means that it can only look for planets with orbital periods shorter than 75 days. This, of course, is too short to find a true Earth analogue. *CoRoT* could conceivably find "hot ocean-planets," though.[4] Such planets, whose existence has been postulated in the literature[5,6] and were discussed briefly in chapter 2, would be rocky planets with dense steam atmospheres, similar to what Venus may have been like early in its history (chapter 6). Finding such planets would be very exciting for astronomers, even if none of them were capable of harboring life.

The mission that astrobiologists are most excited about, though, is NASA's *Kepler* mission, which launched successfully in March 2009.[7] *Kepler* is a 0.95-m diameter telescope that is specifically designed to look for planets with the same size and orbital characteristics as Earth. To do this, *Kepler* will stare continuously at a patch of the Milky Way, where it will measure the brightnesses of ~100,000 stars every half hour (figure 12.4, in color section). This sounds incredible, to be sure, but it is yet another example of what can be accomplished with modern CCD technology. *Kepler* is scheduled to operate for at least 4 years, and it will orbit the Sun rather than the Earth, so it should be capable of finding planets like Earth with 1-year orbital periods.

Let's consider what we might learn if the *Kepler* mission is successful. (There are as yet no results in mid-March 2009.) We saw earlier that the chance of observing a transit of an Earth-like planet orbiting at 1 AU around a Sun-like star is 5×10^{-3}. If every star in the *Kepler* field of view was a G2 star and had one Earth-like planet orbiting around it at 1 AU, then the number of detected Earths should be $5 \times 10^{-3} \times 10^{5} = 5 \times 10^{2}$, or 500. The actual number expected is almost a factor of 10 lower than this, because many of the Kepler target stars are brighter than the Sun, and because much of their habitable real estate lies beyond 1 AU, but this is still enough planets to allow us to compute statistics. The key question is often phrased as: What is the fraction of stars that have at least one Earth-like planet orbiting around them? The astronomers have a special name for this number: η_{Earth} ("eta sub Earth").* Under the idealized assumptions described above, the number of planets that *Kepler* detects should equal approximately $500 \times \eta_{\text{Earth}}$. So η_{Earth} would be just the number of detected planets divided by 500.

The reason this number is important is that it tells us how big a telescope we must build if we want to directly observe an Earth-like planet around another star. NASA hopes to do this with their proposed twin *Terrestrial Planet Finder* (*TPF*) missions, which we discuss in the next chapter. If η_{Earth} is large, then Earth-like planets should exist around many of the nearby stars, and the *TPF* telescopes need not be too large. If η_{Earth} is small, though, then we will need to look at more stars, some of which will necessarily be more distant, and so the telescopes must be made significantly larger.

Note that *Kepler*, even if it is successful, will not provide targets for *TPF*. That's because almost all of the stars in the *Kepler* field of view are very distant, roughly 500 pc (1600 light-years) on average. A few nearby stars may be included in the *Kepler* field, but the odds that one of them has a transiting Earth are, as we have seen, rather low. So *Kepler* can provide us with statistics as to how many Earth-like planets may exist, but it cannot tell us where to look. Even the statistical information, though, would be invaluable.

Observing Exoplanet Atmospheres during Transits

We began this chapter by pointing out that the transit method of finding planets is in a special category, because it lies somewhere in between indirect and direct detection. So far in this chapter we have been discussing indirect detection: the blocking of a star's light as a planet passes

*Note that η_{Earth} is closely related (but not identical) to the third factor in the Drake equation, n_E, which is the number of Earth-like planets per star, for stars that are known to have planets.

directly in front of it. But the transit method also has the potential to provide information about the nature of the planet's atmosphere, and it has recently begun to be used in this manner. How does this process work?

To begin, let's first think about why it is hard to directly observe extrasolar planets in the first place. It is difficult for three reasons: (1) the planet is very close to its parent star, (2) the parent star is much brighter than the planet, and (3) the planet itself is intrinsically dim, i.e., it would be difficult to see even if the parent star were not around. Of these three problems, the first is the most difficult to deal with. Overcoming it requires that the aperture of the telescope be quite large (see next chapter), and it probably also requires that the telescope be located in space, above the perturbing effects of the Earth's atmosphere. Some astronomers believe that it may eventually be possible to find Jupiter-sized planets from the ground by using giant 30-m or even 100-m diameter telescopes, but directly observing Earth-sized planets is likely to remain a space-based endeavor.

A transiting planet, however, can, in principle, be examined without having to physically separate the light from the planet from that of the star. Two different methods are available, one of which works best in the visible and the other in at thermal infrared wavelengths. The first method, called *primary transit spectroscopy*, is illustrated schematically in figure 12.5. As the planet passes in front of the star, some of the light from the star passes through the planet's atmosphere on its way to Earth. If the planet is tiny and has a relatively thin atmosphere, like Earth, then the fraction of the star's light passing through the planet's atmosphere is negligible, and no detectable signal should be expected. But for an extended "hot Jupiter" type of planet, orbiting close to its parent star, the signal should be much larger. Furthermore, if the planet's atmosphere contains gases that absorb at specific wavelengths, then the planet should block out more of the star's light at those wavelengths, and hence it will appear bigger and create a deeper transit. By measuring the planet's effective radius at different wavelengths, it is thus possible to create a crude *spectrum* of the planet's atmosphere. We'll discuss the process of taking spectra in more detail in the next chapter. For now, recall from chapter

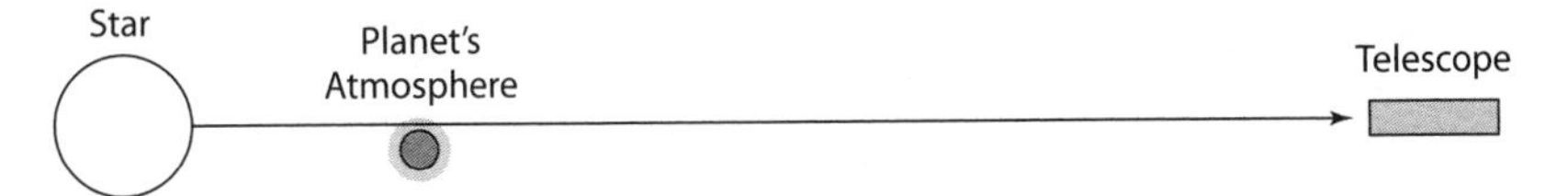

Figure 12.5 Diagram illustrating the technique of primary transit spectroscopy. Some of the light reaching the observer passes through the atmosphere of the planet. The planet appears bigger at wavelengths at which the atmosphere absorbs.

3 that one obtains a spectrum by separating light into its component wavelengths, as occurs in a rainbow.

In 2002, the same group who had measured the light curve for HD 209458 using *HST* (figure 12.3b) showed that they could create a primary transit spectrum from their data.[9] To do this, they measured the depth of the transit at different wavelengths near 590 nm, where the sodium D lines are found. Recall from the last chapter (figure 11.6) that the D lines are among the most prominent features in the spectra of Sun-like stars. The results of the analysis were quite exciting. The planet appeared distinctly bigger, i.e., the transit depth was deeper, inside the D lines than outside them (see figure 12.6), indicating that the element sodium was also present in the atmosphere of this particular giant planet.

Now, to the casual observer this may not appear to be a particularly noteworthy accomplishment. After all, if sodium is present in most Sun-like stars, and in the Earth as well, then it shouldn't be too surprising that it is also present in the atmosphere of a hot Jupiter-type planet. What was remarkable about this measurement, though, was that it demonstrated that we are now capable, if only just barely, of characterizing the atmospheres of extrasolar planets. The hope, of course, is that we will be able to do this much better in the future, using the types of instruments described in the next chapter.

Subsequent to this study, other investigators have used *HST* to observe HD 209458b during transits at a variety of wavelengths. Travis Barman of the Lowell Observatory in Flagstaff, Arizona, looked in the near-infrared, around 970 nm, where H_2O absorbs strongly.[10] He saw evidence for H_2O at approximately solar abundances (see figure 12.7). This again should not be too surprising—after all, H_2O is an abundant component of giant planet atmospheres in our own Solar System, and so

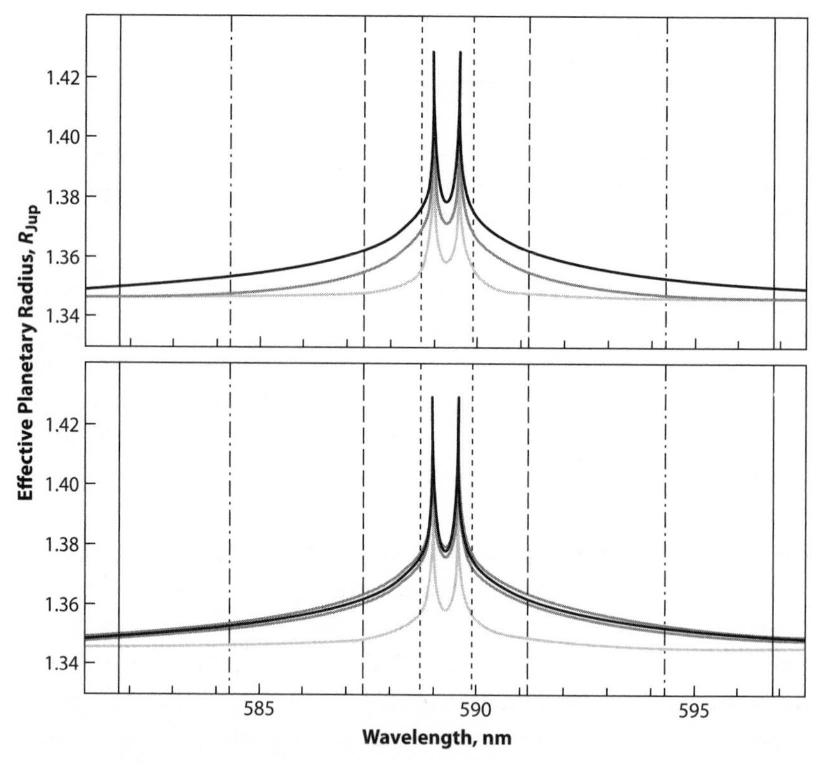

Figure 12.6 Diagram showing the (modeled) apparent radius of transiting planet HD 209458b as a function of wavelength. The planet appears to be larger at the two wavelengths (the D lines) at which sodium absorbs strongly.[9] (*Top*) Variation with cloud height. (*Bottom*) Variation with sodium abundance. (Reproduced by permission of the AAS.)

it might be expected to be present also in an extrasolar giant planet atmosphere. Other types of measurements, though, had yielded different results (see below), and so to many researchers this result was reassuring.

The *HST*/HD 209458b story has one more chapter. In 2003, the French astronomer Alfred Vidal-Madjar and his colleagues used the STIS instrument onboard *HST* to study the planet's transit at very short UV wavelengths.[11] The STIS instrument can measure the flux of Lyman alpha (Ly α) radiation at 121.6 nm. Ly α is a wavelength of UV radiation that can be either emitted or absorbed by atomic hydrogen. The French team

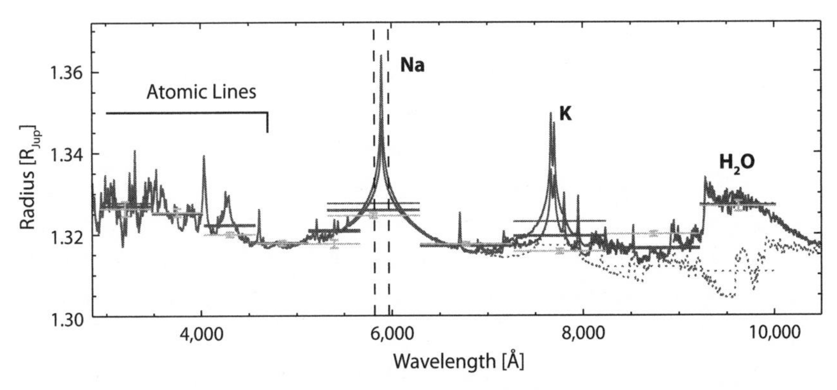

Figure 12.7 Primary transit spectrum (effective planetary radius) of HD 209458b in the visible and near-infrared. The horizontal bars indicate the STIS data. The solid curve is a model spectrum with H_2O included at solar abundances, and the dashed curve is a model spectrum without H_2O.[10] (Reproduced by permission of the AAS.)

found that the depth of the transit in the Ly α line was about 14 percent. By comparison, the transit depth at visible wavelengths was only 1.6 percent (figure 12.3). Hence, the planet appeared to be bigger at Ly α by a factor of $14/1.6 \cong 9$. The amount of light blocked by a planet is proportional to its radius squared, so this meant that the diameter of the planet was effectively 3 times larger at Ly α than in the visible. The authors interpreted this as indicating that the planet was losing an immense amount of hydrogen to space, as might be expected given its hot, hydrogen-rich atmosphere. This result is evidently controversial, and so it may or may not hold up. But it is yet another indication of the power of this technique. Transit data really can provide a wealth of information about the nature of an extrasolar planet.

Secondary Transit Spectroscopy

An entirely different way of obtaining a spectrum of a transiting extrasolar planet is to take advantage of the fact that, if a planet passes in front of its parent star, it must also pass behind it. This alternative approach is called *secondary transit spectroscopy*. The way it works is illustrated in figure 12.8. Suppose that one first takes a spectrum of the planet at a time when the planet is beside the star. This yields a spectrum of the combined star-plus-planet. Next, one takes a second spectrum of the star at a time when the planet is behind it. This, of course, yields only the spectrum of the star itself. If one then subtracts the spectrum of the star from that of the star-plus-planet, one obtains the spectrum of the planet. At least one hopes that is true. In practice, this is very difficult because the star is so much brighter than the planet. At visible wavelengths, the contrast ratio between the star and the planet is overwhelming, and so this technique is not useful. But at thermal–infrared wavelengths where the planet is relatively bright, the secondary transit technique can yield interesting results.

This technique has now been applied to HD 209458 (Charbonneau's original transiting system) by Jeremy Richardson and his colleagues at the NASA Goddard Space Flight Center[13] (see figure 12.9). They used the *Spitzer Space Telescope* to look at the system at thermal-infrared wavelengths. *Spitzer* is a 0.85-m, thermal-infrared space telescope that is in operation at the time of this writing (early 2009). The intriguing conclusion reached by this team was that the hot Jupiter planet orbiting HD 209458 did *not* appear to have much, if any, water. This, admittedly strange, conclusion was reached by comparing the slope of the model spectrum (darker curve, figure 12.9b) with the slope of the data as one moves toward short wavelengths. H_2O has a strong absorption band centered at about 6.3 μm, and the fact that the edges of this band are seen in the model but not in the data suggests that H_2O is absent from the planet's atmosphere. One can see, however, that the data are quite noisy, that is, the error bars are large. Furthermore, the observed spectrum does not extend to short enough wavelengths to make this a definitive measurement of H_2O (or lack thereof). And recent observations of other hot Jupiters[14] suggest that water is present in most such planets. So perhaps one should not make too much of this particular result. Things ought to improve, though, as our observational technology advances; indeed, the next section describes some exciting studies that may be carried out within the next few years.

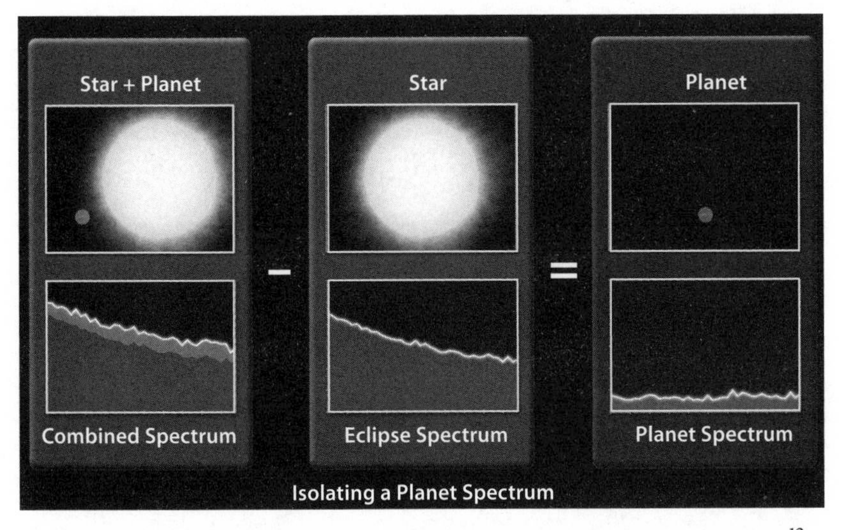

Figure 12.8 Diagram illustrating the technique of secondary transit spectroscopy.[12]

Characterizing Earth-Like Planets around M Stars

Before leaving this chapter, we should point out the potential that these transit techniques, particularly secondary transit spectroscopy, have for the near future. Recall that this technique involves measuring the spectrum of the combined star-plus-planet, and then subtracting the spectrum of the star itself. Thus far, this procedure has been applied only to Jupiter-sized planets, like HD 209458b, orbiting relatively Sun-like stars. It works there—just barely—because Jupiter's diameter is about 1/10 that of the Sun, and so the projected surface area of the planet is about 1 percent that of the star.

There is little chance that this technique could be used to find an Earth-like planet orbiting a star like the Sun. Earth's diameter is only 1/10 that of Jupiter, and so its projected surface area is only 0.01 percent that of the Sun. That is probably too small to produce any kind of measureable secondary transit signal. Consider, though, what would happen if one used this technique on M stars. Recall that M stars are dim, red

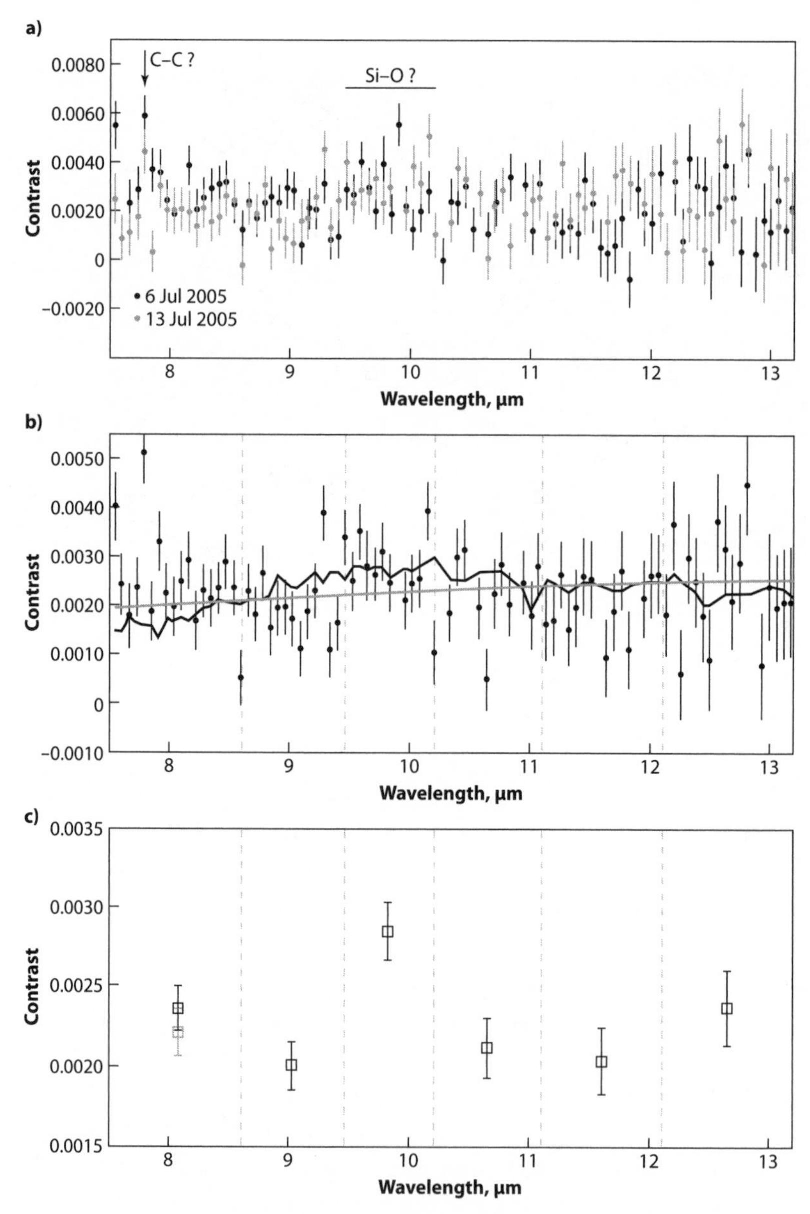

Figure 12.9 Secondary transit spectrum of the planet HD 209458b. (a) Data from two separate secondary eclipses, shown in black and gray. (b) Average data for the two eclipses compared to a model calculation (black curve). (c) Coarse binning of the data showing increased emission from the planet near 9.5μm, possibly caused by silicate vapor in the planet's atmosphere.[13] (Reproduced by permission of *Nature*.)

stars with masses less than 0.5 times that of the Sun. The diameter of an M star is only about 10–30 percent that of the Sun; thus, the smaller (later) M stars are about the same size as Jupiter. An Earth-like planet orbiting a late M star would have a projected surface area that was about 1 percent that of the star—the same ratio as for a Jupiter-sized planet orbiting the Sun. So it is conceivable that secondary transit spectroscopy could be applied to such planets.

This leads to an exciting possibility, which again is being pursued by Dave Charbonneau. (Obviously, he has not yet run out of ideas!) Suppose that one was able to survey nearby M stars to see which of them has planets. If the radial velocity method can be applied successfully to M stars—which has not been possible in the past—then there should be lots of candidates to look at. M stars are, after all, the most numerous types of stars, and so most of Earth's nearest neighbors are M stars. One could then follow up, using ground-based telescopes, and determine which of them has transiting planets. The habitable zone of a typical M star lies at roughly 0.1 AU (figure 10.4), although there is a great deal of variation within this single spectral class. The radius of an M star is about 1/10 that of the Sun, or 7×10^4 km. Hence, the odds of seeing a transiting planet within the habitable zone of an M star are roughly 0.5 percent—the same as *Kepler*'s odds of finding Earth at 1 AU around a star like the Sun. (This can be easily understood, because the habitable zone around an M star is 10 times closer than around the Sun, but the star's diameter is also 10 times smaller.) So, if one surveys several hundred nearby M stars, and if most of these stars have planets within their habitable zones, then one should have a good chance of observing a transit.

What makes this even more exciting is that within the next few years, astronomers should have a wonderful new tool for observing such events. NASA's *James Webb Space Telescope* (*JWST*) is scheduled for launch in 2016. *JWST* is a large telescope, 6.5 m in diameter, which will operate in the thermal infrared. (*HST*, by comparison, is only 2.4 m in diameter.) So *JWST* can be used, like *Spitzer*, to do secondary transit spectroscopy. *Spitzer*, though, has only an 85-cm mirror. *JWST*'s mirror is 8 times larger in diameter, which means that it collects 64 times as much light. More photons mean that one can take a much better spectrum. Hence, *JWST* should be able to obtain much more detailed spectra of extrasolar planets, and the planets that will be accessible to *JWST* should include ones within the habitable zones of their parent M stars. It is thus conceivable that potentially habitable planets could be identified and perhaps even characterized within the next 10 years. Of course, such planets would probably be tidally locked and could be plagued by a host of other problems, as discussed in chapter 10. Still, this promises yet another leap forward in our search for habitable planets.

· Chapter Thirteen ·

Direct Detection of Extrasolar Planets

In the last chapter we heard about the impressive array of techniques that have already been used to find extrasolar planets or that are planned for the very near future. While these methods are all promising, none of them—with the possible exception of *SIM* (the space-based astrometry mission)—is likely to provide information on Earth-size planets orbiting in the habitable zones of Sun-like stars. Being able to do that is the Holy Grail of planet finding. And we don't simply wish to find such planets; we also want to learn something about what they are like. In practice, this means that we want to analyze their atmospheres *spectroscopically*. As has already been mentioned, creating a spectrum of a planet's atmosphere would allow us to identify different gases that it contains and might also provide clues as to whether life exists on the planet's surface.

We saw in the last chapter that it is already possible to take a crude spectrum of the atmosphere of a planet that transits its parent star. This can be done at visible wavelengths using primary transit spectroscopy or in the thermal infrared using secondary transit spectroscopy. But neither of these methods is expected to work well for small, Earth-like planets orbiting large, Sun-like stars, so we are not likely to find a true Earth analogue in this manner. Furthermore, because the probability of transits is relatively low, the planets that we do find by this method are likely to be fairly far away. What we would prefer to do is to observe Earth-size planets around nearby Sun-like stars, without having to rely on transits. This means that we must be able to *spatially* separate the light from the planet from that of the star. And that turns out to be a tall order.

What Wavelength Region Should We Choose?

The first, and most important, question that one must ask is what wavelength region should we use? Planets give off significant amounts of electromagnetic energy in two different spectral regions: the visible/near-infrared and the thermal-infrared. At visible/near-IR wavelengths, the planet is seen in reflected starlight. With our naked eyes, we can see the sunlight reflected from Mercury, Venus, Mars, Jupiter, and Saturn in our own Solar System. In principle, one can do the same thing for planets around other stars, but in practice it is much more difficult for a variety of reasons. At thermal-infrared wavelengths, the planet emits its own radiation. This, after all, is how planets cool themselves: they absorb some of the light given off by their parent star, and they re-emit this energy as thermal-infrared radiation.

From a planet-hunter's point of view, each wavelength region has both its advantages and its disadvantages. To understand why, it is convenient to start with figure 13.1. This diagram shows how the Earth and Sun would appear if viewed from a distance of 10 pc, or 32.6 light-years. The vertical scale shows the flux of radiation from each object, in logarithmic units. The horizontal scale shows the wavelength in microns. (Recall that 1 micron $=1\ \mu\text{m} = 10^{-6}$ m.)

The first thing one notices, not surprisingly, is that the Sun is much brighter than the Earth at all wavelengths. That's because it is both much hotter and much bigger. The Sun's radiation peaks in the visible part of the spectrum, which extends from 0.4 to 0.7 μm. But, as one can see from the figure, the Sun's radiation extends out to much longer wavelengths as well. Technically, the Sun's spectrum resembles that of a black-

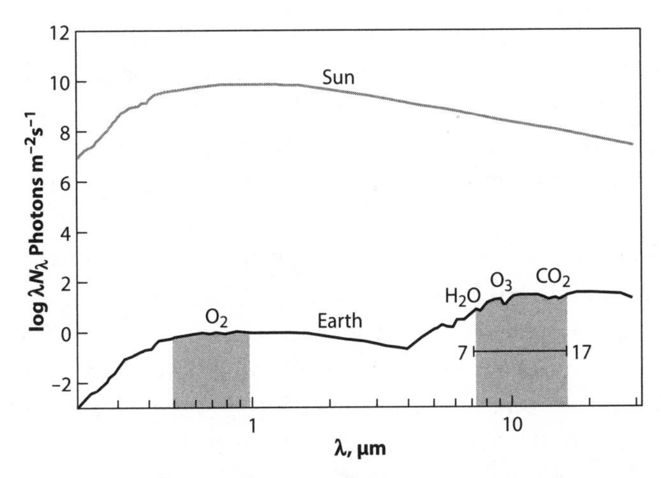

Figure 13.1 Diagram illustrating how the Earth and Sun would appear if viewed from a distance of 10 pc. The shaded areas are the wavelength regions that would be observed by a telescope operating in the visible/near-IR (at left) or in the thermal-IR (at right). (Courtesy of Chas Beichman, NASA/JPL.)

body with an effective temperature of ~5780 K. If one looks at this spectrum in more detail, it can be seen to have hundreds of thousands of absorption lines. (See, for example, figure 11.6.) That is what allows astronomers like Michel Mayor and Geoff Marcy to measure Doppler shifts and to search for planets using the radial velocity method. But at the low resolution of figure 13.1, the Sun's spectrum looks fairly smooth, and it behaves in a nice, predictable manner.

Earth's spectrum, by contrast, is much more complex. One can see clearly that it consists of two different parts. The first part, which extends from the left side of the figure out to about 4 μm, is parallel to the Sun's radiation flux. That's because, at these wavelengths, one is seeing the Earth in reflected sunlight. Earth reflects about 30 percent of the sunlight that hits it, and this light could potentially be observed from afar. The second portion of the Earth's radiation curve, from 4 μm out to very long wavelengths, is the Earth's own thermal-IR radiation. (For convenience, we shall sometimes substitute "IR" for infrared.) Earth's emitted radiation peaks at these wavelengths because the Earth is much cooler than the Sun. Its spectrum is crudely similar to that of a blackbody with a temperature of ~255 K. But there are "wiggles" in the Earth's spectrum that are visible even at this low spectral resolution. The wiggles are caused by the presence of various gases in Earth's atmosphere, and they would be of obvious interest if they were seen in the atmosphere of an extrasolar planet. In the next chapter, we'll discuss what we might learn from them. For now, though, let's bypass this issue and just concentrate on the basic problem of observing the Earth.

At first glance, the choice of which wavelength region to observe at is obvious: relative to the Sun, the Earth is much brighter in the infrared than it is in the visible; hence, one would want to build a telescope that was sensitive in the thermal-IR. The ratio between the Sun's brightness and the Earth's brightness is called the *contrast ratio*. In the thermal-IR, the contrast ratio is about 10^7, which is pretty awful, i.e., the planet is extremely dim compared to the Sun. But, in the visible, the contrast ratio is about 10^{10}, which is 1000 times worse! Hence, it should not be too surprising to learn that when NASA researchers first started studying the idea of constructing a planet-finding telescope, back in the early 1980s, they focused initially on the thermal-IR spectral region.

There is another aspect to this problem, however, that works the other way. To separate out the light from the planet from that of the star, one needs to be able to spatially *resolve* the two objects. A familiar example is that of resolving the binary star Mizar, which is the star at the bend in the handle of the Big Dipper. If you have very good eyesight, and if the night is clear, you may be able to identify Mizar as a binary star simply by looking at it with your naked eyes. If you have bad eyesight, like me, then you need a pair of binoculars to see this. In any case, one needs to be able to see reasonably well to discern that there are indeed two objects, that is, to resolve the binary system.

Physicists have been interested in this problem for a long time, and they long ago demonstrated that there is a theoretical *diffraction limit* to how well one can see with a telescope of a given size. Suppose that two objects, the star and its planet, define an angle, θ (theta), as seen by the observer (figure 13.2). Suppose also that the diameter of the telescope mirror (or lens) is D, and the wavelength of light at which one is observ-

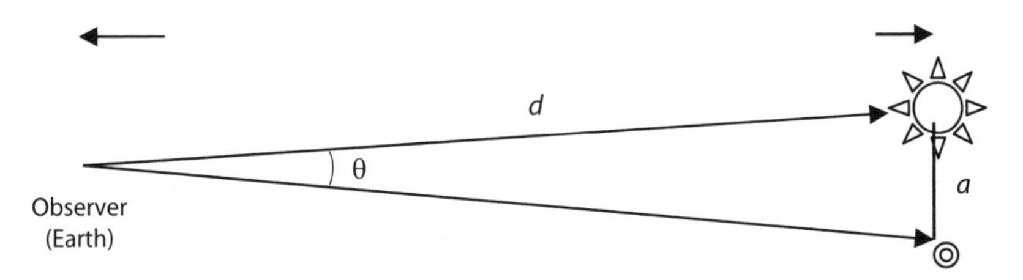

Figure 13.2 Diagram illustrating the geometry involved in observing a planet around a nearby star.

ing is λ (lambda). Then, mathematically one can show that the *smallest* angle that one can theoretically resolve is given by

$$\theta \cong \frac{\lambda}{D}. \tag{13.1}$$

Here, θ is in radians, and λ and *D*, which are both distances, just need to be in the same units. In practice, it is nearly impossible to make observations near this theoretical limit, and so one measures the resolution of an instrument in terms of multiples of the ratio, λ/D.

Already, before one looks at any numbers, equation 13.1 tells us two things that are physically important. First, the smallest angle that can be resolved by a telescope depends on the diameter of its primary mirror, or lens. A bigger mirror allows you to see objects that are closer together. That is why it is easier to resolve *Mizar* with a pair of binoculars than with your unaided eye—one is using the diameter of each lens, typically 35–50 mm, in place of that of the pupil of your eye (~7 mm). Second, the angle that one can resolve also depends on the wavelength of light at which one is observing: the shorter the wavelength, the smaller the angle that one can resolve. This is a critical issue for planet-finding telescopes because it means that if one wants to observe at long wavelengths, where the contrast ratio is more favorable, one needs a bigger telescope.

To see how bad the problem actually is, we need to put in some numbers. We have already seen how to do this when we discussed parallax in chapter 11. To search around a reasonable number of stars, we need to be able to look for planets like Earth out to a distance of about 10 pc, or 32.6

light-years. For simplicity, let's assume that the planet we are observing is located at 1 AU from a solar-type star. This distance is labeled "*a*" in figure 13.2. The star is located at a distance *d* from the Earth. In reality, unless the plane of the planet's orbit is precisely in the plane of the sky, the planet will usually appear closer to the star than shown. However, if we wait until the right moment to observe, then it can be seen at its full separation, as indicated in the drawing. (For this reason, it would be nice to know the planet's orbit *before* trying to observe it. The *SIM* mission could potentially provide this information, and so the *SIM* investigators are quick to point out that their mission could provide significant benefits to *TPF*.)

Recall now that, based on the definition of a parsec, a planet at 1 AU from a star that is 1 pc distant will define an angle of 1 arcsecond with respect to the star. The angle between the planet and the star is inversely proportional to the distance, *d*, so the same planet orbiting a star at 10 pc will form an angle of 0.1 arcsecond.* To use this information in equation 13.1, we need to convert from arcseconds to radians. The conversion factor is 4.848×10^{-6} radians per arcsecond ($= 2\pi/1{,}296{,}000$), so 0.1 arcseconds $\cong 5 \times 10^{-7}$ radians. Let's assume now that we are observing the system in the midvisible at a wavelength of 0.5 μm, or 5×10^{-7} m. If we plug these numbers back into equation 13.1 and solve for the telescope diameter, we obtain $D = \lambda/\theta \cong 1$ m. Hence, a perfect, or "diffraction-limited," telescope with a diameter of 1 m would, in principle, be capable of resolving this planet from its parent star.

We have already noted, though, that for various technical reasons, real telescopes can't make it down to the theoretical resolution limit. A conservative goal—one adopted in the initial design study[1] for NASA's *Terrestrial Planet Finder-Coronagraph* (*TPF-C*)—is to observe at $\theta = 4\lambda/D$, rather than at λ/D itself. If we make this assumption, then we find that the mirror must grow to about 4 m in diameter. If we further require that we be able to observe out to 1 μm, so that we can take a visible/near-IR spectrum, the mirror size doubles to 8 m. Wow, that is big! By com-

*In figure 13.2, the angle θ is related to the distances *a* and *d* by the formula: $\sin\theta = a/d$. Because the angle is very small, however, $\sin\theta \cong \theta$, so we can write $\theta \cong a/d$.

parison, the *Hubble Space Telescope* is only 2.4 m in diameter. Putting a telescope of this size up in space would be an enormous undertaking—although there may well be ways to do it, as discussed below.

But now, suppose we want to observe instead in the thermal-IR, near 10 μm. This wavelength is 10 times longer, and so, according to equation 13.1, the telescope would have to be 10 times bigger. That implies that the telescope mirror would have to be ~80 m in diameter. Ugh! That is really bad news! Putting an 80-m telescope up into space is well beyond our present capabilities. We might be able to do that someday, as technology advances, but the chance that this will happen in our lifetimes is very small. Does this mean, then, that one should abandon the thermal-IR?

Infrared Interferometers: TPF-I and Darwin

The answer to this question is no. One can get around the spatial resolution problem by using a technique that radio astronomers have exploited for a long time. Radio waves have very long wavelengths, meters or longer, and so it has always been a challenge to obtain high spatial resolution using radio telescopes. But radio astronomers solved this problem long ago using a technique called *interferometry*. An interferometer uses 2 or more telescopes separated by a distance that is typically much larger than either telescope's diameter. The advantage of doing this is that the effective diameter that one uses in equation 13.1 becomes the distance between telescopes, rather than the diameter of the individual telescopes. This obviously has huge advantages, because one can build two modest-size telescopes and space them far apart, and thereby obtain the same spatial resolution that could be obtained with one huge telescope. As an extension of this technique, one can build 2-dimensional arrays of telescopes arranged in various patterns, which then provide good spatial resolution in multiple directions. The *Very Large Array* (*VLA*) of radio telescopes in Soccoro, New Mexico, is an outstanding example of this technique. Science fiction buffs may recognize the *VLA* as the tool ostensibly used by Jodie Foster in the movie *Contact* to search for radio signals from intelligent beings elsewhere in the galaxy.

Both NASA and ESA (the European Space Agency) have studied space-based missions that would use interferometry to search for extrasolar planets at thermal-IR wavelengths. NASA's mission is called *Terrestrial Planet Finder-Interferometer* (*TPF-I*), while ESA's mission is called *Darwin*. NASA's original idea for how to build *TPF-I* was to put four 2-m-diameter telescopes on an 80-m boom, as shown in figure 13.3. The telescopes would be arranged in such a way as to produce a *nulling interferometer*. By this, one means that it would null out the light from the star and preserve the light from the planet(s) orbiting around it. For this technique to work, the planet and the star must be lined up with the boom on which the telescopes are mounted. Unfortunately, one wouldn't know ahead of time which direction that was. Hence, the boom would need to be rotated in the plane perpendicular to the line of sight to the star as the observations were being made. As the boom rotated, the interferometer would become sensitive to planets in different positions around the target star.

This boom design for *TPF-I*, sometimes called "TPF-on-a-stick," has its merits, but it also has its disadvantages. Recall from chapter 11 that NASA hopes to fly a much smaller space-based interferometer called *SIM* sometime during the next few years to look for planets using the astrometric method. The current design for *SIM* calls for two 50-cm-diameter visible telescopes on a 6-m boom. Even at this much shorter boom length, the *SIM* design team has had to cope with serious design problems. Booms vibrate, especially since they have to be connected to rapidly spinning "reaction wheels" that store angular momentum and thereby allow the boom itself to rotate. These problems would likely become even worse for an 80-m boom such as that proposed originally for *TPF-I*.

ESA had a different idea for their *Darwin* mission, and this design has also been tentatively adopted now for *TPF-I*. ESA's plan was to fly 3 or 4 telescopes on separate spacecraft, and then to combine the beams from each telescope at yet another spacecraft (figure 13.4). Such a *free-flying interferometer* would have several advantages over an interferometer-on-a-boom. Most importantly, it wouldn't vibrate, because the spacecraft would not be physically connected. One could also adjust the dis-

Figure 13.3 An early NASA concept for *TPF-I* (*Terrestrial Planet Finder–Interferometer*). This space-based instrument could be used to search for extrasolar planets at thermal-IR wavelengths. (Courtesy of NASA/JPL.)

tances between the different spacecraft to be optimal for a given star system. If the planet was close to its parent star, or if the star itself was far away, then the distance could be made larger so as to provide increased spatial resolution. For nearby targets, the distances could be made shorter. One problem that hasn't been mentioned is that, if the resolution is too good, the disk of the star itself becomes visible, and it then becomes more difficult to block out the starlight. Hence, one would want to keep the telescope spacing close enough so that the star appears as a point source.

A free-flying *Darwin* or *TPF-I* mission would have great flexibility and could eventually yield exciting results. Several good spectral features can be found at these wavelengths, as discussed in the next chapter. The main drawbacks to such a mission are that it would require accurate formation-flying in space, and the telescopes would have to be cooled to low temperatures (~50 K) to operate effectively in the thermal-infrared. (Otherwise, the emission from the telescope itself would swamp the signal from the planetary system being observed.) Distances between spacecraft would need to be maintained to approximately 1-cm accuracy even as the array itself was shifted to new formations to look for planets at different positions around the star. The distances between spacecraft would have to be measured much more accurately than this—to a small fraction of the wavelength being used for observations—but this is relatively easy to do by using *laser metrometry*, as indicated by the laser beams in figure 13.4. While all of these things are possible using current technology, everyone agrees that this will end up being an expensive mission, probably costing several billions of dollars. As pointed out in chapter 11, NASA has come up with that type of money for space missions numerous times previously. However, it is not easy to do when government budgets get tight, as they are at the time this book is being written.*

Searching for Planets at Visible Wavelengths: TPF-C

Let's go back now to the other possibility, which is to search for extrasolar planets at visible wavelengths. The problem here, if you recall, is that the contrast ratio between the planet and the star is very large—a factor of 10^{10} for the Earth going around the Sun. So one would need to block out the starlight very effectively. This can be accomplished, in principle, by using an instrument called a *coronagraph*. Coronagraphs get their name from instruments that were used originally to study the Sun. They were first introduced in 1930 by the French astronomer Bernard Lyot. By blocking out the visible disk of the Sun, it is possible to study the hot *corona* that surrounds it.

One must use tricks (e.g., taking advantage of polarization) to observe the Sun's corona from the ground because Earth's atmosphere scatters so much sunlight that the daytime sky is always bright. But if one can block out the Sun's disk from up in space, the effect of a coronagraph is much

*Government spending is a funny thing, though. When the U.S. government really wants to do something, the amount of money available is nothing short of amazing. For example, we are currently spending about $8 billion per month to maintain our troop presence in Iraq. If we could pull out of Iraq for just 2 weeks, the money saved should be enough to do *TPF-I*. Or, to compare to another recent government investment, the cost of *TPF-I* would be about 0.5 percent of the $700 billion Wall Street bailout orchestrated in October 2008.

Figure 13.4 Artist's concept of ESA's proposed *Darwin* mission. This would be a free-flying interferometer operating in the thermal-IR. NASA's proposed *TPF-I* mission has a similar design. (Image courtesy of NASA.)

more pronounced. This happens naturally during total solar eclipses. By an accident of fate, the Moon and the Sun appear to be nearly the same size, when viewed from Earth.* Hence, when the Moon passes in front of the Sun during its orbit around Earth, it makes a nearly ideal coronagraph. In practice, some solar eclipses are "better" than others because the eccentric nature of the Moon's orbit means that it is sometimes closer to the Earth (and hence appears bigger) during an eclipse and sometimes farther away (and hence smaller). When the Moon's size is smaller than that of the Sun, a small ring of the Sun's photosphere remains visible during totality, and it is termed an *annular eclipse.*

*Guillermo Gonzalez and Jay Richards, in their book *The Privileged Planet*, have argued that the fact that the Sun and the Moon appear to be the same size is unlikely to be a coincidence. They see this as an example (one of many) of divine intervention. It should be borne in mind, though, that the Moon is constantly receding from the Earth as a consequence of tidal dissipation; hence, it looked bigger than the Sun in the distant past, and it will appear smaller in the distant future. It may still seem divinely planned to some that the two objects are the same size right now, but to others (including me), this appears to be completely coincidental.

Even if one is observing from space so that scattering from the atmosphere is not a problem, it is still difficult to see a planet close to its much brighter parent star. The reason is that light from the star is *diffracted* from the edges of the telescope mirror and from any irregularities on the mirror surface itself. Recall that light is a type of wave and that waves can be bent as they go around objects in their path. The problem of diffraction is illustrated by the pattern that one observes when light travels through a pinhole in an opaque sheet. If one places a light source on one side of the sheet, and a camera on the other, the picture that is recorded is *not* simply the image of a pinhole. Instead, the pinhole has rings around it, as illustrated in figure 13.5. These rings are called *Airy rings* after the British Royal Astronomer Sir George Airy who first described them mathematically back in the mid-19th century.

When one observes a star with a telescope, the geometry is somewhat different, but the underlying physics is the same. Instead of being diffracted from the edge of a pinhole, light from the star is diffracted by the edge of the telescope mirror. And even if one blocks out the central part of the star's image, a significant fraction of the starlight can remain. Recall that the planet that one is trying to see is 10^{10} times dimmer than the star, so it can obviously be completely swamped if too much starlight leaks through. Overcoming this problem is what the science of coronagraphy is all about.

NASA has been studying coronagraphic planet-finding missions for several years now. The name given to the full-scale version of this mission is *Terrestrial Planet Finder–Coronagraph* (*TPF-C*). Small coronagraphs have been studied as well, but it takes a big one to have a good chance of locating terrestrial planets. The good news, as pointed out earlier, is that a telescope that operates in the visible can be much smaller than one that operates in the thermal infrared: it needs to be only about 8 meters in diameter if it operates at $4\lambda/D$. This is still too large to launch on a single spacecraft if the telescope is round. One can get around that problem, though, by using an elliptical mirror (figure 13.6). The left panel in the figure shows a schematic diagram of an 8×3.5 m elliptical mirror. Such a mirror could be stacked lengthwise in the cargo bay of an existing rocket, and so it could be launched from the ground as a single

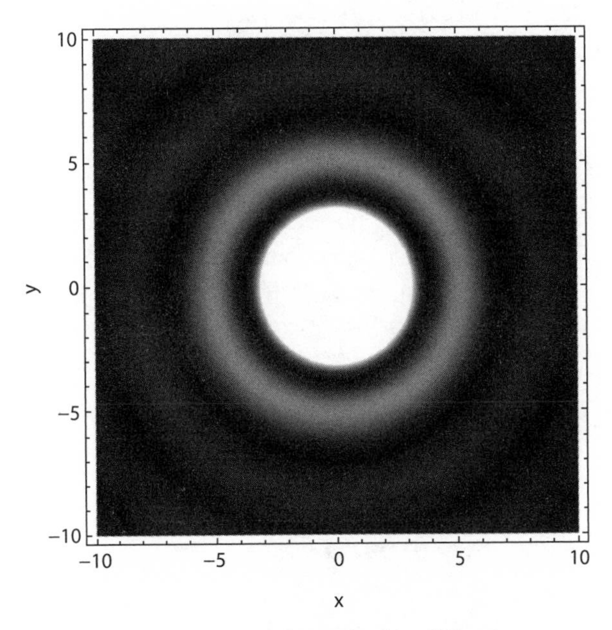

Figure 13.5 Diagram showing the Airy diffraction pattern produced by a circular aperture.[2]

piece. It is not a normal mirror, though, as one can tell by looking at the diagram. Rather, it looks like a mirror that is wearing a Halloween mask. The mask is a critical part of how this particular coronagraph would work. (This design was called the Flight Baseline 1 design from the *TPF-C* study group mentioned in the preface.) By blocking out different parts of the mirror, one can effectively null out the light from the star, allowing planets to be detected. Note how much of the mirror surface is blocked near the edges of the long axis of the ellipse. This effectively reduces the effect of diffraction from those edges, allowing greater starlight suppression in that direction. The panel on the right shows this nulling effect, which is extremely pronounced in the horizontal direction, along the long axis of the mirror. Starlight suppression along the short axis of the mirror is relatively poor. Hence, this telescope, like the boom design for *TPF-I*, would need to be rotated in the plane perpendicular to the line of sight, so that it could look for planets in different locations around the star.

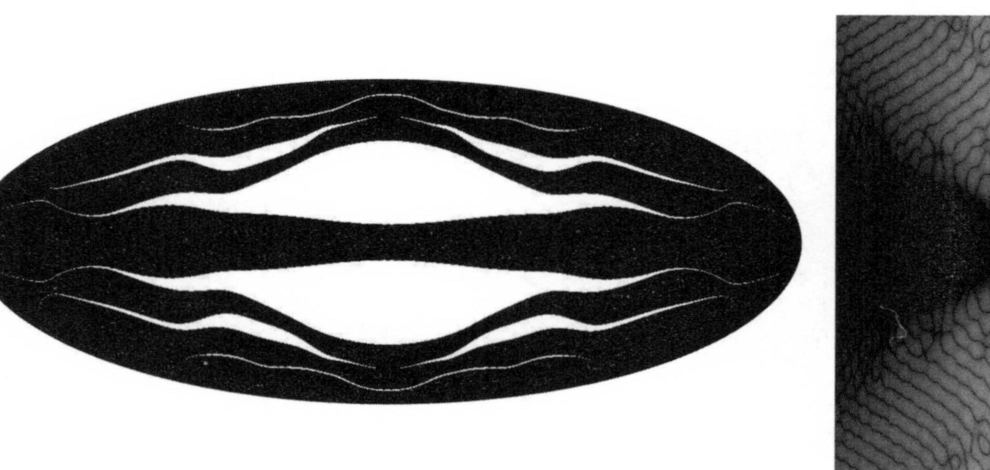

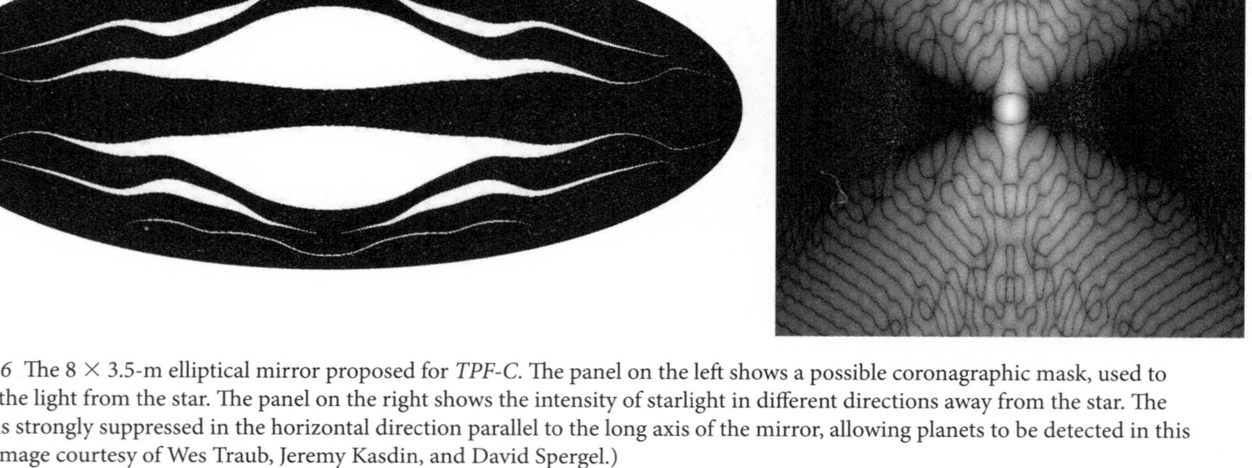

Figure 13.6 The 8×3.5-m elliptical mirror proposed for *TPF-C*. The panel on the left shows a possible coronagraphic mask, used to suppress the light from the star. The panel on the right shows the intensity of starlight in different directions away from the star. The starlight is strongly suppressed in the horizontal direction parallel to the long axis of the mirror, allowing planets to be detected in this region. (Image courtesy of Wes Traub, Jeremy Kasdin, and David Spergel.)

The Visible Occulter: TPF-O

Within the last two years, a new concept has been proposed to look for extrasolar planets at visible wavelengths.[3] This idea comes from an astronomer named Webster Cash at the University of Colorado. As with *TPF-C*, the big problem is to somehow block out enough starlight to overcome the huge contrast ratio between the star and the planet. Instead of using an internal coronagraph, though, Cash has suggested that one fly an occulting disk between the telescope and the system to be observed (figure 13.7). NASA has since picked up on this design, and they have renamed it *Terrestrial Planet Finder–Occulter* (*TPF-O*). The disk, or starshade (which looks more like a flower in Cash's design), would have a diameter of ~50 m and would be flown at a distance of ~70,000 km from the telescope. This makes the starshade approximately the same apparent size, as viewed from the telescope, as the inner part of the planetary system being observed, out to a little less than 1 AU. These specifications are for observing a star like our own Sun at a distance of 10 pc; one would need to adjust the distance accordingly if the star was at a different distance or if it had a different luminosity than the Sun, so that its habitable zone was closer in or farther out. One can think of this technique as being analogous to a solar eclipse, when the Moon passes in front of the Sun. In this case, though, the "Moon" is artificial, and it can be moved from place to place so as to allow the telescope to observe different stars.

The disadvantage of *TPF-O* is that it would require formation flying and multiple spacecraft, like *TPF-I*. It would also be more difficult to move from one target star to the next, as this would involve moving the starshade and/or the telescope considerable distances. This latter problem might be minimized by flying two separate starshades and maneuvering one into position while the other was being used to make observations. But this would make the mission even more complicated and expensive.

The good news about *TPF-O* is that the starlight suppression appears to work exceedingly well, allowing this type of telescope to achieve very high planet-star contrast ratios. In a conventional coronagraph, in which the starlight is blocked within the telescope itself—for example, by a mask

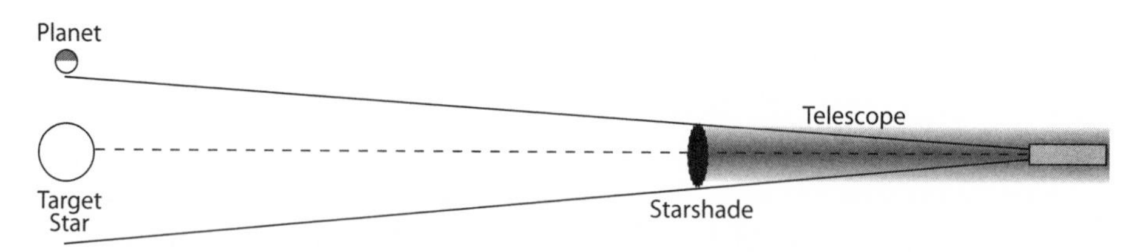

Figure 13.7 The visible occulter proposed by Webster Cash (University of Colorado) and adopted by NASA as its proposed *TPF-O* mission. The telescope and the occulting disk would be separated by a distance of ~70,000 km.[4]

like the one shown in figure 13.6—the diffracted starlight is still somewhere within the system. Hence, if one is not extremely careful, it can get back into the light path and can potentially overwhelm the light from the planet. With the occulting disk, however, most of the starlight diffracted from the edge of the disk is completely gone—lost somewhere in space. Hence, the light that does get through to the telescope is almost all from the surrounding planets. The planets are still very dim, and so one still needs a big telescope (~4 m diameter) to see them effectively and to collect enough light to form a spectrum. But the optical properties of the telescope mirror, and the overall quality of the system, do not need to be any better than the existing *Hubble Space Telescope*. So we could probably go out and build this instrument tomorrow, given sufficient money. By contrast, both *TPF-C* and *TPF-I* still require some technical development in order to achieve a good contrast ratio. Hence, *TPF-O* may be the way to go if we want to observe Earth-like extrasolar planets within the next two decades.

Nearby Target Stars

Thus far, we have discussed how to look for extrasolar Earth-like planets, but we have not discussed where to look. Astronomers, however, have already put a great deal of effort into figuring out which stars would be best to look at. No one has done more work on this topic than Robert

TABLE 13.1
Top 10 Target Stars for *TPF-C*

Rank	*Name*	*Constellation*	*Distance (ly)*	*Spectral type*
1	Alpha Centauri A	Cen	4.3	G2V
2	Alpha Centauri B	Cen	4.3	K1V
3	Tau Ceti	Cet	12	G8V
4	Achird	Cas	19	G3V
5	Beta Hydri	Hyi	24	G2V
6	Delta Pavonis	Pav	20	G8V
7	Tabit	Ori	26	F6V
8	Gamma Leporis	Lep	29	F7V
9	Epsilon Eridani	Eri	10	K2V
10	Keid	Eri	16	K1V

Data from http://en.wikipedia.org/wiki/Terrestrial_Planet_Finder.
Based on work by R. Brown, Space Telescope Science Institute, Baltimore, MD.

(Bob) Brown at the Space Telescope Institute in Baltimore, Maryland, who was one of the members of our *TPF-C* study committee. A list of the top 100 target stars for *TPF-C* can be found on his website.[5] These have been reproduced in simplified form on Wikipedia.[6] The first 10 of them are listed in table 13.1.

The number one star on Brown's list is, not surprisingly, Alpha Centauri (α Cen). At 4.3 light-years distance, α Cen is (almost) the closest star to the Sun. (Proxima Centauri, as we have seen, is slightly closer.) It is also a near twin of the Sun, as it has the same spectral classification, G2V. As Bob would probably tell you, though, α Cen may *not* be the most probable star around which to find an Earth-like planet. It is easy to observe, because it is nearby and relatively bright; however, it is part of a triple star system. Planets may or may not form in such a system, and even if they do, they may not have stable, circular orbits, and they may not have sufficient amounts of volatiles. None of this is impossible;

however, this system is so different from our own Solar System that we cannot confidently extrapolate based on anything that we observe here.

Many of the other stars in Brown's list are good candidates, however. Tau Ceti and Epsilon Eridani are two nearby, single, relatively Sun-like stars whose (postulated) planetary systems have appeared often in science fiction books as targets for future human colonization. And there are roughly 100 other solar-type stars within 60 light-years (or ~20 pc) that are at least partly observable by either *TPF-C* or *TPF-I/Darwin*. Indeed, several thousand stars can actually be found within this distance. Most of these, however, are red dwarf stars, which are much less luminous than the Sun. These stars are less likely to harbor habitable planets, for reasons discussed in chapter 10. Furthermore, most of them are very dim, even though they are close, making them difficult to observe. That is why both *TPF* missions are primarily focused on stars not too different from the Sun.

There is another important reason for focusing on Sun-like stars. *TPF-C*, in particular, has trouble observing stars that are either much brighter or much dimmer than our Sun. Bright blue A stars, for example, have habitable zones that are far away from the star, and hence that are easy to resolve spatially with a telescope of modest size. However, the contrast ratio between the planet and the star is much worse than for the Earth around the Sun.* For dim red M stars, the problem is just the opposite: the contrast ratio is favorable, because the planet is again the same brightness as Earth, whereas the star is much dimmer than our Sun. However, the habitable zone is very close to the star, and so a very large telescope would be needed to resolve the planet from the star.

TPF-I suffers from the same problem for bright blue stars: the planet-star contrast ratio is poor. For dim red stars, however, the free-flying version of *TPF-I* is able to cope much more effectively, because the spacecraft can be flown at a greater separation distance. Thus, *TPF-I* (or *Darwin*) may be the best tool for investigating planets around M stars. We

*One can understand this problem by noting that although the A star is much brighter than the Sun, the planet must receive approximately the same amount of starlight as does Earth; otherwise, it could not be habitable.

shall see in the next chapter that the ideal solution is to fly both *TPF-C* and *TPF-I*. The reasons for doing that, though, have less to do with finding planets, and more to do with understanding what we find. For that is where this whole planet-finding business really begins to become exciting.

· Chapter Fourteen ·

The Spectroscopic Search for Life

In the last chapter, we discussed several possible space-based missions that have been proposed to look for Earth-like planets around other stars. Some of these (*TPF-C* and *TPF-O*) would operate in the visible/near-IR, while others (*TPF-I* and *Darwin*) would operate at thermal-IR wavelengths. The primary goal of any such mission would be to look for small, rocky planets in the habitable zones of their parent stars. But we wouldn't just want to *find* the planets—indeed, that may already have been done beforehand if the *SIM* mission is flown first. An equally important goal of the *TPF* missions is to *characterize* the atmospheres and surfaces of such planets by studying them spectroscopically.[1] Some of the gases in Earth's atmosphere, including oxygen and ozone (O_2 and O_3), absorb electromagnetic radiation at either visible or infrared wavelengths. Most of Earth's O_2 comes from photosynthesis (chapter 4), and all of Earth's O_3 comes from O_2 (chapter 9); hence, the presence of either one of these gases would be at least suggestive of life. But would this evidence be definitive? Or, to put it another way, are there possible *false positives* for life that might lead us to think that a planet was inhabited when it really was not? If and when a *TPF*-type mission is flown, these issues are certain to be raised. So let's first discuss what types of absorption features might be seen in each wavelength region and then tackle the more difficult question of what does it really mean if we see them.

Spectral Resolution

Before looking at graphs of planetary spectra, we should first address the question of *spectral resolution*. This term refers to how finely we divide up the electromagnetic spectrum. At very high resolution, one subdivides the spectrum into literally millions of different wavelengths. Astronomers define spectral resolution, R, mathematically by the equation $R = \lambda/\Delta\lambda$. Here, λ is the wavelength at which one is observing, and $\Delta\lambda$ is the smallest wavelength difference that one can resolve. A typical high-resolution spectrometer for a ground-based telescope might have an R value of 200,000. Hence, if it is used to make observations at wavelengths near 1000 nm (1 μm), the width of each subinterval in the spectrum would be 1000 nm/200,000 = 0.005 nm. Some examples of high-resolution spectrometers, which we discussed previously in chapter 11, are the instruments used by Michel Mayor and Geoff Marcy to measure precise radial velocities of nearby stars. There, high resolution is required because they are measuring tiny Doppler shifts caused by the planet's gravitational tug on the star.

For *TPF* or *Darwin*, however, it will be virtually impossible to make high spectral resolution measurements. The reason is straightforward—extrasolar planets are very faint, and so there simply are not enough available photons. One can think of the problem this way: if one wants to make spectral measurements at an R value of 200,000, one needs at least 200,000 photons (one in each wavelength bin) just to make a small portion of a spectrum near a particular wavelength. Indeed, one needs many more photons than that to compute accurate statistics over a range of wavelengths. It is possible to obtain lots of photons for the ground-based radial velocity measurements. There, investigators use large telescopes, as much as 10 meters in diameter for the Keck telescopes on Mauna Kea, to look at relatively bright, nearby stars. Even so, they are limited to observing bright stars, because the dimmer, more distant stars do not yield enough photons.

In looking for extrasolar planets, the photon-counting problem is much more severe. Recall that in the visible the planet is likely to be 10^{10}

times dimmer than its parent star. Hence, even if the star itself is fairly bright, any planet orbiting around it will be extremely dim. A friend on the *TPF-C* committee described the problem this way: If one were to keep track of the incoming photons by hand, instead of with a CCD, one could basically stand there and count them off on one's fingers as they arrived. Of course, some targets are better than others, and so more photons will be available in some cases. The bottom line, though, is that for an initial planet-finding mission, the spectral resolution is likely to be low: $R = 100$, or less. Achieving higher spectral resolution is not impossible, but it would require telescopes that are much larger and more expensive than the ones described in the previous chapter. We may build such telescopes someday—indeed, we'll speculate about this possibility in the next chapter—but for an initial *TPF* mission, we'll likely do only what is necessary to get a first-order answer to our questions.

The Visible / Near-IR Region: TPF-C or -O

Let's begin by considering the visible/near-IR spectral region. This is the wavelength region that could potentially be observed by *TPF-C* or *TPF-O* (the coronagraph or the occulter). Let's begin also by looking at Earth. We have a pretty good idea of what the Earth looks like from space because we've had satellites circling around it for the last 40 years or so. Surprisingly, though, our best idea of what a complete "whole-Earth" spectrum would look like does not come from satellites. That's because most satellites orbit relatively close to the Earth, so when an instrument on board is used to look down at the Earth, it sees only a small portion.

Would-be planet finders agonized about this problem for a number of years. Some suggested putting a space telescope at the Earth-Sun L1 Lagrange point and using it to take observations of the Earth from a large distance. Recall that L1 is the Lagrange point that lies in between Earth and the Sun (figure 7.2). Such a telescope has actually been proposed for other reasons. A possible NASA mission called *DSCOVR* (formerly *Triana*) would place an Earth-observing telescope at L1 to monitor global change—again, trying to capture the "whole-Earth," or *disk-averaged*, view.

Other astronomers, though, came up with a clever (and much cheaper!) idea for obtaining a whole-Earth spectrum. They could accomplish the same goal by using *Earthshine* data from the Moon.[2,3] You may have noticed that if you look at the Moon when it is in crescent phase, you can see not only the brightly lit crescent, but also a dim glow from the unlit portion. Why is the dark side of the Moon glowing? The bright side glows, obviously enough, because it reflects light from the Sun. The dark side glows because it reflects light from the Earth (which, of course, came originally from the Sun). That means that the light from the lunar dark side, termed "Earthshine," can be used to create a spectrum of the Earth. The trick is to subtract the spectrum of the dayside, thereby removing the features caused by the Moon's surface and leaving only the signature of the Earth. Such a spectrum represents an average over a large fraction of the Earth's surface, because the reflected light comes from all parts of the Earth that are sunlit, as seen from the Moon.

Although it appears dim to the naked eye, Earthshine is actually fairly bright, and so it contains lots of photons. Hence, one can easily subdivide it to create a medium-resolution ($R = 600$) spectrum (figure 14.1). In the figure, the wiggly curve at the top represents the Earthshine data. The smooth curve running through the data, which matches the "clear sky" curve from lower down, is a model fit to the data. From this, one can see that three different gases can be detected in Earth's atmosphere: O_2, O_3, and H_2O. They are all seen as absorption bands: the brightness within these bands is less than it is elsewhere in the spectrum because the gases are absorbing some of the incident sunlight. O_2 itself has three different absorption bands that can be seen at this spectral resolution. The brightest of these is the so-called O_2 "A" band at 7600 Å, or 760 nm. (Recall that the visible portion of the spectrum extends from 400 to 700 nm. This band is just beyond that region in the near-infrared.) The A band is easy to observe and was singled out almost 30 years ago as a possible indicator of life on extrasolar planets.[4] The O_2 "B" band at 6900 Å is also easy to pick out, as are the three H_2O absorption bands at 7200, 8200, and 9400 Å.

O_3 has one broad band in the visible that extends all the way from 5000 to 7000 Å. O_3 absorbs even more strongly at shorter UV wavelengths,* between 2000 and 3000 Å, but one cannot see that wavelength region in this particular spectrum. The O_3 band in the visible is considered less easy to observe on an extrasolar planet than are the O_2 bands because it is easily masked by clouds[5] and because at wavelengths below 6000 Å its effects can be confused with those of Rayleigh scattering (labeled "ray" in the model curves near the bottom of the figure). As discussed in chapter 7, Rayleigh scattering—scattering by air molecules—increases toward shorter wavelengths, making our sky appear blue. The Earthshine spectrum in figure 14.1 is also very "blue," as shown by the increase in intensity at shorter wavelengths.

One other possible bioindicator may be present in the Earthshine spectrum shown in figure 14.1. If one looks carefully at the wavelength region between 7000 and 7500 Å, one can see an increase in the amount of reflected light toward longer wavelengths. (In looking for this, try to mentally remove the effect of the H_2O band at 7200 Å.) This increase may reflect the so-called *red edge* of chlorophyll. The leaves of land plants reflect sunlight much more efficiently in the near-IR ($\lambda > 7000$ Å) than they do in the visible; hence, there is a significant increase, or "edge," in the reflection spectrum of plants at this particular wavelength. Marine plants and algae do this as well, although the effect is not as pronounced. This red edge is easy to pick out if one looks directly at a leaf (figure 14.2), or if one looks down from space at a patch of densely vegetated land, such as the Amazon rainforest.[6] It is less visible in figure 14.1 because most of the Earthshine light used to make this spectrum was reflected originally from the Pacific ocean, where the chlorophyll signal is not as strong.

The red edge arises from a combination of effects. Chlorophyll—the green pigment in plant cells—absorbs visible sunlight and funnels this energy to other parts of the cell where it is used for photosynthesis. Near-IR photons with wavelengths > 7300 Å are not useful for photosynthesis,[7] and so plants have evolved in such a way that they are less strongly absorbed. Indeed, tiny vesicles (bubbles) within leaf cell walls may have developed in this way to help keep the leaf cool and thereby minimize water loss.[7]

*The short-wavelength ozone band is the one that protects Earth's surface from biologically harmful UV radiation.

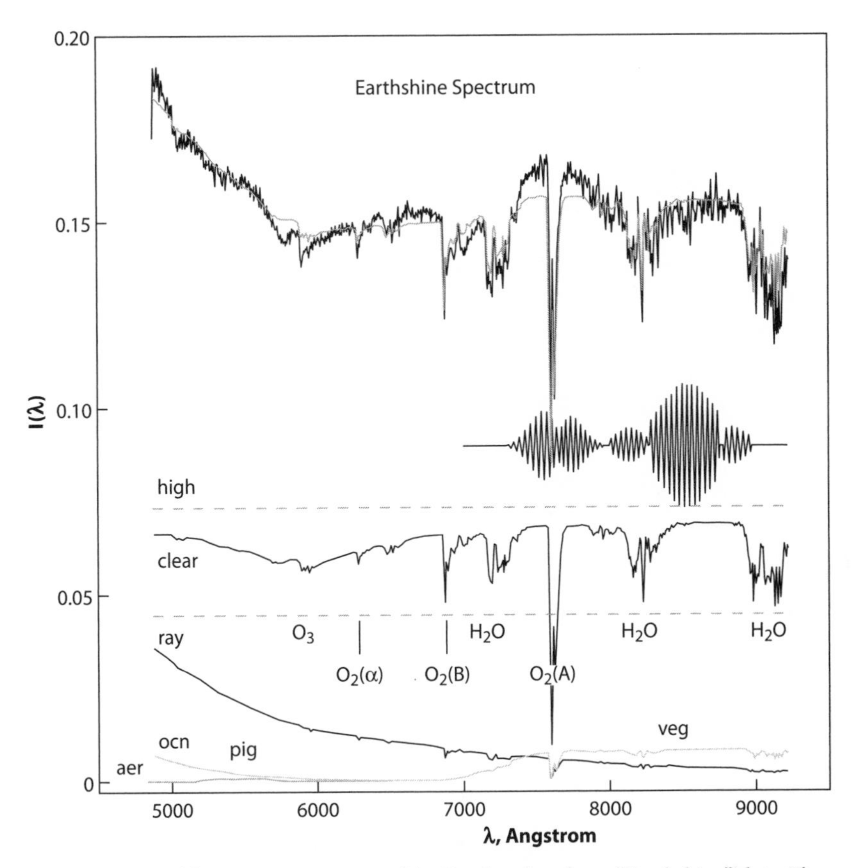

Figure 14.1 Visible/near-IR spectrum of the Earth, taken from "Earthshine" data. The wiggly curve at the top shows the actual data. The smooth curve running through it shows the model fit (from below).[3] (Reproduced by permission of the AAS.)

The red edge is an interesting phenomenon and has been extensively discussed in the literature,[6,7,9–12] but it may not be a reliable bioindicator. One reason is that it is only marginally apparent in disk-averaged spectra such as that shown in figure 14.1. It is also not clear that the red edge would exist for an alien type of vegetation or, if it did, that it would occur at the same wavelength as it does on our planet. Would plants on other planets use chlorophyll to harvest sunlight, as on Earth, or might they evolve a different type of biochemical antennae system? Would they require visible-light photons to perform oxygenic photosynthesis, or might

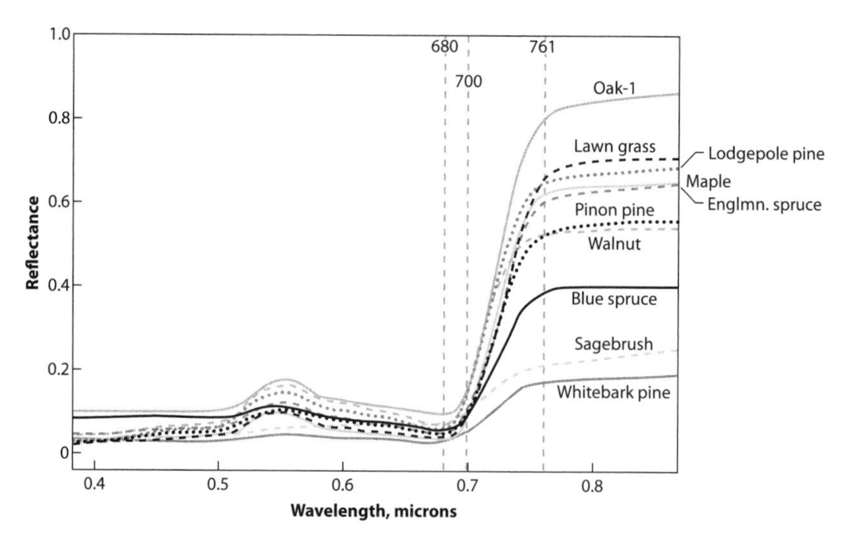

Figure 14.2 Laboratory reflectance spectra of different land plants. (From Kiang et al. 2007,[7] based on data from Woolf et al. 2002.[3] Reproduced by permission of Mary Ann Liebert Publishing.)

they develop systems that could utilize lower-energy, near-IR photons?[10,12] These questions may not be answerable without physically examining the plant life of interest. And, as that cannot be done without actually visiting the alien planet, the red-edge phenomenon may not prove to be a practical tool for searching for extraterrestrial life. But you can be sure that if we were to obtain a visible/near-IR spectrum of an extrasolar terrestrial planet, it would be scrutinized carefully to see if a red edge was present.

What would we be able to say if we were to see a spectrum like that shown in figure 14.1 on a planet orbiting another star? It is actually premature to ask this question, because the spectral resolution of this figure is considerably higher than we are likely to achieve with *TPF-C* or *TPF-O*. A more realistic, $R = 70$, visible/near-IR spectrum of Earth is shown in figure 14.3.* The figure also shows $R = 70$ spectra of Venus, Neptune, and Saturn's moon, Titan. All of these curves are so-called *synthetic spectra*, that is, they were generated from a computer model, rather than being based on real data. That should not invalidate our analysis, however, as

*The scale is now in μm, so one has to divide the wavelengths shown in figure 14.1 by 10^4 to convert values.

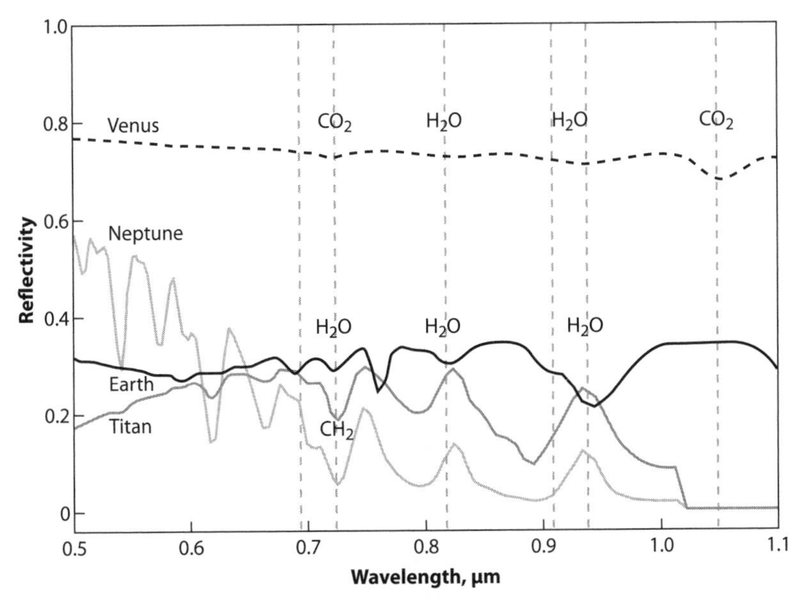

Figure 14.3 Venus, Earth, Neptune, and Titan at visible/near-IR wavelengths and at $R=70$ spectral resolution, based on a theoretical model calculation.[13] (Courtesy of Vikki Meadows, University of Washington. Reproduced by permission of Cambridge University Press.)

we saw from figure 14.1 that such models are capable of accurately simulating real planetary spectra.

If one examines the "Earth" curve in figure 14.3, one can see that most of the features seen in the Earthshine data can still be observed. This includes the A and B bands of O_2 and the three H_2O bands mentioned above. The broad O_3 band can also be seen, although it is unclear whether this would remain true on a planet that had clouds.[13] (The synthetic spectra shown in figure 14.3 are for cloud-free atmospheres.) By contrast, the spectrum of Venus in the visible is almost flat until one gets out beyond 1 μm (10,000 Å), where CO_2 absorption bands can be seen. CO_2 is relatively easy to observe on Venus because Venus' atmosphere contains about 90 bars of it (chapter 6). Earth's atmosphere also contains CO_2, but its concentration is so low, ~380 ppm, that it is impossible to see at this spectral resolution. (However, one can see CO_2 easily in the thermal-IR, as discussed below.) CO_2 is also difficult to observe for Mars (not shown) at these wavelengths because, even though Mars' atmosphere is rich in CO_2, the pressure is too low. Titan and Neptune also show strong absorption features at these wavelengths. The bands at wavelengths longer than 0.6 μm are all caused by CH_4, which is present in both atmospheres. The shorter-wavelength absorption features on Neptune are mostly from NH_3. Neither of these gases is visible in Earth's spectrum because both are too scarce. (CH_4 is present at 1.7 ppm, and NH_3 is present at ppb levels and is not uniformly mixed.)

Although this analysis has been brief, we can draw several conclusions from it:

1. If even a low-resolution ($R = 70$) spectrum can be obtained, it should be possible to distinguish a planet like modern Earth from other planets like those in our Solar System.
2. H_2O is reasonably easy to observe in the near-IR. These bands, though, are coming from atmospheric water vapor, and so this does not guarantee that liquid water is present at the planet's surface (which, of course, is our assumed prerequisite for life). We'll return to this issue below, as there may be other ways of looking for surface liquid water.
3. O_2 should also be observable in an extrasolar planet atmosphere at near-IR wavelengths if it is present at high concentrations, as it is on the modern Earth. As already mentioned, most of Earth's O_2 comes from photosynthesis, which is carried out by plants, algae, and cyanobacteria. Predicted atmospheric O_2 concentrations prior to the origin of photosynthesis are small—too low to produce a detectable spectroscopic signal.[14] Hence, the observation of O_2 in an extrasolar planet atmosphere would provide strong evidence for the existence of life. That said, there may be some "false positives" that we would need to watch out for, so we shall return to this topic as well.

The Thermal-IR Region: TPF-I or Darwin

What about spectroscopic observations in the thermal-IR? This is the wavelength region that would be observed by either *TPF-I* or *Darwin* (the two of which, one hopes, might eventually be combined into a single mission). Is the information available at these wavelengths any different from what we might learn in the visible?

The answer to this question is an emphatic yes! Figure 14.4 shows thermal-IR spectra of Venus, Earth, and Mars. Looking first at Venus and Mars, one can see that at this relatively low spectral resolution only a single feature is clearly visible: the 15-μm band of CO_2. This band is created by the strong, "bending" mode of the CO_2 molecule, and it is the primary reason why CO_2 is a strong greenhouse gas.

If we look at Earth's spectrum, the 15-μm CO_2 band is also clearly visible, even though Earth's CO_2 concentration is relatively low. Thus, the thermal-IR is an excellent place to look for this gas. It is also a good way of distinguishing terrestrial (rocky) planets like Venus, Earth, and Mars from gas giant planets like those in the outer Solar System, because the gas giants lack CO_2. But Earth's thermal-IR spectrum contains additional information. The absorption at short wavelengths (<8 μm) is caused by H_2O, as is the absorption at long wavelengths (>17 μm). The short-wavelength feature is the 6.3-μm band mentioned in chapter 12, and the long-wavelength feature is the H_2O *rotation band*, caused when H_2O molecules are made to rotate. The latter band extends all the way out to the microwave region, where it allows you to heat food rapidly in your microwave oven (because the food contains H_2O). So, as in the visible/near-IR, it should be possible to determine whether a planet has water vapor in its atmosphere, but not necessarily on its surface.

Even more interesting is the strong band of O_3 centered at 9.6 μm. This band is clearly visible, even though ozone is only a trace constituent of Earth's atmosphere. Ozone resides up in the stratosphere, however, and so it is readily observable when one looks down at the planet from space. Suppose that one were to see this feature in the atmosphere of an extrasolar planet. What would that tell us? Well, the ozone is formed photochemically from O_2, and so we would know that the planet must have O_2 in its atmosphere. So the basic implication is the same as if one observed O_2 directly in the near-IR: the planet appears to have photosynthesis, and thus it is probably inhabited. But one also finds that O_3 can be detected even if only small amounts of O_2 are present.[16] This can be seen in figure 14.5a, which shows the strength of the 9.6-μm band as a function of the O_2 concentration for an Earth-like planet orbiting our own Sun. These synthetic spectra were calculated from the model atmospheres shown in figure 9.2. Surprisingly, as the O_2 concentration

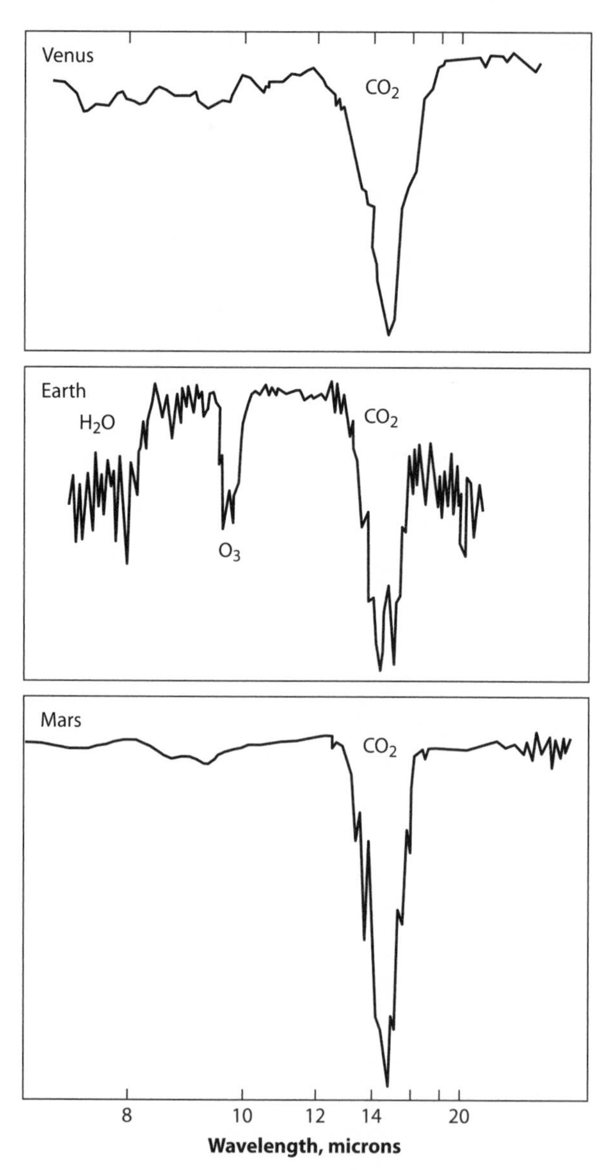

Figure 14.4 Thermal-IR spectra of Venus, Earth, and Mars.[15] (Courtesy of R. Hanel, NASA/Goddard.)

is reduced by a factor of 10, or even 100, the strength of the 9.6-μm band remains almost the same. Not until O_2 is reduced by a factor of 1000 or more does the band start to disappear. The reasons have to do partly with the nonlinear nature of the ozone photochemistry (chapter 9) and partly with the fact that the stratospheric temperature drops as the amount of ozone decreases (figure 9.2b), making the absorption feature appear stronger. Thus, O_3 is, in some ways, an even more sensitive indicator of O_2 than is O_2 itself. By comparison, the O_2 A band becomes difficult to see below about 10 percent of Earth's O_2 abundance (figure 14.5b).

Looking for Life on Early Earth-Type Planets

Thus far, we have focused on O_2 and O_3 as potential indicators for life. But this is a fairly narrow strategy—one predicated on the assumption that other inhabited planets will resemble the modern Earth. However, we already know of an example of a planet that is, or was, inhabited and that would not exhibit the signature of either of these two gases, namely, the early Earth! As discussed in chapter 4, O_2 did not become abundant in Earth's atmosphere until about 2.4 Ga, based on sulfur isotopes and other geologic evidence. O_3 should have been scarce as well, because it is formed from O_2. But life has probably existed on Earth since at least 3.5 Ga, and possibly earlier. Would we be able to tell this if we were able to observe an early-Earth-type planet with some flavor of *TPF* telescope?

This question is a difficult one, as we obviously know less about Earth's early atmosphere than we do about its present one. However, our models for how Earth's atmosphere evolved, discussed in chapters 3 and 4, can give us some ideas. The most obvious gas to look for is methane, CH_4. Methane was probably relatively scarce prior to the origin of life, as the prebiotic atmosphere is thought to have been dominated by N_2 and CO_2.[17-19] Questions remain as to whether this *weakly reduced** prebiotic

*A weakly reduced atmosphere is one in which O_2 is absent, but the remaining gases are still relatively oxidized. Carbon, for example, is present mostly as CO_2. A strongly reduced atmosphere is one rich in hydrogen-bearing gases such as CH_4 and NH_3.

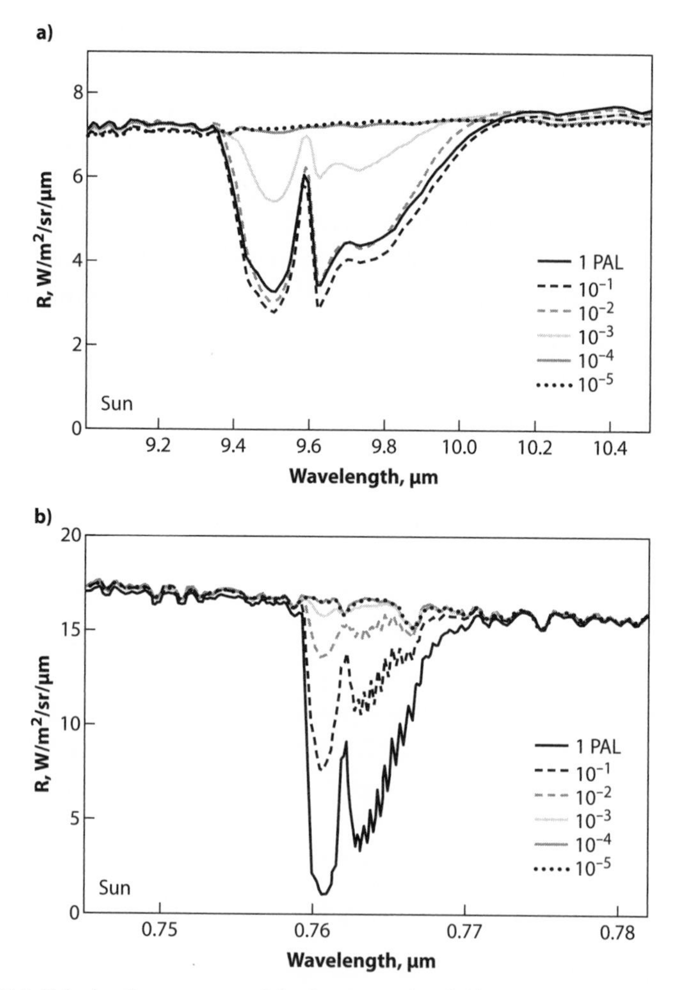

Figure 14.5 Calculated appearance of the O_3 9.6-μm band (a) and the O_2 0.76-μm band (b) as a function of atmospheric O_2 concentration. "PAL" means "times the present atmospheric level."[5] (Reproduced by permission of Mary Ann Liebert Publishing.)

atmosphere model is correct, as older models supported more *strongly reduced* mixtures,[20,21] and some recent authors have resurrected these older ideas.[22] Most workers would agree, however, that atmospheric CH_4 concentrations probably increased dramatically once methanogenic bacteria evolved and began generating CH_4 biologically. Today's CH_4 concentration is relatively low, 1.7 ppm, but prior to the rise of O_2 the methane lifetime would have been longer and its concentration could have been of the order of 1000 ppm or more.[23,24] This amount of CH_4 would be easy to see in the thermal-IR by looking for the strong 7.7-μm band[25] (see also figure 14.8 later in this chapter), and it might be observable in the visible/near-IR as well, although this would be somewhat more difficult.[26] CH_4-rich atmospheres can also generate Titan-like organic hazes (chapter 4), which might also be observable spectroscopically. So, if planets like the postbiotic Archean Earth exist elsewhere, a mission such as *TPF-I* might well record the spectrum of a CH_4-rich terrestrial planet. Whether or not this would be interpreted as a sign of extraterrestrial life is not clear. It would certainly generate a huge amount of debate.

Possible False Positives for Life

This brings us back to the issue of possible false positives for life. CH_4 is a possible false positive if abiotic terrestrial planets are capable of sustaining high atmospheric CH_4 concentrations. But that is something that is widely recognized. What about O_2 and O_3? Are these gases really reliable bioindicators, or are there ways in which O_2 might accumulate in a planet's atmosphere abiotically?

The answer to this question is almost certainly yes: it is easy to conceive of abiotic methods of producing O_2. The most obvious one is by way of a Venus-like runaway greenhouse.[27] Think of Venus itself: suppose for the sake of argument that Venus started out with as much water as is present in Earth's oceans today, 1.4×10^{21} liters. This assumption may not be strictly correct, but the argument remains valid even if its initial water endowment was much smaller. Suppose also that Venus lost most of its water within the first few hundred million years of its history by the mechanism discussed in chapter 6: photodissociation of H_2O by solar ultraviolet light, followed by escape of hydrogen to space. That process should have left an enormous amount of oxygen behind—enough to produce about 240 bars of surface pressure if it was all converted to O_2.[28] Eventually, all of this O_2 might have been consumed by reactions with the planet's surface. But such reactions are slow, and indeed some authors have questioned whether it is really possible to dispose of an entire ocean's worth of oxygen in this way.[29] So perhaps a model in which Venus started out with only 10 percent of Earth's water is more realistic. Either way, the analysis implies that for at least some time interval following the loss of water, perhaps as long as a billion years, O_2 could have been an abundant component of Venus' atmosphere. O_3 would likely have been present as well, and so the spectroscopic signals of O_2 and O_3 should have been readily observable to anyone who was around to look for them.

Would a situation like this fool us if we were to see such an early Venus-type planet around another star? Probably not, because we would have several other clues as to what had happened. First, the planet would be near or inside the inner edge of the habitable zone. Second, its H_2O bands should be weak or nonexistent, depending on which stage of the process we were seeing. If we were unfortunate enough to observe this process as it was only partway to completion, the H_2O might still be there. But it would be up in the planet's stratosphere, as well as down in its troposphere, and so the odds of finding a well-developed ozone layer would be small. That's because the by-products of H_2O photolysis, if present in high abundances, can catalyze ozone destruction, just as do the chlorine-containing molecules produced by photolysis of freons. Hence, if we had measurements in both the visible and thermal-IR wavelength regions, we could probably resolve this question one way or the other.

A second false positive for life that is easy to identify ahead of time would be a frozen planet like Mars near or outside the outer edge of the habitable zone.[27] Mars itself has about 0.13 percent O_2 in its atmosphere.[30] This is too little to see spectroscopically because Mars' atmosphere is very thin; however, the process that produces it is universal and could thus operate on other planets as well. Mars' O_2 does not come from photosynthesis; rather, it comes from photodissociation of H_2O, followed by escape of hydrogen to space—the same process just discussed for early Venus.* On early Venus, the rate of abiotic O_2 produc-

*One might think that Mars' O_2 comes from photodissociation of CO_2, followed by recombination of O atoms with each other to form O_2. The net reaction, then, would be $2CO_2 \longrightarrow 2CO + O_2$. But this cannot be the whole story because the observed $CO:O_2$ ratio in the martian atmosphere is 0.6, rather than 2. Furthermore, the time required to replenish Mars' O_2 on Mars by photodissociation of H_2O followed by hydrogen escape is only about 10^5 years. Hence, it is this latter process that controls the martian O_2 concentration.

tion would have been large because the stratosphere was wet, and the hydrogen escape rate was high. That is not the case for a cold planet like Mars. But Mars lacks appreciable O_2 sinks on its surface. It is small enough to have lost much of its internal heat, and so volcanism today is either sporadic or entirely absent. Consequently, there are few reduced gases, like hydrogen, with which O_2 might react. Furthermore, Mars' surface is cold and dry, and so the rate of surface oxidation is presumably very slow. Mars' surface *is* highly oxidized, to be sure—that is why it is called the "red planet"—but most of this oxidation probably happened a long time ago. In the absence of liquid water, surface erosion is relatively slow, so fresh, unoxidized rocks that might serve as an O_2 sink are exposed only infrequently.

What, then, limits the O_2 concentration in Mars' atmosphere? The key factor is that Mars is so small ($\sim$0.1 Earth mass) that oxygen can escape from Mars' atmosphere at a slow rate.[31] This happens by ion recombination reactions in Mars' ionosphere, for example:

$$O_2^+ + e \longrightarrow O + O. \qquad \text{(R1)}$$

Here, an O_2^+ ion (a molecule that has lost an electron) recombines with an electron, *e*, to form two atomic oxygen atoms. These O atoms have extra energy, which means that they are moving rapidly. If one of them is headed upward, which must always be the case, it is possible for the O atom to escape from Mars' gravity. Mars also loses O atoms through sputtering by the solar wind, as discussed in chapter 9.

A second false positive, then, might be a planet like Mars that had limited surface sinks for O_2. If the planet was just a little bigger than Mars, say 0.2 Earth masses, it should be able to hold onto its oxygen, because reaction R1 would not provide enough energy for the O atoms to escape. It might also be warm enough in its core to keep its magnetic field in place for an extended period of time. This combination of circumstances could, in principle, allow O_2 to accumulate indefinitely in such a planet's atmosphere. If the planet were too big, then this phenomenon would be less likely to occur, because then it would presumably be volcanically active, like Earth, and O_2 would be consumed by reaction with volcanic hydrogen. But there could be a range of planet masses for which such "super-Mars" false positives might occur.

Finally, other types of false positives might occur if we were to mistake the absorption bands of one gas for those of another gas that is considered to be a bioindicator. For example, the French astronomer Franck Selsis and his colleagues have pointed out that CO_2 has weak absorption bands at 9.4 and 10.4 μm that could conceivably be mistaken for the 9.6-μm O_3 band in a dense, CO_2-rich atmosphere.[32] Such an ambiguity could eventually be resolved, though, either by going to higher spectral resolution, so that the CO_2 bands would split, or by using visible/near-IR spectra to look directly for the presence of O_2.

I apologize to the reader if the above discussion seems overly detailed. However, I have envisioned many times the debate that is sure to erupt if and when O_2 is detected in an extrasolar planet atmosphere. I'm also afraid that this event may occur far enough in the future that I may not be around to comment on it when it happens. Thus, the preceding discussion is an attempt to insert my two cents' worth into the debate and, at the same time, to caution others that one needs to think carefully about what the detection of O_2 on an extrasolar planet implies. It may indicate the presence of extraterrestrial life, but one can think of situations where it may not. So we'll want to examine all aspects of a planet carefully if we ever make such a detection, and we'll almost certainly want to follow up with more detailed observations to be sure that our interpretations are correct.

Polarization Measurements: Looking for the Glint of Surface Water

How might we differentiate between habitable planets like Earth and frozen Mars-like planets like the one described above? The key difference, other than planetary size, is whether liquid water is present at the planet's surface. Is it possible to look for liquid water directly?

Although conventional spectroscopic measurements are sensitive only to water vapor, it may be possible to use *polarization* to determine whether surface water is present. The term "polarization" refers to the direction in which the electric field points in an electromagnetic wave, such as light. In unpolarized light, the electric field can point in any direction. In linearly polarized light coming toward an observer, the electric field points either straight up-and-down or straight right-to-left. Polarized sunglasses take advantage of this effect to eliminate glare and thereby make it easier to see in bright sunlight. What makes them work is the fact that the up-and-down component of the reflected sunlight—technically, the component *parallel* to the plane in which the reflection takes place—is not as bright as the right-to-left, or *perpendicular*, component. The effect is especially pronounced near certain angles (see below). Hence, if one blocks the right-to-left component by using an up-and-down polarized lens, the light that gets through will include less reflected light, and hence less glare.

What makes this technique potentially useful for studying extrasolar planets is the fact that the parallel component of the reflected light vanishes entirely when the angle of incidence of the sunlight is near the *Brewster angle*, which for liquid water is 53.1°. (The angle of incidence, θ, is half the angle between the incident and reflected ray. See figure 14.6.) One would not get this effect if the surface was solid ground because it would not be smooth enough to create a polarization signal. But, if the planet has an ocean, the *polarization fraction* of the starlight reflected from the planet's surface should peak when the angle formed by the star, the planet, and the observer (i.e., the Earth) is equal to twice the Brewster angle, or 106.2°. An extrasolar planet going around its star will pass through the Brewster angle at almost the same time as it is farthest from its parent star and, hence, is easiest to see.[33] One has to be careful here because Rayleigh scattering by the atmosphere also causes polarization, and that polarization peaks when the star-planet-observer angle is 90°. We should be able, though, to distinguish between these two angles. So, if we can design a planet-finding telescope that includes a *polarimeter* for measuring polarization, it may be possible to answer the question of

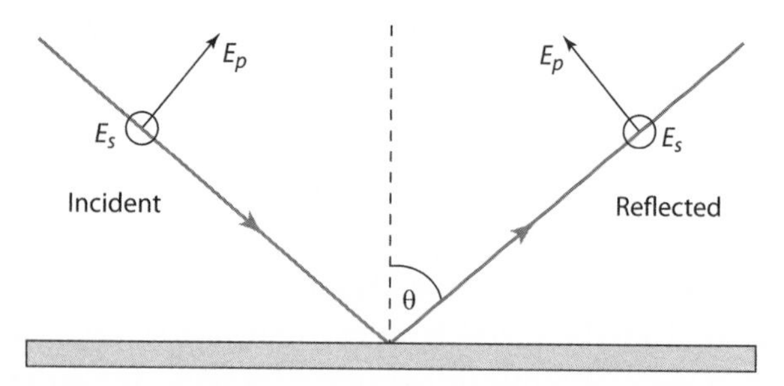

Figure 14.6 Diagram illustrating the reflection of polarized light. E_s represents the perpendicular component of the light (electric field in and out of the paper); E_p represents the parallel component of the light (electric field parallel to the plane defined by the incident and reflected rays). The parallel component of the reflected light vanishes when the angle between the two rays, divided by two, is equal to the Brewster angle (53.1° for water). (Modified from http://en.wikipedia.org/wiki/Polarization.[34])

whether an extrasolar planet has liquid water on its surface. This might not happen on a first-generation *TPF-C* or *TPF-O* mission, but it could be done on a follow-up mission, if a potentially habitable planet had already been identified.

The Holy Grail: Simultaneous Detection of O_2 and Reduced Gases

Suppose that we found an extrasolar planet well within the habitable zone of its parent star, that we were able to prove definitively that it had liquid water on its surface, and that we also saw a strong spectroscopic signal of O_2 or O_3 in its atmosphere. Would we then be convinced that the planet was inhabited?

I personally would be 99 percent sure that life was present if I saw this combination of planetary characteristics. But I am also pretty sure that skeptics would still remain. Indeed, short of visiting the planet and finding things crawling around on its surface, it is likely that no data bearing on extrasolar life will ever be considered definitive by all observers. That is simply to be expected, though. There are still people who think that the Earth is flat and others who believe that global warming is not caused by humans. One cannot hope to convince everyone about anything.

That said, it is possible to significantly increase our confidence in remote life detection by looking for additional biogenic gases besides O_2 (or its photochemical by-product, O_3). Indeed, the idea for how to do so is an old one. It was suggested nearly simultaneously by two different authors, Joshua Lederberg and James Lovelock, writing in the journal *Nature* back in 1965.[35,36] Both authors were thinking about detecting life on planets in our own Solar System, Mars in particular, rather than on planets around other stars. Lederberg's paper came out first and was the more general. He suggested that the best type of remote evidence for life would be the observation of extreme thermodynamic disequilibrium in a planet's atmosphere. Earth's atmosphere is in such a state, largely because of the presence of gases given off as a by-product of metabolism. One must be careful about applying Lederberg's argument, though, because all planetary atmospheres are out of thermodynamic equilibrium to some degree. In part, this is because in relatively cool atmospheres, like that of Earth, the kinetic reactions between molecules are too slow to maintain equilibrium. Furthermore, Earth's atmosphere is also influenced by UV and X-ray photons from the Sun, which cause photochemical reactions that would not otherwise take place. These photons come partly from the Sun's surface, which has a temperature of ~6000 K, and partly from the chromosphere and corona, which are even hotter. Hence, the radiation field in Earth's atmosphere is not in thermodynamic equilibrium with the atmosphere itself and would not be so even in the absence of life. Lederberg's criterion for remote life detection was the identification of *extreme* thermodynamic equilibrium, and so it may still be considered correct, but it requires refinement to be used effectively.

Lovelock's argument was based on the same principle, but it was more specific. Lovelock's thinking about this topic was spurred by the fact that he was involved, along with Carl Sagan, in planning for NASA's *Viking* missions to Mars. Recall that the two *Viking* spacecraft were launched in 1976, 11 years after Lovelock's paper was written. They both contained probes that landed successfully on the surface of Mars and that searched unsuccessfully for life in the uppermost few centimeters of martian soil. During the planning for these missions, Lovelock argued that NASA didn't need to spend billions of dollars to send probes to Mars to search for life. Instead, they could look for the simultaneous presence of trace gases that were out of equilibrium with each other. In particular, the presence of O_2, along with a reduced gas such as CH_4 (or N_2O) that reacts with it, would be strong evidence for the existence of life. In Earth's present atmosphere, O_2 (at 21 percent by volume) and CH_4 (at 1.7 ppm by volume) are out of thermodynamic equilibrium by a factor of roughly 10.[20] At the time, neither of these gases had been detected in Mars' atmosphere, and so Lovelock argued that the planet must be uninhabited.

NASA did not buy Lovelock's argument, and so they sent the *Viking* probes anyway. They were right do so, because Lovelock's argument about martian life was actually backward from what it should have been. The *absence* of O_2 and CH_4 does not prove that life is absent. After all, alien organisms need not necessarily produce the same by-products that we see on modern Earth (although thermodynamics suggests that it would make sense energetically for them to do so). More importantly, these gases might be generated in such small quantities that it is difficult to see them. We know now, for example, that Mars' atmosphere *does* contain O_2 (0.13 percent), and there are hints that it contains minute concentrations of CH_4 as well (chapter 8). The O_2 is produced photochemically, as we saw earlier, and the CH_4 (if it exists) may be produced abiotically as well by interactions between water and warm rocks. But a biotic source for CH_4 cannot be ruled out, and so NASA has plans to investigate this issue further.

If one inverts the argument, however, Lovelock's criterion is a good one: the simultaneous detection of O_2 and CH_4, or O_2 and N_2O, would be strong evidence for the existence of life. Based on the example just discussed, one needs to modify this statement further by requiring that both gases be present in abundances that cannot be explained by abiotic processes. This would rule out the type of water–rock interactions that have been suggested as a source of martian methane. Making proper use of this criterion requires that one have a good understanding both of atmospheric photochemistry and of the rate of abiotic production of the gases under consideration.

Would it be possible to apply this "Holy Grail" of remote life detection to a planet like modern Earth? In one of the last scientific papers that he wrote before his death, Carl Sagan and his colleagues attempted to answer this question using data collected from NASA's *Galileo* spacecraft.[6] *Galileo* was a mission to the planet Jupiter, not to Earth. But, to save fuel and reduce the flight time out to Jupiter, Galileo swung by Venus once and Earth twice on its way out to the outer Solar System. If one does this in just the right way, it is possible to obtain a gravitational boost from the planet as one goes by it. On the second pass by Earth, Sagan asked the mission controllers to activate the scientific instruments on board and use them to observe the Earth. One of those instruments was called NIMS, the Near Infrared Mapping Spectrometer. As it flew by, the NIMS instrument acquired the spectra shown in figure 14.7. The panel at the left shows the short-wavelength part of the NIMS spectrum, just beyond the visible. In it, one can clearly see the O_2 A band at 0.76 μm—the same band seen in the Earthshine data (figure 14.1) and in the synthetic spectrum of Earth (figure 14.3). The two right-hand panels show the longer-wavelength portion of the NIMS spectrum, which extended out to about 5 μm. Here, one can see absorption bands of both CH_4 and N_2O. They are tiny because the concentrations of these gases are low (1.7 ppmv for CH_4, 0.3 ppmv for N_2O). But, from Sagan's perspective, the answer to the Holy Grail question was a resounding "yes." One can indeed confirm that Earth is inhabited simply by observing its atmosphere, provided that one is capable of looking closely enough.

Could we do the same thing for an Earth-like planet orbiting another star? Probably not at the present time—at least, not with the types of telescopes being studied for the *TPF* and *Darwin* missions. Indeed, none of these missions is being designed to look at the mid-IR wavelengths shown in the right-hand panels of figure 14.7. In part, this is because there are fewer photons available at these wavelengths to make the required observations (figure 13.1), and in part this is because the CCD detectors that would be used at shorter wavelengths for *TPF-C* do not operate beyond about 1.1 μm. If one went to larger telescopes and used other types of detectors, one could in theory overcome these problems. Both O_2 and CH_4 have now been detected in Earthshine data obtained

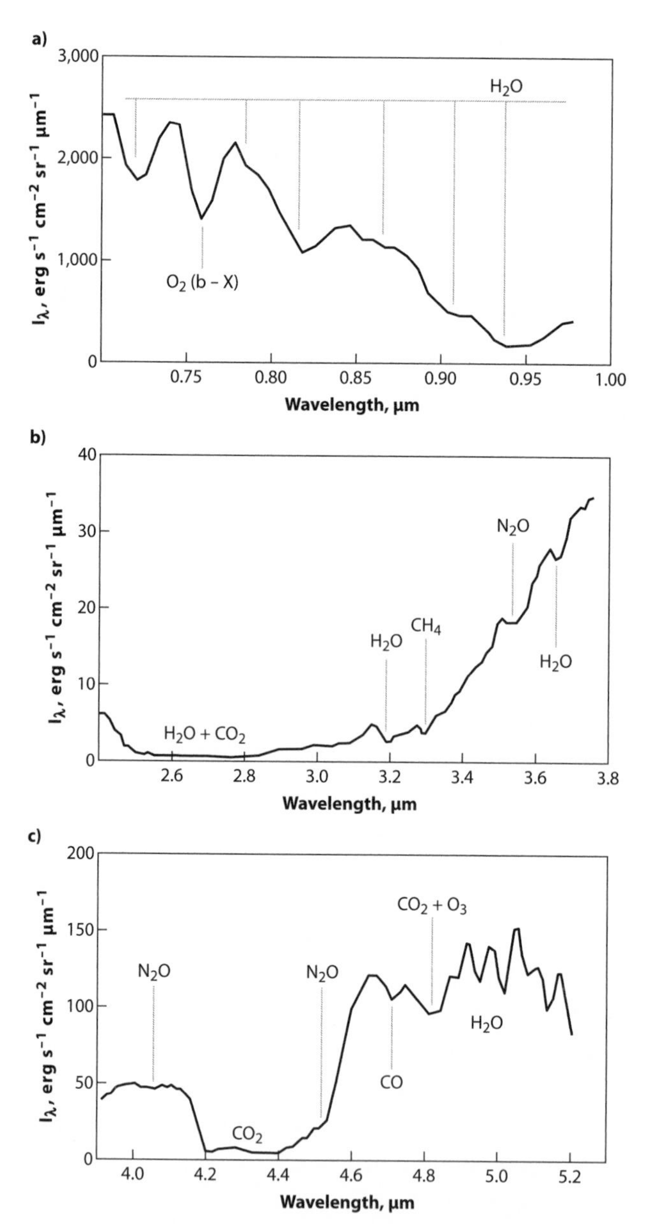

Figure 14.7 A near-infrared spectrum of Earth taken from the spacecraft Galileo while on its way out to Jupiter. (a) The shorter infrared wavelengths, including the O_2 A band at 0.76 µm (here identified as "O_2(b-X)"). (b, c) The longer wavelength region, which includes absorption bands of CH_4 and N_2O.[6] (Reproduced by permission of *Nature*.)

from ground-based telescopes,[37] so it should be possible in principle to do the same thing for extrasolar planets. Perhaps this will be done on the proposed *Life Finder* mission discussed in the next chapter. However, it is highly unlikely that any space-based project of this magnitude will be launched within the next 20–30 years. Indeed, we should probably count ourselves fortunate if a *TPF*-class mission is launched during this time frame.

One can, however, think of possible ways in which the simultaneous measurement of O_2 plus CH_4 might be achieved on an Earth-like extrasolar planet. One possibility would be if we were to observe a planet similar to the Proterozoic Earth, between about 2.0 and 0.8 Ga, if significant amounts of CH_4 were generated in marine sediments.[38] It might also be possible to do this for a modern Earth-like planet orbiting an M star.[39] At wavelengths < 200 nm, active M stars, which have lots of flares, emit more UV radiation than the Sun. UV fluxes from two active M stars, AD Leo and GJ (Gliese) 643, are shown in figure 14.8a, along with the fluxes from F, G, and K stars. (The latter are the same as shown previously in figure 10.3.) This short-wavelength UV radiation can dissociate O_2, and so an M-star planet that has O_2 in its atmosphere should also have plenty of O_3. However, the near-ultraviolet flux from M stars is much lower than that from the Sun, especially in the 200- to 300-nm range, where ozone absorbs (figure 14.8a). The lack of near-UV radiation on the M-star planet changes the atmospheric photochemistry in such a way that the lifetime of methane is very much longer than on Earth. So, if methane was being produced by methanogenic bacteria at the same rate as on modern Earth, the concentration of methane in the planet's atmosphere could be as high as 500 ppm,[39] enough to be readily observed by *TPF-I* or *Darwin* (see figure 14.8b). So perhaps M-star planets, if they can indeed harbor life, will end up being the place where extraterrestrial life is first definitively detected. If so, that would further dilute our present geocentric view of the inhabited universe.

In summary, there are many potential spectroscopic indicators of life, some of which are more definitive than others. Even the Holy Grail of remote life detection—the simultaneous measurement of O_2 (or O_3) and

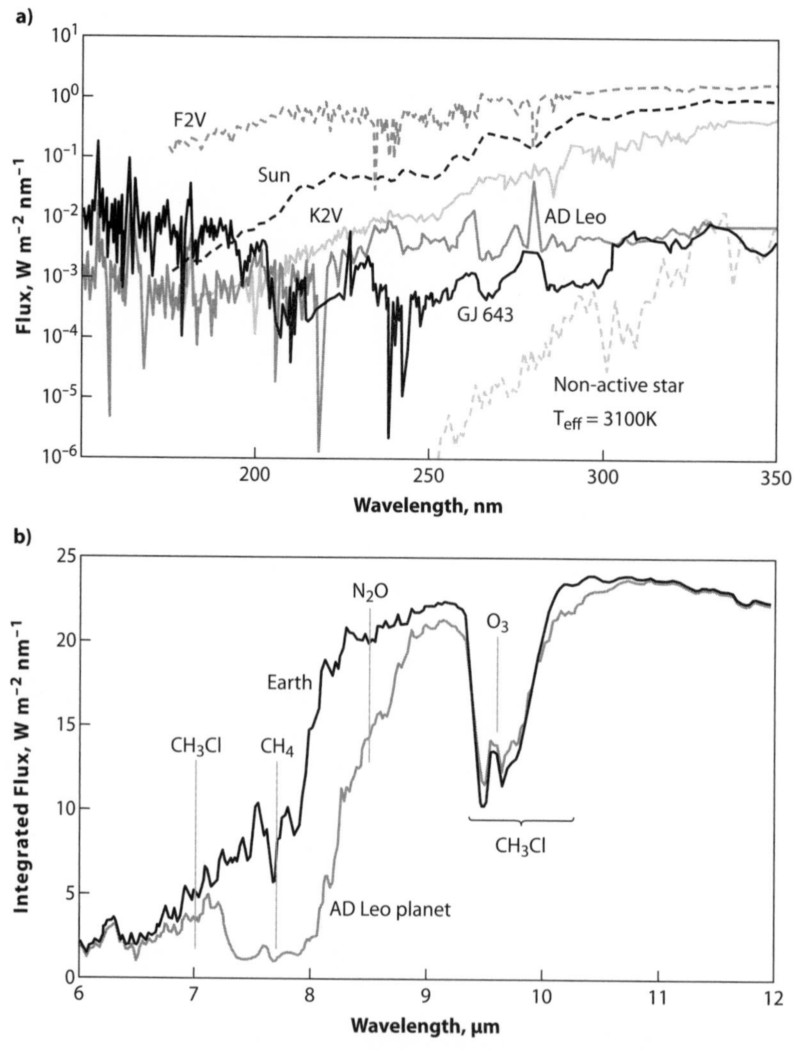

Figure 14.8 (a) Calculated ultraviolet fluxes from different types of stars for an Earth-like planet orbiting at an equivalent distance of 1 AU. AD Leo and GJ 643 are M stars, as is the "nonactive" star at the lower right. (b) Thermal-IR spectra of Earth and of a hypothetical Earth-like planet orbiting AD Leo. The AD Leo planet's atmosphere contains 500 ppm of CH_4.[39] (Reproduced by permission of *Nature*.)

CH_4 (or N_2O)—would no doubt leave some people unconvinced. But, remember, we are talking here about first-generation *TPF*-type missions that will hopefully occur within the next 20–30 years. What might happen beyond that time frame? We'll look at that question in the next and final chapter.

· Chapter Fifteen ·

Prospects for the More Distant Future

The two *TPF* missions described in the previous chapter could go a long way to answering some of our fundamental questions about whether other Earth-like planets exist and whether some of them might be inhabited. We saw, though, that neither *TPF-C* nor *TPF-I/Darwin* is likely to be able to simultaneously detect O_2 or O_3 and a reduced gas such as CH_4 or N_2O on a planet that is a duplicate of the modern Earth. We might be able to accomplish this on a Proterozoic Earth-type planet or on an Earth-like planet orbiting an M star, but these are special types of planets that may or may not exist in the local solar neighborhood. How then might we proceed if we saw a planet with an atmosphere that looked like that of present Earth, i.e., that was rich in O_2 and O_3, but we were not sure whether or not the oxygen was actually produced biogenically?

NASA's Life Finder Mission

The obvious answer to this question is to scale up the size of the *TPF* missions so that one can obtain more photons, along with improved spatial resolution. NASA has already started thinking about this possibility, even though they have not yet begun building *TPF* itself. In the space game, visions of what might be done generally precede the actual doing of them by long time intervals. Indeed, *TPF* itself was already being discussed within NASA way back in 1980 as part of a program called TOPS (Towards Other Planetary Systems). I remember this well because the researcher in the office next door to mine at NASA Ames Research Center, David Black, was involved in this effort. It takes the patience of Job, combined with the lifetime of Methuselah, to see some planned NASA missions actually take place.

The post-*TPF* mission that has been discussed by NASA is referred to as *Life Finder*. Some concepts for a *Life Finder*–type mission have already been proposed. The French astronomer Antoine Labeyrie has probably done more thinking on this topic than has anyone else. He has proposed a *hypertelescope*,[3] an example of which is shown in figure 15.1. This particular concept consists of 37 small, free-flying optical telescopes, which would be used as an interferometer. This design was actually proposed by Labeyrie and his collaborators from Boeing-SVS as a possible design for the *TPF* mission itself,[4] and it was given the name *Exo-Earth Detector*. However, the fact that it involves 37 separate telescopes suggests that it is really a second-generation *TPF* concept, and it has not been pursued seriously by NASA for that reason.

Labeyrie's imagination does not stop with *Exo-Earth Detector*. He has also described a much larger space-based hypertelescope that he calls *Exo-Earth Imager (EEI)*.[3] This instrument would consist of 100 or more free-flying telescopes, each 3 m in diameter, arranged in three concentric rings of up to 150 km in diameter. Such an unabashedly futuristic telescope array should, in principle, have two big advantages over a first-generation *TPF* mission. First, the total collecting area would be of the order of 100 times larger than *TPF*. The number of photons collected would be correspondingly larger, and these could be used to create a much higher-resolution spectrum. Although it has not been explicitly demonstrated, such an instrument would likely be capable of simultaneously observing O_2 and CH_4 in Earth's present atmosphere, thereby satisfying Lovelock's criterion for remote life detection.

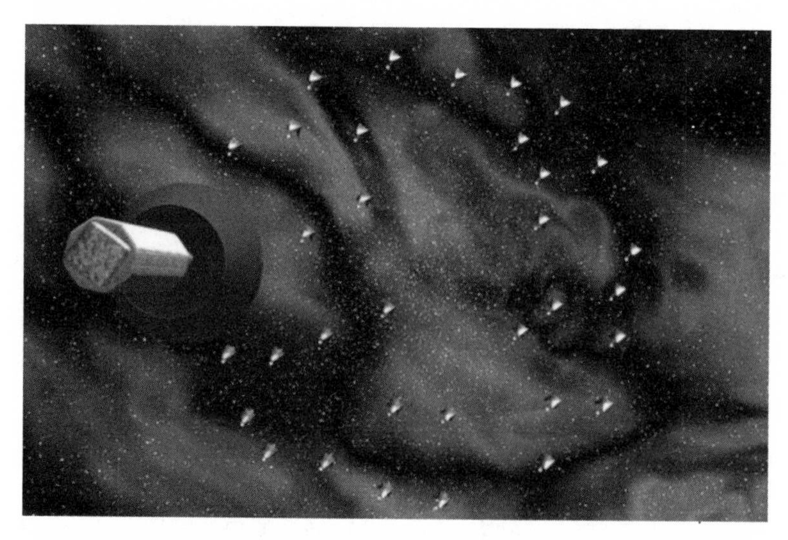

Figure 15.1 Artist's rendition of *Exo-Earth Detector*, a hypertelescope that could be used as a follow-up mission to *TPF*.[3]

The second advantage of the *EEI* array would be its large effective diameter. Recall from chapter 13 that the angular resolution, θ, of a telescope or an interferometer is proportional to λ/D, where λ is the wavelength at which one is observing and D is the effective diameter of the telescope. *EEI* would have an effective diameter of 150 km, or 1.5×10^5 m. Suppose that it was being used to make observations in the mid-visible at $\lambda = 500$ nm (5×10^{-7} m). Then, θ (in radians) would be roughly 5×10^{-7} m/1.5×10^5 m $\cong 3 \times 10^{-12}$. Suppose further that the planet that one was observing was located at a distance of 3 pc ($\cong 10^{14}$ km) from Earth. Then, the theoretical spatial resolution should be of the order of 3×10^{-12} ($\times 10^{14}$ km) $= 300$ km. That means that one should, in principle, be able to take a multipixel image of a planet like Earth! Indeed, Labeyrie has simulated what the Earth might look like at this distance, based on a 30-minute exposure with *EEI* (figure 15.2, in color section). The picture is a little fuzzy, to be sure, but one can still make out the continents of North and South America, along with the distinction between rocky, mountainous areas and vegetated land. If Earth-like planets do indeed exist within a distance of 3 pc, perhaps someday we'll be studying them at this resolution.

Using the Sun as a Gravitational Lens

I was fortunate to have been seated next to Frank Drake at a banquet dinner at a NASA Astrobiology meeting several years ago, and the topic of conversation got around to post-*TPF* extrasolar planet missions. Drake, of course, is the radio astronomer who, along with Carl Sagan, came up with the now-famous Drake equation discussed in chapter 1. I had heard a little of Labeyrie's ideas and was relating them to Frank. He looked at me and said: "You realize, don't you, that there is a much cheaper way to do this?" I had no idea what he meant, so I asked him to explain. "All you have to do," he said, "is to send a spacecraft out away from the Solar System in the exact opposite direction from the star in which you are interested. Once the spacecraft gets to ~600 AU, you can use the Sun as a gravitational lens. The light from the stellar system will be amplified by a factor of roughly a million, and you will have plenty of photons with which to do whatever you want."

Drake was at least partly correct in his assertion. Indeed, I learned later that he had written an article[5] describing how this phenomenon might be used to facilitate interstellar communication. (He was at that time more interested in radio waves than in visible light waves.) The idea itself had been proposed sometime previously[6] by another radio astronomer, Von Eshleman, from Stanford University, and is based on even earlier work by Einstein[7] and Liebes.[8] As is true for many things these days, you can now work out the basic physics for yourself by looking it up on Wikipedia. Recall from chapter 11 that a star will focus the light from a source directly behind it to form an Einstein ring around it (see figure 11.8). For this phenomenon to work, the ring radius must be larger than the radius of the Sun, $\sim 7 \times 10^5$ km. If one works out the math,[9] one finds that for this to be true, the distance, D_L, of the spacecraft from the Sun has to be at least 8.3×10^{13} m, or ~550 AU. In practice, one might need to go several times farther than this to avoid light coming from the Sun's corona. Eshleman[6] assumed that any observations would be made from 2200 AU, which would make the ring radius equal to twice the radius of the Sun.

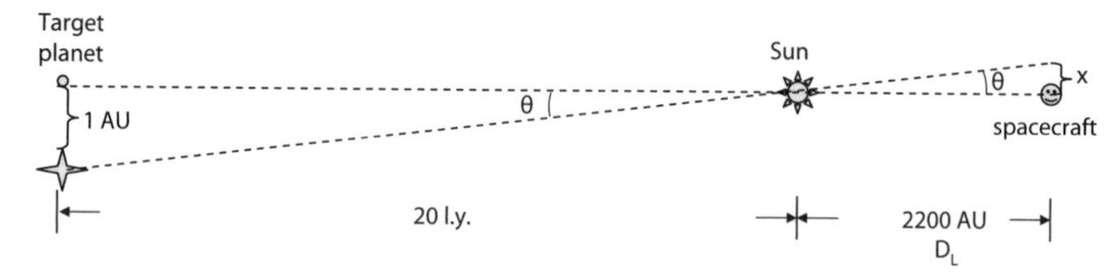

Figure 15.3 Slightly oversimplified diagram of the use of the Sun to lens another planetary system. In reality, the light rays from the target system would be curved by the Sun's gravity. The spacecraft used for the observations would be lined up with the target planet and the Sun; hence, the parent star of the planet would appear slightly off-axis. Distances taken from Eshleman 1979.[6]

One might think that 2200 AU is a long distance, and indeed it is, although it is still much less than the distance to the nearest star (about 4 light-years, or 250,000 AU). The Solar System itself extends only out to about 30 AU (Neptune's orbit), so this distance would be nearly 70 times further. Several spacecraft, including Pioneers 10 and 11 (launched in 1972/73) and Voyagers 1 and 2 (launched in 1977) have already made it out well beyond Neptune's orbit. The one that is most distant, Voyager 1, is currently about 105 AU from the Sun and is drifting away at ~3.6 AU per year.[10] If it were headed toward Alpha Centauri (which it is not), it would get there in about 70,000 years. Voyager 1 will reach 2200 AU distance in a little less than 600 years.

Science fiction authors, though, have imagined propulsion schemes that might allow a spacecraft to reach such distances much more quickly. Perhaps the best idea is that proposed by Larry Niven and Jerry Pournelle in their novel *The Mote in God's Eye*.[11] In the novel, an alien spacecraft from another planetary system is seen approaching Earth. The spacecraft is propelled by a solar sail pushed by a powerful laser located in the aliens' home planetary system. This provides continuous acceleration even when the spacecraft is far from any stellar energy source. With a large sail of this type and a laser operating from, for example, the Moon, a spacecraft could achieve high velocities without having to pack along a lot of (very heavy) fuel. The spacecraft would not be able to slow down effectively, so it would reach 2200 AU and keep right on going, but that is OK because the gravitational focusing by the Sun would continue to operate out to far greater distances, essentially to infinity. Thus, at first glance, the idea of using the Sun as a gravitational lens appears to be an intriguing possibility.

On further examination, though, this idea is fraught with difficulties. First, one would need a good coronagraph on board the spacecraft to block out the light from our own extremely bright Sun. That by itself should not be a particularly difficult problem for a civilization advanced enough to contemplate such a mission, although one would need to worry about the Sun's corona, as pointed out above. Second, the Sun is continually moving around the center of mass of the Solar System (figure 11.2), so aiming the spacecraft would be difficult: it would remain aligned with its target star system for only a relatively short time unless one could "tack" horizontally to follow the Sun's motion. This might be achieved, for example, by making the solar sail partially reflective and tilting it so that beamed light was reflected at an angle. Third, and most importantly, the light from the planet of interest would be mixed with that of its parent star, which, of course, is much brighter. This problem would be insurmountable but for the fact that the gravitational focusing by the Sun is sufficiently precise that it should readily distinguish the planet from its parent star, provided that one knew ahead of time exactly where to look for the planet.* That is doable, in principle, because one would already know the details of the planet's orbit from previous astrometric or direct imaging studies. The idea then would be to focus the gravitational lens on a particular part of the planet's orbit at a time when it was expected to be there. The planet would then be imaged uniformly into the Einstein ring, which for Eshleman's parameters (star at 20 light-years, observer at 2200 AU) would have an angular radius of 0.88 arcsec.[7] The ring would be sufficiently large that it could be resolved with a

*One can verify this by calculating the parameter X in Eshleman (1979) (see his figure 2). X is the nondimensional distance of the observer from the line defined by the target planet (or star) and the Sun, $X = x/[4GM_L/c^2]^{0.5}$, where x is the distance shown in figure 15.3. X would be close to zero for the planet observation because one would line the spacecraft up with the planet and the Sun. This gives an amplification factor of about 2×10^5 for the planet, given Eshleman's observational parameters (see text). The star would be slightly off-axis, and hence would be less strongly amplified. For a star-planet separation of 1 AU, the star's X value would be ~ 0.18, resulting in an amplification factor of unity or less.

modest-sized telescope less than a meter in diameter. Some of the parent star's light would still be present, and it should still outshine the planet by a factor of $\sim 10^5$; however, its light would not be uniformly distributed around the Einstein ring, and hence it could, in principle, be separated out. I am not a good enough gravitational physicist to solve this problem myself, but there may well be someone reading this paragraph who could do so.

Considering the list of potential problems just enumerated, it is not obvious that this technique for extrasolar planet characterization is worth pursuing. It might well make more sense to simply build Labeyrie's *Exo-Earth Imager*. Such an instrument, while costly, would have the huge advantage of being able to be used on more than one target. But if NASA by this time is still being run by someone who is interested in better/faster/cheaper missions, and if the logistical difficulties discussed above can be overcome, then perhaps the idea of using the Sun as a gravitational lens might eventually bear fruit.

The Drake Equation Revisited: The Search for Extraterrestrial Intelligence

So, then, where does all of this lead? Of course, one can certainly venture even further into the realm of science fiction and propose interstellar missions manned by robots or multiple generations of human colonists. But I do not particularly want to go there, as those who actually write science fiction have already done so much more effectively than I could. I myself grew up reading science fiction novels by Asimov and Heinlein, and thus I have always had great respect for writers who can stimulate the imaginations of young people and help steer them into actual science. No, for me, the problem comes full circle. I suspect that if we discover that life does exist elsewhere, and especially if we see evidence for O_2-rich worlds similar to our own, we will eventually return to the question that motivated Frank Drake and Carl Sagan to organize their meeting at Green Bank Observatory more than 40 years ago: Does *intelligent* life exist elsewhere in the galaxy? After all, we don't just want to find out if we are the only form of life in the universe; we want also to find out if there is anyone out there with whom we might converse.

Let's go back then to the Drake equation, which was introduced in chapter 1, and see if we can say anthing more about it, based on the information discussed in this book. Recall that the number of advanced, communicating civilizations in the galaxy, N, is given by the product of 7 numbers:

$$N = N_g f_p n_e f_l f_i f_c f_L. \tag{15.1}$$

What will we have learned about the solution to this equation once *TPF* and *Lifefinder* have flown? We already saw in chapter 1 that the first term in the Drake equation, N_g (the number of stars in the galaxy), is ~400 billion (4×10^{11}). This number comes from doing star counts in representative parts of the sky and then multiplying by the volume of the Milky Way galaxy. It is uncertain by at least a factor of 2, but that still is much less than the uncertainties in other terms in the formula.

The second term, f_p, the fraction of stars that have planets, is not known nearly as well. However, as a result of the planet-finding studies discussed in chapters 11 and 12, we are now able to place at least a lower bound on it. At the time of this writing, early 2008, the percentage of nearby single, solar-type stars that have known planets is approximately 5 percent of those that have been observed. Another 7 percent of stars observed by the radial velocity technique have long-term drifts in velocity that are likely caused by planets in distant orbits that have not yet had time to complete an entire orbit.* So we can say with some certainty that the fraction of single, solar-type stars that have planets is at least 0.12. At least half of the stars in the galaxy are members of binary or multiple star systems, for which the existence of planets is less clear. This may be compensated, though, by the fact that we have probably missed a lot of

*Suppose that one has observed a star only during one-half of a planet's orbital period, during which time the planet has been moving toward the observer (on Earth). Then, the star will appear to have been drifting slightly away from the observer during this time. But if one keeps observing for another half orbital period, then the star will drift back toward the observer, thereby demonstrating that its motion is being driven by an orbiting planet.

low-mass planets around single stars because of the limited detection sensitivity of the RV method and the limited amount of time during which RV measurements have been made. Hence, it seems safe to conclude that f_p is of order 0.1 or greater.

The third factor, n_E, is the number of Earth-like planets per planetary system. Here, again, we have enough information to say something meaningful. One can estimate this factor by comparing the average spacing of planets in our own inner Solar System with the width of the continuously habitable zone. (Recall that we already did this in chapter 10.) Our own system has four planets between the orbits of Mercury and Mars, 0.4–1.52 AU, so the average interplanetary spacing is ~0.35 AU. The 4.6-billion-year CHZ extends from at least 0.95–1.4 AU, and so its width is at least 0.45 AU. Hence, if planets are evenly spaced, or even if they are spaced logarithmically (which seems more likely), the odds are good that at least one of them within each system should be within the CHZ. So, optimistically, n_E could be equal to 1, or even slightly higher.

More realistically, this number could be too high for a number of reasons. First, only about 20 percent of nearby single stars are F-G-K stars similar to the Sun. Most of the others are dim, red M stars, which pose a variety of problems for habitable planets, as discussed in chapter 10. So, if we wish to be conservative, we should rule these stars out as parent stars for habitable planets. Furthermore, not all of the rocky planets within the CHZs of F-G-K stars are necessarily habitable. As we discussed in chapter 2, the delivery of volatiles (including water) to terrestrial planets is a stochastic process. This means that some of these planets may be too dry to support life and others may be too wet. Water planets do not necessarily pose a problem for life itself, but advanced life like humans requires dry land on which to evolve. Hence, several of Ward and Brownlee's "Rare Earth" arguments may well apply. Similarly, large obliquity variations on Earth-like planets that lack large moons could potentially pose a problem for advanced life, even though we argued in chapter 9 that this situation might occur less frequently than Ward and Brownlee suggested. Taking all of these factors into consideration, and with the goal of being conservative, we should probably reduce our estimate for n_e by at least a factor of 10, i.e., $n_e \cong 0.1$, with a great deal of included uncertainty.

Fortunately, we may soon be able to determine this third term in the Drake equation observationally, as we have done, or are in the process of doing, with the first and second terms. The first step should be taken by NASA's upcoming *Kepler* mission, which we discussed in chapter 12. By looking for planetary transits around ~100,000 not-so-nearby stars, *Kepler* should tell us what fraction of stars have planets orbiting within their habitable zones. The second part of the required information—whether or not these planets have volatiles—will have to wait for *TPF* or *Darwin*. The obliquity question is even tougher, but it could potentially be answered by making prolonged observations with an instrument similar to the *Life Finder* array discussed above. So, although we cannot evaluate n_e at present, we may be able to do so in the more distant future.

The fourth factor in the Drake equation, f_l, is the fraction of Earth-like planets on which life evolves. This one sounds tough but could actually prove to be relatively easy. As we have seen, *TPF* or *Darwin* may provide a preliminary answer to this question within the next 20–30 years by identifying planets whose atmospheres contain O_2 or O_3. And *Life Finder*, as its name implies, should eventually do much better, both in terms of number of planets searched and in numbers of biogenic gases observed. So, again, this term should eventually be evaluated with some degree of confidence.

What about f_i, the fraction of inhabited planets on which intelligent life evolves? This factor is almost impossible to evaluate observationally by itself. After all, humans evolved close to 2 million years ago, and for most of the time since then we would have been extremely difficult to detect remotely. Only within the last 50–100 years have we reached the stage where we are modifying our planetary environment in ways that might be observed from afar. So, from a practical standpoint, this factor should probably be lumped together with f_c, the fraction of intelligent species that develop a technical civilization capable of radio communication. We already know, in principle, how to detect such civilizations. The most straightforward method is to use radio telescopes like the giant Arecibo telescope described in chapter 1. The Arecibo telescope is sufficiently powerful that it could, in principle, communicate with a similar telescope located on the opposite side of the Milky Way galaxy, were the two instruments to be pointed at each other. The time lag would be long, however, of the order of 50,000 years, so one would need to be extremely patient if one wished to have a two-way conversation.

In spite of the daunting distances and time lags involved in interstellar communication, the SETI Institute[12] has been carrying out radio-wave searches for extraterrestrial intelligence from Arecibo and other telescopes for many years now. The goal at present is not to talk back and forth, but rather to listen and see if anyone else is out there. Optical laser searches are also being performed, notably by researchers at the Columbus Optical Search for Extraterrestrial Intelligence Observatory (COSETI)[13] near Columbus, Ohio. Optical SETI looks for pulses of monochromatic (laser) light that might be intentionally emitted as beacons by extraterrestrial civilizations.

Even though SETI searches have been carried out for several decades now, researchers have just begun to search effectively for intelligent life in our corner of the galaxy. That's because the distance to which we can *eavesdrop* at radio wavelengths is relatively small—far enough to reach only the very closest stars. As its name implies, "eavesdropping" means listening for radio signals that were not beamed at us intentionally. Such signals might include broadcasts from radio or TV stations, as well as the more powerful pulses emitted by military radar, for example, the U.S. DEW (Distant Early Warning) line[14] in northern Canada and Alaska. Eavesdropping mode, as such search strategies are termed, is considered by many researchers to have a much greater likelihood of success than does looking for beacons or beamed signals, because it does not rely on the idea that extraterrestrials would intentionally try to make contact with us. That said, eavesdropping has its own associated uncertainties. The intensity of Earth's broadcast radio signals could conceivably decrease significantly in the future as long-distance communications are increasingly carried over cable and satellite networks. So some SETI skeptics have suggested that a truly advanced civilization may actually emit little or no leakage energy that might be detected at radio wavelengths.

Not all researchers are this pessimistic, however. More ambitious radio searches for extraterrestrial life are planned, and some are presently being carried out. The prime instrument that is likely to be used in the near future is the Allen Telescope Array (ATA), currently under construction at Hat Creek Radio Observatory about 300 miles northeast of San Francisco (figure 15.4, in color section). The ATA project was initially started with a generous private donation from Paul Allen, the co-founder (with Bill Gates) of Microsoft Corporation. While Allen's Family Foundation continues to put money into the project, the ATA now attracts funding from more conventional sources as well, including University of California, Berkeley, the SETI Institute, and the National Science Foundation. After all, it is not just a tool for looking for extraterrestrial life—it is a powerful radio telescope that has lots of conventional astronomical applications. When it is eventually completed, the ATA will consist of 350 telescopes, each 6.1 m ($\sim$20 ft) in diameter, yielding a total collecting area of about one hectare (10,000 m^2). That's a factor of 7 less than Arecibo, which has a diameter of 305 m, so this by itself is not that spectacular. However, ATA can also be used easily at different frequencies without changing receivers and it has a much wider field of view than the Arecibo telescope, making it an ideal tool for performing all-sky surveys. The ATA is also explicitly designed to allow *piggybacking*, so that SETI searches can be carried out while the array is being used for other purposes. At the time of this writing, 42 telescopes are in operation. The others will be added as funding permits.

Even larger radio telescopes are now being planned. The one that is closest to being built is called the Square Kilometer Array (SKA).[16] As its name suggests, the SKA will have a combined collecting area of approximately one square kilometer, or 10^6 m^2—a factor of 100 greater than the ATA. Its spatial extent will also be much greater. About half of the projected 300 radio dishes will be concentrated within a core area of 5 square kilometers. But the other 150 dishes will be spread out over continental distances. For example, for one of the possible core sites near Perth in Western Australia, the array would include stations as far away as New Zealand. Site selection for the SKA is scheduled to be completed by 2011, and construction could begin within a year following that, provided that the required funding, more than $1.6 billion U.S., comes through from an international group of over 30 institutes in 15 countries. The SKA should, in principle, have SETI capabilities far beyond those of the Allen Telescope Array. Whether it will actually be used for this purpose remains to be seen. Unlike the ATA, its primary focus will be on other, more conventional astronomical topics, and the various groups who contributed money to its construction will no doubt have first say on how it is used.

Let's return now to the Drake equation. The product $f_i f_c$ may eventually be determined observationally using radio telescopes like the SKA and perhaps even larger ones that have yet to be imagined. But that prospect is a long way off yet, and it is perhaps more interesting to see what we can say about the Drake equation at the present time. So let's step back for a moment and put in the numbers that we think we know something about, along with some conjectures about those that are more uncertain.

Based on the discussion above, we have for the first three terms: $N_g f_p n_e \cong (4 \times 10^{11})\,(0.1)\,(0.1) = 4 \times 10^9$. For the next three factors, we shall abrogate all responsibility and simply use Carl Sagan's estimate from *Cosmos*[17]: $f_l f_i f_c \cong 1/300$. Sagan, of course, was an optimist, and so this number could be too high. We are not likely to improve significantly on it, though, in the space of a few short pages here. If we adopt this value and then round off our arithmetic, we obtain

$$N \cong 10^7 f_L. \tag{15.2}$$

Here, f_L is the fraction of a planet's lifetime during which it supports a technical civilization.

We have left f_L in the equation because it is the most uncertain factor of all—so much so that it dominates virtually every discussion of the Drake equation. This is where optimists and pessimists really sort themselves out. A pessimist would point out that we have been capable of radio communication for only a little over 100 years, and we have built big radio telescopes for less than half that time. Furthermore, our technical civilization could conceivably come to an abrupt end within the next few hundred years if we launch an all-out nuclear war or if we succumb to human-induced environmental problems such as global warming. I personally do not believe that global warming itself could destroy our technical civilization, but the possible atmospheric CO_2 increases are quite large—a factor of 6–8 higher than preindustrial values[18,19]—and the associated climate warming could trigger massive societal upheavals that might be more difficult to deal with than would climate change itself. Thus, the lifetime of a technical civilization could potentially be as low as 100–1000 years. Main-sequence stars like the Sun live for roughly 10 billion years, and so a pessimist might predict that f_l is between 10^{-8} and 10^{-7}. If so, then $N \leq 1$, and we could conceivably be the only communicating civilization in the entire galaxy. Michael Hart would be right, even if his climate models were wrong.

I myself am more of an optimist, having been influenced during my formative years by Carl Sagan. A true optimist would suggest that humans might survive as long as the Earth itself remains habitable, assuming that we resolve our present environmental issues in some amicable way. I myself used to think that the upper limit on the lifetime of our civilization was ~500 million years—the time until atmospheric CO_2 drops below the limit for C_3 photosynthesis (chapter 7). Having become aware of the possibility of constructing a solar shield, however (also discussed in chapter 7), I see no particular reason why humankind could not survive for the remainder of the Sun's main sequence lifetime, about 4–5 billion years. We would likely evolve biologically during that time, and so we might not qualify as "humans," but we could conceivably maintain technological capabilities for this entire time span. So, if we take 5 billion years as an upper bound for the lifetime of our (or any) technical civilization, then $f_l \cong 0.5$, and the number of communicating civilizations could be as high as 5×10^6, i.e., 5 million. If we further assume[20] that most of the stars in the galaxy are contained within a disk of radius 15 kpc and thickness 300 pc, the average distance between civilizations should be about 75 pc, or ~250 light-years. Interstellar communication with alien civilizations would still be slow, but the presence of such civilizations might at least be detectable.

This entire analysis presumes, of course, that interstellar travel is impossible. If that assumption breaks down, and if colonization of other planetary systems is feasible, then technological civilizations could conceivably be widespread throughout the galaxy. But then we would have to wonder, as did the famous Italian physicist Enrico Fermi,[21] why we have not yet been contacted by aliens. The writers of Star Trek have already figured this out, of course: it's the Prime Directive, a variant of the more general *zoo hypothesis*,[22] which postulates that alien races choose to leave us alone so as not to corrupt our developing technical civilization. Carl Sagan himself was fond of such optimistic speculation, and who knows—it might actually be right. We shall leave that for the movies and for the writers of science fiction, however. Such ideas are not currently testable and thus do not fall into the realm of science, which is what this book is about.

I would not wish to end this narrative on a purely speculative note. My own scientific interests are firmly grounded in what might be accomplished in the field of extrasolar planet exploration during the next 20–30 years—the time frame during which I myself hope to be around to see what happens. If things go well, that is, if we avoid large-scale warfare and resolve our ongoing economic and environmental problems, then we should be able to make great progress in the study of exoplanets. In particular, we should be able to answer the question of whether Earth-like planets exist around nearby stars, and we may also get some strong hints as to whether life exists elsewhere in the galaxy. These questions are of the same fundamental significance as Copernicus' proposition that the Earth orbits the Sun, rather than vice versa. If we can answer them, we will have done our part in extending the Copernican revolution. This is fascinating from a purely scientific standpoint, but it has philosophical implications as well. Each time we broaden our horizons, we learn a little more about where humans fit into this vast universe, and we are generally humbled by the experience. My own guess is that, just as we learned that the Sun is an ordinary star, we will find that Earth is an ordinary planet and that life itself is a commonplace phenomenon that exists on most, or all, such planets. But this is simply a guess. The challenge is to determine empirically whether these speculations are correct. Let us hope that we can muster the resources and the technical skills to answer these questions within the next generation.

Afterword to the Paperback Edition

Since the publication of the hardcover edition of this book two years ago, several important developments have occurred in the exoplanet field, some of which are discouraging but most of which are quite promising. First, the bad news (or part of it): NASA's *Space Interferometry Mission* (*SIM*), which was mentioned in chapter 11, has been canceled indefinitely. *SIM* was an expensive mission that would have looked for Earth-mass planets around 60 or more nearby stars. These are for the most part the same stars that might eventually be targeted by a *Terrestrial Planet Finder* mission, and so *SIM* could potentially have given us a target list for *TPF*. *SIM*, along with its smaller cousin, *SIM Lite*, failed to impress the 2010 Astronomy and Astrophysics Decadal Survey committee, however,[1] and so, after more than half a billion dollars' investment by NASA over almost 20 years, it has now disappeared for the foreseeable future. I will not say "disappeared for good," because such mission concepts have a history of being resurrected, albeit sometimes in different form. Space-based astrometry is still the only technique available for determining the absolute masses of nontransiting exoplanets. So, if we were to find an apparently Earth-like planet using some other technique, such as direct imaging (i.e., *TPF*), then we might build a *SIM*-like telescope anyway to figure out its mass.

Now the good news, which comes from two different sources. First, NASA's *Kepler* mission, mentioned in chapter 12, launched successfully

[1] Astro2010: The Astronomy and Astrophysics Decadal Survey. http://sites.nationalacademies.org/bpa/BPA_049810.

in March 2009, and has now collected almost 2.5 years of data. Recall that *Kepler* looks for exoplanets using the transit method, that is, by measuring the slight reduction in a star's light that occurs when a planet passes directly in front of it. A major public data release from *Kepler* was made in December 2011, just as this afterword was being written. The results are summarized in table A.1. *Kepler* has now identified some 2326 "planet candidates" orbiting a somewhat smaller number of stars. They are called "planet candidates" rather than "planets" because most of them have not been confirmed by other techniques, such as ground-based radial velocity (RV) measurements, and because other possible sources of interference, such as background eclipsing binary stars, remain to be ruled out for many of them.[2] The planet candidates are divided into five different size categories based on their radius, R_p, relative to Earth's radius, R_E. The majority of the planets found so far are classified as Neptune-sized, although this category, 2–6 R_E, is relatively broad and may include planets that are considerably different in structure and composition from Uranus and Neptune. By comparison, Neptune and Uranus each have a radius of about 4 R_E. Almost 900 planet candidates fall into the Earth and super-Earth size categories, with radii $< 1.25\ R_E$ and $< 2.0\ R_E$, respectively. 2.0 R_E is chosen as the upper limit for super-Earths because this is about the size above which a planet is expected to start capturing significant amounts of gas from the protostellar nebula during accretion, converting it from a rocky planet into a gas or ice giant. Because a planet's volume scales as its radius cubed, and its mass scales nearly linearly with its volume (for a given composition), a 2-R_E rocky planet would have a mass of roughly 8–10 Earth masses, putting it close to the theoretical upper limit for rocky planets discussed in chapter 2.

While all these planet discoveries by *Kepler* are interesting, the most exciting results pertain to planets within the habitable zones of their parent stars. *Kepler* has found several dozen of these already, although many of them are ice or gas giants, and thus are not likely to be habit-

[2] Most *Kepler* target stars are too dim to allow RV measurements to be made accurately; furthermore, many planet candidates are too small to provide a strong RV signal even around a relatively bright star.

TABLE A.1
"Planet Candidates" Identified by *Kepler* as of December 2011

Candidate label	*Candidate size* (R_E)	*Number of candidates*
Earth-size	$R_p < 1.25$	207
Super-Earths	$1.25 < R_p < 2.0$	680
Neptune-size	$2.0 < R_p < 6.0$	1181
Jupiter-size	$6.0 < R_p < 15$	203
Very large size	$15 < R_p < 22.4$	55
TOTAL		2326

able. The one that has received the greatest amount of attention so far is Kepler-22b. This is a 2.4-R_E object in a 290-day orbit around a late G star, some 600 light years from Earth. Although the planet's radius puts it in the Neptune size category, it is significantly smaller than Neptune and could conceivably be a rocky planet. The amount of starlight hitting this planet is about 10 percent higher than that received by Earth, putting it near the inner edge of its star's habitable zone, but most likely within it.

The real goal of *Kepler*, of course, is not to identify specific Earth-like planets, but rather to calculate the statistics of Earth-like planets around other stars. Planets like Kepler-22b are so far away that they would be difficult to observe directly even if we had a *TPF* mission of the sort described in chapter 13. Transit spectroscopy could in principle be performed (because, by definition, all planets identified by *Kepler* must transit), but this, too, is difficult, because the *Kepler* target stars are relatively dim, so not many photons are available. This should not be a big worry, however. What we are really hoping to get from *Kepler* is a measurement of the parameter η_{Earth}—the fraction of Sun-like stars that have at least one rocky planet within their habitable zone. The *Kepler* team themselves have been very cautious about this. Their dataset is only just beginning to become long enough to measure η_{Earth} directly, and so the *Kepler* scientists prefer to hold off on their estimates and wait for more data to roll in.[3] Other astronomers, however, are more daring. Two different estimates of η_{Earth} have already been published, both of them based on the earlier, February 2011, data release. This dataset contained only the first 4 months of *Kepler* data and so was arguably complete only for orbital periods ≤ 42 days. The authors of the first paper used the entire dataset (which included orbital periods up to 100 days) and derived an estimate of (2±1) percent for η_{Earth}.[4] The author of the second paper used only the data for periods < 42 days and calculated that η_{Earth} was (34±14) percent.[5] This, of course, involves a bold extrapolation, as none of these short-period planets actually orbit within their star's habitable zone. As the two published estimates for η_{Earth} differ by more than an order of magnitude, one is tempted to conclude that the *Kepler* team is right in withholding judgment on this issue. However, given that these earlier data were known to be incomplete for the longer orbital periods, it seems likely that the more optimistic 34 percent estimate is closer to the truth. And that would indeed be good news if it holds up, because it means that we might be able to get away with building a smaller, 4-m-class *TPF* telescope and still have a good chance of being able to image a planet like Earth.

The other good news on the exoplanet front comes from ground-based radial velocity searches. Both U.S.-based research groups[6] and European groups[7] have released updated RV survey results within the last year. The European group, headed by Michel Mayor at the Geneva

[3] Think about how this analysis works: Normally, three transits are required to identify a planet. The first one tells you that something has happened; the second tells you that it was not some random fluctuation; and the third, if it occurs at the same interval, tells you that you really might be observing a planet. So, if one were observing our own Solar System with *Kepler*, one would need at least three years of data before one could determine that planet Earth existed. At this time, *Kepler* has been operating for only 2.5 years, so it should not yet have detected Earth.

[4] Catanzarite J and Shao M. 2011. The occurrence rate of Earth analog planets orbiting Sunlike stars. *Astrophys. J.* 738. DOI: 10.1088/0004-637X/738/2/151.

[5] Traub WA. 2011. Terrestrial habitable zone frequency from Kepler. http://arxiv.org/abs/1109.4682v1, accepted by *Astrophys. J.*

[6] Howard AW, Marcy GW, Johnson JA, Fischer DA, Wright JT, et al. 2010. The occurrence and mass distribution of close-in super-Earths, Neptunes, and Jupiters. *Science* 330: 653–655.

[7] Mayor M, Marmier M, Lovis C, et al. 2011. The HARPS search for southern extra-solar planets XXXIV. Occurrence, mass distribution and orbital properties of super-Earths and Neptune-mass planets. http://arxiv.org/abs/1109.2497v1.

Observatory, currently has the greatest sensitivity to small planets, so I shall cite their results here. They find, remarkably, that more than 50 percent of solar-type (FGK) stars have a planet of any mass with an orbital period < 100 days. This figure increases to almost 70 percent for the brighter F and G stars. (The smaller K stars appear to have somewhat fewer planets.) Their data mostly include planets with masses > 10 Earth masses, but, when corrected for the known bias against detecting small planets, this implies that the frequency of Earths and super-Earths is probably quite high. Thus, these RV results are entirely consistent with the new results from *Kepler*. Both observational techniques suggest that Earth-sized planets orbiting within stellar habitable zones may be relatively common in our part of the galaxy.

Finally, to return briefly to space-science politics, I must admit to being discouraged about the prospects for flying *TPF* within the next 20 years. Even if η_{Earth} is relatively high, and so we can build a relatively small, 4-m *TPF*, such a space telescope is still a multibillion-dollar mission. Budgets are tight at the moment, both for NASA and for the United States as a whole, partly as a result of the economic downturn and partly because the politicians in Washington cannot agree on how to balance our federal budget. The imbalance there is of the order of several trillions of dollars over the next decade, and the problems that need to be addressed—Social Security, Medicare, Medicaid, taxes—are much broader than the issues that are the focus of this book. One might think that space science could still progress normally in this budget climate, as the amounts of money that are needed are small compared with the money needed for these other megaprograms, but that perception is almost certainly wrong. In reality, all discretionary federal spending—which includes research agencies like NASA, NSF, and DOE—is in danger of undergoing severe cuts over the next few years. To add to these larger budget problems, NASA's astrophysics program has its own major issue: its *James Webb Space Telescope* (*JWST*), which is currently under development, is over cost and behind schedule. *JWST* was originally projected to cost ~$2.5 billion and to launch by 2014. Its estimated cost is now over $8 billion, and the launch date has been moved back to 2018. Even that schedule may prove impossible to meet if NASA fails to come up with the extra money it needs to complete the telescope. Before all this happened, many of us had hoped that *TPF* might be picked as the top flagship-class astronomy mission by the 2020 Decadal Survey and that it could be constructed and launched before 2030. Indeed, the 2010 Decadal Survey was strongly supportive of this goal. But for this to happen, NASA would need to invest several hundred million dollars in the current decade to bring at least one of the technologies discussed in chapter 13 up to snuff. The prospects for getting that technology-development money in time for it to do any good seem slim, so getting started on *TPF* early in the next decade now looks less and less likely. If we have to wait until 2030 to get started, then the mission will not launch until close to 2040, and this author—who was born in 1953—may or may not be around to see it happen. So, if any of you who are reading this book have ideas on how to speed things up, please give me a ring! There is no reason to wait another 30 years for this mission to proceed. We are on the verge right now of being able to determine whether there is life on planets around other stars, and we should take the initiative to do so.

Notes

CHAPTER 1

1. Aristotle, in Guthrie WKC trans., 1953, of Aristotle's *On the Heavens*. Cambridge, UK: Cambridge University Press.

2. Epicurus, in Bailey C, ed. and trans. 1926. *Epicurus: The Extant Remains*. Oxford, UK: Clarendon Press.

3. From: http://commons.wikimedia.org/wiki/File:GiordanoBrunoStatueCampoDeFiori.jpg

4. Shapley H, ed. 1953. *Climatic Change : Evidence, Causes, and Effects*. Cambridge, MA: Harvard University Press, 318 pp.

5. Shklovskii IS, Sagan C. 1966. *Intelligent Life in the Universe*. San Francisco: Holden-Day, 509 pp.

6. From: http://www.naic.edu/public/about/photos/hires/aoviews.html.

7. Sagan C, Agel J. 1973. *Cosmic Connection: An Extraterrestrial Perspective*. Garden City, NY: Anchor Press, 301 pp.

8. Ward PD, Brownlee D. 2000. *Rare Earth: Why Complex Life Is Uncommon in the Universe*. New York: Copernicus.

9. http://www.paleothea.com/Majors.html

10. Lovelock JE. 1979. *Gaia: A New Look at Life on Earth*. Oxford, UK: Oxford University Press.

11. Lovelock JE 1988. *The Ages of Gaia*. New York: WW Norton.

12. Lovelock JE 1991. *Gaia: The Practical Science of Planetary Medicine*. London: Gaia Books.

13. Lovelock JE 2006. *The Revenge of Gaia: Earth's Climate in Crisis and the Fate of Humanity*. New York: Basic Books, 176 pp.

CHAPTER 2

1. Lewis JS, Prinn RG. 1984. *Planets and Their Atmospheres: Origin and Evolution*. Orlando, FL: Academic Press.

2. Ringwood AE. 1966. Chemical evolution of the terrestrial planets. *Geochim. Cosmochim. Acta* 30: 41–104.

3. http://meteorites.asu.edu/met-info/. Photo by D. Ball, Arizona State University.

4. Prinn RG et al. 1989. Solar nebula chemistry: origin of planetary, satellite and cometary volatiles. In: Atreya SK, et al., eds. *Origin and Evolution of Planetary and Satellite Atmospheres.* Tucson, AZ: University of Arizona Press, pp. 78–136.

5. Safronov VS. 1972. Evolution of the Protoplanetary Cloud and Formation of the Earth and Planets. NASA report.

6. Wetherill GW. 1985. Giant impacts during the growth of the terrestrial planets. *Science* 228: 877–79.

7. Wetherill GW, et al. 1986. Accumulation of the terrestrial planets and implications concerning lunar origin. In: *Origin of the Moon.* Houston, TX: Lunar and Planetary Inst., pp. 519–50.

8. Wetherill GW. 1991. Occurrence of Earth-like bodies in planetary systems. *Science* 253: 535–38.

9. http://www.harmsy.freeuk.com/oimages/oort_cloud.jpg. Reproduced by permission of the illustrator, Donald Yeomans, of NASA's Jet Propulsion Laboratory.

10. Chyba CF. 1987. The cometary contribution to the oceans of primitive Earth. *Nature* 330: 632–35.

11. Owen T, Bar-Nun A, Kleinfeld I. 1992. Possible cometary origin of heavy noble gases in the atmospheres of Venus, Earth, and Mars. *Nature* 358: 43–46.

12. Oro J. 1961. Comets and the formation of biochemical compounds on the primitive Earth. *Nature* 190: 389–90.

13. Chyba CF, Thomas JJ, Brookshaw L, Sagan C. 1990. Cometary delivery of organic molecules to the early Earth. *Science* 249: 366–73.

14. Robert F. 2001. *Science* 293:1056–8.

15. Weissman PR, Carusi A, Valsecchi GB. 1985. Dynamical evolution of the Oort cloud. In: *Dynamics of Comets: Their Origin and Evolution.* Dordrecht: D Reidel, pp. 87–96.

16. Raymond SN, Quinn T, Lunine JI, 2006. High-resolution simulations of the final assembly of Earth-like planets, I: terrestrial accretion and dynamics. *Icarus* 183: 265–82.

17. Raymond SN, Quinn T, Lunine JI, 2004. Making other earths: dynamical simulations of terrestrial planet formation and water delivery. *Icarus* 168: 1–17.

Chapter 3

1. Gough, DO. 1981. Solar interior structure and luminosity variations. *Solar Phys.* 74: 21–34.

2. Chaisson E, McMillan S. 2008. *Astronomy Today,* 6th ed. Upper Saddle River, NJ: Pearson/Addison-Wesley.

3. Willson LA, et al. 1987. Mass loss on the main sequence. *Comments Astrophys.* 12: 17–34.

4. Boothroyd AI, et al. 1991. Our Sun, II: early mass loss of 0.1 Mo and the case of the missing lithium. *Astrophys. J.* 377: 318–29.

5. Wood BE, et al. 2002. Measured mass loss rates of solar-like stars as a function of age and activity. *Astrophys. J.* 574: 412–25.

6. Wood BE, et al. 2005. New mass-loss measurements from astrospheric Ly alpha absorption. *Astrophys. J.* 628: L143–L146.

7. Sagan C, Mullen G. 1972. Earth and Mars: evolution of atmospheres and surface temperatures. *Science* 177: 52–56.

8. Kasting JF, et al. 1988. How climate evolved on the terrestrial planets. *Sci. Am.* 256: 90–97.

9. Valley JW, Peck WH, King EM, Wilde SA. 2002. A cool early Earth. *Geology* 30: 351–54.

10. Walker JCG, Klein C, Schidlowski M, Schopf JW, Stevenson DJ, et al. 1983. Environmental evolution of the Archean–Early Proterozoic Earth. In: *Earth's Earliest Biosphere: Its Origin and Evolution.* Princeton, NJ: Princeton University Press, pp. 260–90.

11. Svensmark H. 2007. Cosmoclimatology: a new theory emerges. *Astron. Geophys.* 48: 18–24.

12. Kuhn WR, Atreya SK. 1979. Ammonia photolysis and the greenhouse effect in the primordial atmosphere of the Earth. *Icarus* 37: 207–13.

13. Sagan C, Chyba C. 1997. The early faint Sun paradox: organic shielding of ultraviolet-labile greenhouse gases. *Science* 276: 1217–21.

14. Holland HD. 1978. *The Chemistry of the Atmosphere and Oceans.* New York: Wiley.

15. Lovelock JE. 1979. *Gaia: A New Look at Life on Earth.* Oxford, UK: Oxford University Press.

16. Walker JCG, Kasting JF. 1992. Effects of fuel and forest conservation on predicted levels of atmospheric carbon dioxide. *Global Planet. Change* 97: 151–89.

17. Archer D. 2005. Fate of fossil fuel CO_2 in geologic time. *J. Geophys. Res. Oceans* 110, C9, C09505.

18. Hyde WT, Crowley TJ, Baum SK, Peltier WR. 2000. Neoproterozoic 'snowball Earth' simulations with a coupled climate/ice-sheet model. *Nature* 405: 425–29.

19. Pierrehumbert RT. 2004. High levels of atmospheric carbon dioxide necessary for the termination of global glaciation. *Nature* 429: 646–49.

20. Caldeira K, Kasting JF. 1992. Susceptibility of the early Earth to irreversible glaciation caused by carbon dioxide clouds. *Nature* 359: 226–28.

21. Walker JCG, Hays PB, Kasting JF. 1981. A negative feedback mechanism for the long-term stabilization of Earth's surface temperature. *J. Geophys. Res.* 86: 9776–82.

22. Lovelock J. 1988. *The Ages of Gaia.* New York: WW Norton.

CHAPTER 4

1. Berner RA. 2006. GEOCARBSULF: a combined model for Phanerozoic atmospheric O_2 and CO_2. *Geochim. Cosmochim. Acta* 70: 5653–64.

2. Raymo ME, Ruddiman WF. 1993. Tectonic forcing of Late Cenozoic climate. *Nature* 361: 117–22.

3. Pavlov AA, Toon OB, Pavlov AK, Bally J, Pollard D. 2005. Passing through a giant molecular cloud: "Snowball" glaciations produced by interstellar dust. *Geophys. Res. Lett.* 32: Art. No. L03705, Feb. 4.

4. Thomas BC, Melott AL, Jackman CH, Laird CM, Medvedev MV, Stolarski RS, Gehrels N, Cannizzo JK, Hogan DP, Ejzak LM. 2005. Gamma-ray bursts and the earth: exploration of atmospheric, biological, climatic, and biogeochemical effects. *Astrophys. J.* 634: 509–33.

5. Trotter JA, Williams IS, Barnes CR, Lecuyer C, Nicoll RS. 2008. Did cooling oceans trigger Ordovician biodiversification? Evidence from conodont thermometry. *Science* 321: 550-54.

6. Young GM, von Brunn V, Gold DJC, Minter WEL. 1998. Earth's oldest reported glaciation; physical and chemical evidence from the Archean Mozaan Group (~2.9 Ga) of South Africa. *J. Geol.* 106: 523–38.

7. Lowe DR, Tice MM. 2007. Tectonic controls on atmospheric, climatic, and biological evolution 3.5–2.4 Ga. *Precambrian Res.* 158: 177–97.

8. Evans DA, et al. 1997. Low-latitude glaciation in the Proterozoic era. *Nature* 386: 262–66.

9. Cloud PE. 1972. A working model of the primitive Earth. *Am. J. Sci.* 272: 537–48.

10. Holland HD. 1994. Early Proterozoic atmospheric change, *Early Life on Earth*. New York: Columbia University Press. Redrafted by Y. Watanabe.

11. Courtesy of JW Schopf, University of California, Los Angeles.

12. Farquhar J, et al. 2000. Atmospheric influence of Earth's earliest sulfur cycle. *Science* 289: 756–58.

13. Widdel F, Schnell S, Heising S, Ehrenreich A, Assmus B, Schink B. 1993. Ferrous iron oxidation by anoxygenic phototrophic bacteria. *Nature* 362: 834–36.

14. Canfield D, et al. 2006. Early anaerobic metabolisms. *Philos. Trans. R. Soc. B* 361: 1819–36.

15. http://microbes.arc.nasa.gov/gallery/lightms.html

16. Brocks JJ, et al. 1999. Archean molecular fossils and the early rise of eukaryotes. *Science* 285: 1033–36.

17. Summons JR, et al. 1999. Methylhopanoids as biomarkers for cyanobacterial oxygenic photosynthesis. *Nature* 400: 554–57.

18. Rashby SE, Sessions AL, Summons RE, Newman DK. 2007. Biosynthesis of 2-

methylbacteriohopanepolyols by an anoxygenic phototroph. *Pro. Nat. Acad. Sci USA* 104: 15099–104.

19. Kopp RE, Kirshvink JL, Hilburn I., Nash CZ. 2005. The Paleoproterozoic snowball Earth: climate disaster triggered by the evolution of photosynthesis. *Proc. Natl. Acad. Sci. USA* 102: 11131–36.

20. Kelley DS, et al. 2005. A serpentinite-hosted ecosystem: the Lost City hydrothermal vent field. *Science* 307: 1428–34.

21. Keppler F, et al. 2006. Methane emissions from terrestrial plants under aerobic conditions. *Nature* 439: 187–91.

22. Ueno Y, Yamada K, Yoshida M, Maruyama S, Isozaki U. 2006. Evidence from fluid inclusions for microbial mathanogenesis in the early Archean era. *Nature* 440: 516–19.

23. Fu Q, Lollar BS, Horita J, Lacrampe-Couloume G, Seyfried WE. 2007. Abiotic formation of hydrocarbons under hydrothermal conditions: constraints from chemical and isotope data. *Geochim. Cosmochim. Acta* 71: 1982–98.

24. Woese CR, Fox GE. 1977. Phylogenetic structure of the prokaryotic domain: the primary kingdoms. *Proc. Natl. Acad. Sci. USA* 74: 5088–90.

25. Courtesy of Norman Pace, University of Colorado.

26. Knoll AH. 2003. *Life on a Young Planet*. Princeton, NJ: Princeton University Press, pp. 122–26.

27. Schimel D, et al. 1996. Radiative forcing of climate change. In: Houghton JT, et al., eds. *Climate Change 1995: The Science of Climate Change*. Cambridge, UK: Cambridge University Press, pp. 65–131.

28. Pavlov AA, et al. 2001. UV-shielding of NH_3 and O_2 by organic hazes in the Archean atmosphere. *J. Geophys. Res.* 106: 2326–87.

29. Kharecha P, et al. 2005. A coupled atmosphere-ecosystem model of the early Archean Earth. *Geobiology* 3: 53–76.

30. Haqq-Misra JD, et al. 2008. A revised, hazy methane greenhouse for the early Earth. *Astrobiology* 8: 1127–37.

31. McKay CP, et al. 1991. The greenhouse and antigreenhouse effects on Titan. *Science* 253: 1118–21.

32. Trainer MG, Pavlov AA, DeWitt HL, Jimenez JL, McKay CP, Toon OB, Tolbert MA. 2006. Organic haze on Titan and the early Earth. *Proc. Nat. Acad. Sci. USA* 103: 18035–42.

33. Young GM. 1991. *Stratigraphy, Sedimentology, and Tectonic Setting of the Huronian Supergroup*, Field Trip B5 guidebook, joint meeting Geol. Assoc. Canada, Mineral Assoc. Canada, Soc. Econ. Geol., Toronto.

34. Roscoe SM. 1973. The Huronian Supergroup: a Paleophebian succession showing evidence of atmospheric evolution. *Geol. Soc. Can. Spec. Pap.* 12: 31–48.

35. Rasmussen B, Fletcher IR, Brocks JJ, Kilburn MR. 2008. Reassessing the first appearance of eukaryotes and cyanobacteria. *Nature* 455: 1101–4.

Chapter 5

1. Imbrie J, Imbrie KP. 1979. *Ice Ages: Solving the Mystery*, Hillside, NJ: Enslow, 224 pp.

2. Redrawn following ref. 4, figure 14.5.

3. Petit JR, Jouzel J, Raynaud D, Barkov NI, Barnola JM, et al. 1999. Climate and atmospheric history of the past 420,000 years from the Vostok ice core, Antarctica. *Nature* 399: 429–36.

4. Kump LR, Kasting JF, Crane RG. 2004. *The Earth System*, 2nd ed. Upper Saddle River, NJ: Pearson Education, chapter 14, pp. 270–88.

5. Eriksson E. 1968. Air–ocean–icecap interactions in relation to climatic fluctuations and glaciation cycles. *Meteorol. Monogr.* 8: 68–92.

6. Budyko MI. 1969. The effect of solar radiation variations on the climate of the Earth. *Tellus* 21: 611–19.

7. Sellers WD. 1969. A global climatic model based on the energy balance of the Earth–atmosphere system. *J. Appl. Meteorol.* 8: 392–400.

8. Holland HD. 1978. *The Chemistry of the Atmosphere and Oceans*. New York: Wiley.

9. http://www.snowballearth.org/week7.html

10. Caldeira K, Kasting JF. 1992. Susceptibility of the early Earth to irreversible glaciation caused by carbon dioxide clouds. *Nature* 359: 226–28.

11. Harland WB, Rudwick MJS. 1964. *Sci. Am.* 211 (2): 28–36.

12. Hoffman PF, Kaufman AJ, Halverson GP, Schrag DP. 1998. A Neoproterozoic Snowball Earth. *Science* 281: 1342–46.

13. Hyde WT, Crowley TJ, Baum SK, Peltier WR. 2000. Neoproterozoic 'snowball Earth' simulations with a coupled climate/ice-sheet model. *Nature* 405: 425–29.

14. Courtesy of Adam Maloof, Princeton University.

15. Williams GE. 1975. Late Precambrian glacial climate and the Earth's obliquity. *Geol. Mag.* 112: 441–65.

16. Bills BG. 1994. Obliquity–oblateness feedback: are climatically sensitive values of the obliquity dynamically unstable? *Geophys. Res. Lett.* 21: 177–80.

17. Williams DM, Kasting JF, Frakes LA. 1988. Low-latitude glaciation and rapid changes in the Earth's obliquity explained by obliquity–oblateness feedback. *Nature* 396: 453–55.

18. Levrard B, Laskar J. Climate friction and the Earth's obliquity. *Geophys. J. Int.* 154 (3): 970–90.

19. Kirshvink JL, Schopf JW, Klein C. 1992. In: *A Paleogeographic Model for Vendian and Cambrian Time*. Cambridge, UK: Cambridge University Press, pp. 569–81.

20. Hoffman PF, Schrag DP. 2002. The Snowball Earth hypothesis: testing the limits of global change. *Terra Nova* 14: 129–55.

21. McKay CP. 2000. Thickness of tropical ice and photosynthesis on a snowball Earth. *Geophys. Res. Lett.* 27: 2153–56.

22. Goodman JC, Pierrehumbert RT. 2003. Glacial flow of floating marine ice in 'Snowball Earth'. *J. Geophys. Res.* 108: 3308.

23. Pollard D, Kasting JF. 2005. Snowball Earth: a thin-ice model with flowing sea glaciers. *J. Geophys. Res.* 110: C7, C07010.

Chapter 6

1. Meadows VS, Crisp D. 1996. Ground-based near-infrared observations of the Venus nightside: the thermal structure and water abundance near the surface. *J. Geophys. Res.* 101: 4595–622.

2. Donahue TM, Hoffman JH, Hodges RR Jr. 1982. Venus was wet: a measurement of the ratio of deuterium to hydrogen. *Science* 216: 630–33.

3. McElroy MB, Prather MJ, Rodriguez JM. 1982. Escape of hydrogen from Venus. *Science* 215: 1614–15.

4. Debergh C, Bezard B, Owen T, Crisp D, Maillard JP, Lutz BL. 1991. Deuterium on Venus: observations from Earth. *Science* 251: 547–49.

5. Rasool SI, DeBergh C. 1970. The runaway greenhouse and the accumulation of CO_2 in the Venus atmosphere. *Nature* 226: 1037–39.

6. Gurwell M. 1995. Evolution of deuterium on Venus. *Nature* 378: 22–23.

7. Grinspoon DH. 1993. Implications of the high D/H ratio for the sources of water in Venus' atmosphere. *Nature* 363: 428–31.

8. Robert F. 2001. Isotope geochemistry—the origin of water on Earth. *Science* 293: 1056–58.

9. Goody RM, Walker JCG. 1972. *Atmospheres*. Englewood Cliffs, NJ: Prentice Hall.

10. Lange MA, Ahrens TJ. 1982. The evolution of an impact generated atmosphere. *Icarus* 51: 96–120.

11. Matsui T, Abe Y. 1986. Evolution of an impact-induced atmosphere and magma ocean on the accreting Earth. *Nature* 319: 303–5.

12. Matsui T, Abe Y. 1986. Impact-induced atmospheres and oceans on Earth and Venus. *Nature* 322: 526–28.

13. Zahnle KJ, Kasting JF, Pollack JB. 1988. Evolution of a steam atmosphere during Earth's accretion. *Icarus* 74: 62–97.

14. Pollack JB. 1971. A nongrey calculation of the runaway greenhouse: implications for Venus' past and present. *Icarus* 14: 295–306.

15. Ingersoll AP. 1969. The runaway greenhouse: a history of water on Venus. *J. Atmos. Sci.* 26: 1191–98.

16. Kasting JF. 1988. Runaway and moist greenhouse atmospheres and the evolution of Earth and Venus. *Icarus* 74: 472–94.

17. Hunten DM. 1973. The escape of light gases from planetary atmospheres. *J. Atmos. Sci.* 30: 1481–94.

18. Kasting JF, Pollack JB. 1983. Loss of water from Venus, I: hydrodynamic escape of hydrogen. *Icarus* 53: 479–508.

19. Bullock MA, Grinspoon DH. 1996. The stability of climate on Venus. *J. Geophys. Res.* 101: 7521–29.

20. Hashimoto GL, Abe Y. 2005. Climate control on Venus: comparison of the carbonate and pyrite models. *Planet. Space Sci.* 53: 839–48.

21. Bullock MA, Grinspoon DH. 2001. The recent evolution of climate on venus. *Icarus* 150: 19–37.

22. Lewis, JSPRG. 1984. *Planets and Their Atmospheres: Origin and Evolution.* Orlando, FL: Academic Press, pp. 143ff.

Chapter 7

1. Walker JCG, Kasting JF. 1992. Effects of fuel and forest conservation on predicted levels of atmospheric carbon dioxide. *Global Planet. Change* 97: 151–89.

2. Archer D. 2005. Fate of fossil fuel CO_2 in geologic time. *J. Geophys. Res. Oceans* 110: C9, C09505.

3. Bala G, Caldeira K, Mirin A, Wickett M, Delire C. 2005. Multicentury changes to the global climate and carbon cycle: results from a coupled climate and carbon cycle model. *J. Climate* 18: 4531–44.

4. Kasting JF, Ackerman TP. 1986. Climatic consequences of very high CO_2 levels in the Earth's early atmosphere. *Science* 234: 1383–85.

5. Schwartzman DW, Shore SN, Volk T, McMenamin M. 1994. Self-organization of the Earth's biosphere: geochemical or geophysiological. *Origins Life Evol. Biosphere* 24: 435–50.

6. Ward PD, Brownlee D. 2000. *Rare Earth: Why Complex Life Is Uncommon in the Universe.* New York: Copernicus.

7. Lowe DR, Tice MM. 2007. Tectonic controls on atmospheric, climatic, and biological evolution 3.5–2.4 Ga. *Precambrian Res.* 158: 177–97.

8. Baross JA, Schrenk MO, Huber JA. 2007. Limits of carbon life on Earth and elsewhere. In: Sullivan WTI, Baross JA, eds. *Planets and Life: The Emerging Science of Astrobiology*, Cambridge, UK: Cambridge University Press, pp. 275–91.

9. Caldeira K, Kasting JF. 1992. The life span of the biosphere revisited. *Nature* 360: 721–23.

10. Lovelock JE, Whitfield M. 1982. Life span of the biosphere. *Nature* 296: 561–63.

11. Kasting JF, Pollack JB. 1983. Loss of water from Venus, I: hydrodynamic escape of hydrogen. *Icarus* 53: 479–508.

12. Angel R. 2006. Feasibility of cooling the Earth with a cloud of small spacecraft near the inner Lagrange point (L1). *Proc. Nat. Acad. Sci USA* 103: 17184–89.

13. Early JT. 1989. The space based solar shield to offset greenhouse effect. *J. Br. Interplanet. Soc.* 42: 567–69.

14. http://en.wikipedia.org/wiki/Lagrange_Point_Colonization

Chapter 8

1. Kahn R. 1985. The evolution of CO_2 on Mars. *Icarus* 62: 175–90.

2. Malin MC, Edgett KS, Posiolova LV, McColley SM, Noe Dobrea EZ. 2006. Present-day impact cratering rate and contemporary gully activity on Mars. *Science* 314: 1573–77.

3. Ward WR. 1974. Climatic variations on Mars, 1: Astronomical theory of insolation. *J. Geophys. Res.* 79: 3375–86.

4. Ward WR, Rudy DJ. 1991. Resonant obliquity of Mars. *Icarus* 94: 160–64.

5. Laskar J, Robutel P. 1993. The chaotic obliquity of the planets. *Nature* 361: 608–14.

6. Touma J, Wisdom J. 1993. The chaotic obliquity of Mars. *Science* 259: 1294–97.

7. Mumma MJ, Villanueva GL, Novak RE, Hewagama T, Bonev BP, et al. 2009. Strong release of methane on Mars in northern summer 2003. *Science* 323: 1041–45.

8. Krasnopolsky VA, Maillard JP, Owen TC. 2004. Detection of methane in the martian atmosphere: evidence for life? *Icarus* 172: 537–47.

9. Formisano V, Atreya S, Encrenaz T, Ignatiev N, Giuranna M. 2004. Detection of methane in the atmosphere of Mars. *Science* 306: 1758–61.

10. Malde HE. 1968. The catastrophic late Pleistocene Bonneville flood in the Snake River plain, Idaho. U.S. Geol. Survey Prof. paper 596.

11. Malin MC, Carr MH. 1999. Groundwater formation of martian valleys. *Nature* 397: 589–91.

12. Segura TL, Toon OB, Colaprete A, Zahnle K. 2002. Environmental effects of large impacts on Mars. *Science* 298: 1977–80.

13. Hartmann WK, Ryder G, Dones L, Grinspoon D, Canup RM, Righter K. 2000. In: *The Time-dependent Intense Bombardment of the Primordial Earth–Moon System*. Tucson: University of Arizona Press.

14. Stoffler D, Ryder G. 2001. Stratigraphy and isotope ages of lunar geologic units: chronological standard for the inner solar system. *Space Sci. Rev.* 96: 9–54.

15. Baldwin RB. 2006. Was there ever a terminal lunar cataclysm? With lunar viscosity arguments. *Icarus* 184: 308–18.

16. Gomes R, Levison HF, Tsiganis K, Morbidelli A. 2005. Origin of the cataclysmic Late Heavy Bombardment period of the terrestrial planets. *Nature* 435: 466–69.

17. Tsiganis K, Gomes R, Morbidelli A, Levison HF. 2005. Origin of the orbital architecture of the giant planets of the Solar System. *Nature* 435: 459–61.

18. Wallace D, Sagan C. 1979. Evaporation of ice in planetary atmospheres: ice-covered rivers on Mars. *Icarus* 39: 385–400.

19. Hoefen TM, Clark RN, Bandfield JL, Smith MD, Pearl JC, Christensen PR. 2003. Discovery of olivine in the Nili Fossae region of Mars. *Science* 302: 627–30.

20. Christensen PR, Bandfield JL, Bell JF, Gorelick N, Hamilton VE, et al. 2003. Morphology and composition of the surface of Mars: Mars Odyssey THEMIS results. *Science* 300: 2056–61.

21. Squyres SW, et al. 2004. The Opportunity rover's Athena science investigation at Meridiani Planum, Mars. *Science* 306: 1698–703.

22. Clifford SM. 1991. The role of thermal vapor diffusion in the subsurface hydrologic evolution of Mars. *Geophys. Res. Lett.* 18: 2055–58.

23. Melosh HJ, Vickery AM. 1989. Impact erosion of the primordial atmosphere of Mars. *Nature* 338: 487–89.

24. Pollack JB, Kasting JF, Richardson SM, Poliakoff K. 1987. The case for a wet, warm climate on early Mars. *Icarus* 71: 203–24.

25. Kasting JF. 1991. CO_2 condensation and the climate of early Mars. *Icarus* 94: 1–13.

26. Fogg MJ. 1995. *Terraforming: Engineering Planetary Environments*. Warrendale, PA: Society of Automotive Engineers, 544 pp.

27. McKay CP, Toon OB, Kasting JF. 1991. Making Mars habitable. *Nature* 352: 489–96.

28. Forget F, Pierrehumbert RT. 1997. Warming early Mars with carbon dioxide clouds that scatter infrared radiation. *Science* 278: 1273–76.

29. Halevy I, Zuber MT, Schrag DP. 2007. A sulfur dioxide climate feedback on early Mars. *Science* 318: 1903–7.

30. Postawko SE, Kuhn WR. 1986. Effect of the greenhouse gases (CO_2, H_2O, SO_2) on Martian paleoclimate. *J. Geophys. Res.* 91: D431–D438.

31. Kasting JF, Zahnle KJ, Pinto JP, Young AT. 1989. Sulfur, ultraviolet radiation, and the early evolution of life. *Origins Life* 19: 95–108.

32. Bandfield JL, Glotch TD, Christensen PR. 2003. Spectroscopic identification of carbonate minerals in the martian dust. *Science* 301: 1084–87.

33. Fairen AG, Fernandez-Remolar D, Dohm JM, Baker VR, Amils R. 2004. Inhibition of carbonate synthesis in acidic oceans on early Mars. *Nature* 431: 423–26.

34. Ming DW, Mittlefehldt DW, Morris RV, Golden DC, Gellert R, et al. 2006. Geochemical and mineralogical indicators for aqueous processes in the Columbia Hills of Gusev crater, Mars. *J. Geophys. Res. Planets* 111: E2, E02512.

35. Hurowitz JA, McLennan SM, Tosca NJ, Arvidson RE, Michalski JR, et al. 2006. In situ and experimental evidence for acidic weathering of rocks and soils on Mars. *J. Geophys. Res. Planets* 111: E2, E02519.

36. Robinson, KS. 1992–1996. *The Mars Trilogy: Red Mars, Green Mars, Blue Mars.* Tega Cay, SC: Spectra.

Chapter 9

1. Kasting JF. 2001. Essay review: Peter Ward and Donald Brownlee's "Rare Earth." *Perspect. Biol. Med.* 44: 117–31.

2. Labrosse S, Hernlund JW, Coltice N. 2007. A crystallizing dense magma ocean at the base of the Earth/'s mantle. *Nature* 450: 866–69.

3. http://www.nrc.gov/reading-rm/basic-ref/glossary/exposure.html

4. Jakosky BM, Pepin RO, Johnson RE, Fox JL. 1994. Mars atmospheric loss and isotopic fractionation by solar-wind-induced sputtering and photochemical escape. *Icarus* 111: 271–88.

5. http://space.rice.edu/IMAGE/livefrom/5_magnetosphere.jpg

6. Dole SH. 1964. *Habitable Planets for Man*. New York: Blaisdell, 158 pp.

7. Berkner LV, Marshall LL. 1964. The history of oxygenic concentration in the Earth's atmosphere. *Disc. Faraday Soc.* 34: 122–41.

8. Ratner MI, Walker JCG. 1972. Atmospheric ozone and the history of life. *J. Atmos. Sci.* 29: 803–8.

9. Levine JS, Hays PB, Walker JCG. 1979. The evolution and variability of atmospheric ozone over geologic time. *Icarus* 39: 295–309.

10. Kasting JF, Donahue TM. 1980. The evolution of atmospheric ozone. *J. Geophys. Res.* 85: 3255–63.

11. Segura A, Krelove K, Kasting JF, Sommerlatt D, Meadows V, et al. 2003. Ozone concentrations and ultraviolet fluxes on Earth-like planets around other stars. *Astrobiology* 3: 689–708.

12. Ingersoll AP. 1969. The runaway greenhouse: a history of water on Venus. *J. Atmos. Sci.* 26: 1191–98.

13. Watson A, Lovelock JE, Margulis L. 1978. Methanogenesis, fires and the regulation of atmospheric oxygen. *Biosystems* 10: 293–98.

14. Lovelock JE. 1991. *Gaia: The Practical Science of Planetary Medicine*. London: Gaia Books.

15. Kasting JF, Whitmire DP, Reynolds RT. 1993. Habitable zones around main sequence stars. *Icarus* 101: 108–28.

16. Saunders RS. 1999. Venus. In: *The New Solar System*, ed. Beatty JK, Peterson CC, Chaikin A, eds. Cambridge, MA: Sky, pp. 97–110.

17. http://photojournal.jpl.nasa.gov/catalog/PIA00158

18. http://sos.noaa.gov/gallery/

19. Schaber GG, et al. 1992. Geology and distribution of impact craters on Venus: what are they telling us? *J. Geophys. Res.* 97: 13257–301.

20. Turcotte DL. 1993. An episodic hypothesis for Venusian tectonics. *J. Geophys. Res.* 98: 17061–178.

21. Wetherill G. 1994. Possible consequences of absence of Jupiters in planetary systems. *Astrophys. Space Sci.* 212: 23–32.

22. Hut P, Alvarez W, Elder WP, Hansen T, Kauffman EG, et al. 1987. Comet showers as a cause of mass extinctions. *Nature* 329: 118–26.

23. Alvarez LW, Alvarez W, Asaro F, Michel HV. 1980. Extraterrestrial cause for the Cretaceous-Tertiary extinction: experimental results and theoretical interpretation. *Science* 208: 1095–108.

24. Sleep NH, Zahnle KJ, Kasting JF, Morowitz HJ. 1989. Annihilation of ecosystems by large asteroid impacts on the early Earth. *Nature* 342: 139–42.

25. Laskar J, Robutel P. 1993. The chaotic obliquity of the planets. *Nature* 361: 608–14.

26. Laskar J, Joutel F, Robutel P. 1993. Stabilization of the Earth's obliquity by the Moon. *Nature* 361: 615–17.

27. Wood JA, Hartmann WK, Phillips RJ, Taylor CJ. 1986. In: *Moon over Mauna Loa: A Review of Hypotheses of Formation of Earth's Moon.* Houston, TX: Lunar and Planetary Inst., pp. 17–55.

28. Canup RM, Agnor CB, 2000. Accretion of the terrestrial planets and the Earth–Moon system. In: *Origin of the Earth and Moon*, Canup RM, Righter K, eds. Tucson: University of Arizona Press, pp. 113–29.

29. Williams DM, Pollard D. 2003. Extraordinary climates of Earth-like planets: three-dimensional climate simulations at high obliquity. *Int. J. Astrobiol.* 2: 1–19.

CHAPTER 10

1. Strughold H. 1953. *The Green and Red Planet.* Albuquerque: University of New Mexico Press.

2. Strughold H. 1955. The ecosphere of the Sun. *Aviat. Med.* 26: 323–28.

3. Huang SS. 1959. Occurrence of life in the universe. *Am. Sci.* 47: 397–402.

4. Huang SS. 1960. Life outside the solar system. *Sci. Am.* 202: 55–63.

5. Dole SH. 1964. *Habitable Planets for Man.* New York: Blaisdell, 158 pp.

6. Hart MH. 1978. The evolution of the atmosphere of the Earth. *Icarus* 33: 23–39.

7. Hart MH. 1979. Habitable zones around main sequence stars. *Icarus* 37: 351–57.

8. Kasting JF, Whitmire DP, Reynolds RT. 1993. Habitable zones around main sequence stars. *Icarus* 101: 108–28.

9. Forget F, Pierrehumbert RT. 1997. Warming early Mars with carbon dioxide clouds that scatter infrared radiation. *Science* 278: 1273–76.

10. Mischna MM, Kasting JF, Pavlov AA, Freedman R. 2000. Influence of carbon dioxide clouds on early martian climate. *Icarus* 145: 546–54.

11. Chaisson E, McMillan S, eds. 2008. *Astronomy Today*, 6th ed. Boston, MA: Pearson/Addison-Wesley.

12. Segura A, Krelove K, Kasting JF, Sommerlatt D, Meadows V, et al. 2003. Ozone concentrations and ultraviolet fluxes on Earth-like planets around other stars. *Astrobiology* 3: 689–708.

13. Gladman B. 1993. Dynamics of systems of 2 close planets. *Icarus* 106: 247–63.

14. Chambers JE, Wetherill GW, Boss AP. 1996. The stability of multi-planet systems. *Icarus* 119: 261-68.

15. Margulis L, Walker JCG, Rambler MB. 1976. Reassessment of roles of oxygen and ultraviolet light in Precambrian evolution. *Nature* 264: 620–24.

16. Rambler M, Margulis L. 1980. Bacterial resistance to ultraviolet irradiation under anaerobiosis: implications for pre-Phanerozoic evolution. *Science* 210: 638–40.

17. van Baalen C, O'Donnell R. 1972. Action spectra for ultraviolet killing and photoreactivation in the bluegreen alga *Agmenellum quadruplicatum*. *Photochem. Photobiol.* 15: 269–74.

18. Joshi M. 2003. Climate model studies of synchronously rotating planets. *Astrobiology* 3: 415–27.

19. Joshi MM, Haberle RM, Reynolds RT. 1997. Simulations of the atmospheres of synchronously rotating terrestrial planets orbiting M dwarfs: conditions for atmospheric collapse and the implications for habitability. *Icarus* 129: 450–65.

20. Griessmeier JM, Stadelmann A, Motschmann U, Belisheva NK, Lammer H, Biernat HK. 2005. Cosmic ray impact on extrasolar earth-like planets in close-in habitable zones. *Astrobiology* 5: 587–603.

21. Lissauer JJ. 2007. Planets formed in habitable zones of M dwarf stars probably are deficient in volatiles. *Astrophys. J.* 660: L149–L152.

22. Melosh HJ, Vickery AM. 1989. Impact erosion of the primordial atmosphere of Mars. *Nature* 338: 487–89.

23. Franck S, Kossacki K, Bounama C. 1999. Modelling the global carbon cycle for the past and future evolution of the Earth system. *Chem. Geo.* 159: 305–17.

24. Franck S, Block A, von Bloh W, Bounama C, Schellnhuber HJ, Svirezhev Y. 2000. Habitable zone for Earth-like planets in the solar system. *Planet. Space Sci.* 48: 1099–105.

25. Franck S, Block A, Von Bloh W, Bounama C, Schellnhuber HJ, Svirezhev Y. 2000. Reduction of biosphere life span as a consequence of geodynamics. *Tellus, Series B* 52: 94–107.

26. Ward PDBD. 2000. *Rare Earth: Why Complex Life Is Uncommon in the Universe.* New York: Copernicus.

27. Gonzalez G, Brownlee D, Ward P. 2001. The galactic habitable zone: galactic chemical evolution. *Icarus* 152: 185–200.

28. Lineweaver CH, Fenner Y, Gibson BK. 2004. The Galactic habitable zone and the age distribution of complex life in the Milky Way. *Science* 303: 59–62.

29. Gonzalez G. 1999. Is the Sun anomalous? *Astron. Geophys.* 40: 25–29.

30. Boone RH, King JR, Soderblom DR. 2006. Metallicity in the solar neighborhood out to 60 pc. *New Astron. Rev.* 50: 526–29.

Chapter 11

1. http://www.solstation.com/stars/barnards.htm

2. http://en.wikipedia.org/wiki/Center_of_mass

3. http://en.wikipedia.org/wiki/Parsec

4. http://planetquest.jpl.nasa.gov/Navigator/material/sim_material.cfm

5. Perryman MAC. 2002. GAIA: an astrometric and photometric survey of our Galaxy. *Astrophys. Space Sci.* 280: 1–10.

6. Unwin SC, Shao M, Tanner AM, Allen RJ, Beichman CA, et al. 2008. Taking the measure of the universe: precision astrometry with SIM PlanetQuest. *Publ. Astron. Soc. Pacific* 120: 38–88.

7. Wolszczan A, Frail DA. 1992. A planetary system around the millisecond pulsar PSR1257 + 12. *Nature* 355: 145–47.

8. http://en.wikipedia.org/wiki/Crab_Nebula

9. http://space-art.co.uk/index.html

10. http://exoplanet.eu/catalog-pulsar.php

11. http://www.glenbrook.k12.il.us/GBSSCI/PHYS/CLASS/sound/u11l1c.html

12. http://exoplanet.eu/catalog.php

13. http://www.coseti.org/images/ospect_1.jpg

14. Mayor M, Queloz D. 1995. A Jupiter-mass companion to a solar-type star. *Nature* 378: 355–59.

15. Marcy GW, Butler RP, Williams E, Bildsten L, Graham JR, et al. 1997. The planet around 51 Pegasi. *Astrophys. J.* 481: 926–35.

16. Lin DNC, Bodenheimer P, Richardson DC. 1996. Orbital migration of the planetary companion of 51 Pegasi to its present location. *Nature* 380: 606–7.

17. Rasio FA, Ford EB. 1996. Dynamical instabilities and the formation of extrasolar planetary systems. *Science* 274: 954–56.

18. Goldreich P, Tremaine S. 1980. Disk–satellite interactions. *Astrophys. J.* 241: 425–41.

19. Lin DNC, Papaloizou J. 1986. On the tidal interaction between protoplanets and the protoplanetary disk, 3: orbital migration of protoplanets. *Astrophys. J.* 309: 846–57.

20. Murray N, Hansen B, Holman M, Tremaine S. 1998. Migrating planets. *Science* 279: 69–72.

21. Armitage PJ, Rice WKM. 2005. Planetary migration. In: *A Decade Of Extrasolar Planets Around Normal Stars*. Baltimore, MD: Space Telescope Science Institute.

22. Talk by Geoff Marcy, circa 2002.

23. Udry S, Bonfils X, Delfosse X, Forveille T, Mayor M, et al. 2007. The HARPS search for southern extra-solar planets, XI: super-Earths (5 and 8 M-circle plus) in a 3-planet system. *Astron. Astrophys.* 469: L43–L47.

24. von Bloh W, Bounama C, Cuntz M, Franck S. 2007. The habitability of super-Earths in Gliese 581. *Astron. Astrophys.* 476: 1365–71.

25. Selsis F, Kasting JF, Paillet J, Levrard B, Delfosse X. 2007. Habitable planets around the star Gl581. *Astron. Astrophys.* 476: 1373–87.

26. http://www.astro.cornell.edu/academics/courses/astro201/microlensing.htm

27. http://www.iam.ubc.ca/~newbury/lenses/microlensing.html

28. Bond IA, Udalski A, Jaroszynski M, Rattenbury NJ, Paczynski B, et al. 2004. OGLE 2003-BLG-235/MOA 2003-BLG-53: a planetary microlensing event. *Astrophys. J.* 606: L155–L158.

29. http://exoplanet.eu/catalog-microlensing.php

30. Einstein A. 1936. Lens-like action of a star by the deviation of light in the gravitation field. *Science* 84: 506–7.

31. http://en.wikipedia.org/wiki/Einstein_ring

Chapter 12

1. http://sunearth.gsfc.nasa.gov/eclipse/transit/venus0412.html

2. Charbonneau D, Brown TM, Latham DW, Mayor M. 2000. Detection of planetary transits across a sun-like star. *Astrophys. J.* 529: L45–L49.

3. Brown TM, Charbonneau D, Gilliland RL, Noyes RW, Burrows A. 2001. Hubble Space Telescope time-series photometry of the transiting planet of HD 209458. *Astrophys. J.* 552: 699–709.

4. Selsis F, Chazelas B, Borde P, Ollivier M, Brachet F, et al. 2007. Could we identify hot ocean-planets with CoRoT, Kepler and Doppler velocimetry? *Icarus* 191: 453–68.

5. Kuchner MJ. 2003. Volatile-rich earth-mass planets in the habitable zone. *Astrophys. J.* 596: L105–L108.

6. Leger A, Selsis F, Sotin C, Guillot T, Despois D, et al. 2004. A new family of planets? "Ocean-Planets." *Icarus* 169: 499–504.

7. http://kepler.nasa.gov/

8. http://kepler.nasa.gov/sci/basis/fov-milkyway.html

9. Charbonneau D, Brown TM, Noyes RW, Gilliland RL. 2002. Detection of an extrasolar planet atmosphere. *Astrophys. J.* 568: 377–84.

10. Barman T. 2007. Identification of absorption features in an extrasolar planet atmosphere. *Astrophys. J.* 661: L191–L194.

11. Vidal-Madjar A, des Etangs AL, Desert JM, Ballester GE, Ferlet R, et al. 2003. An extended upper atmosphere around the extrasolar planet HD209458b. *Nature* 422: 143–46.

12. http://www.nasa.gov/mission_pages/spitzer/news/070221/index.html

13. Richardson LJ, Deming D, Horning K, Seager S, Harrington J. 2007. A spectrum of an extrasolar planet. *Nature* 445: 892–95.

14. Tinetti G, Vidal-Madjar A, Liang M-C, Beaulieu J-P, Yung Y, et al. 2007. Water vapour in the atmosphere of a transiting extrasolar planet. *Nature* 448: 169–71.

Chapter 13

1. http://planetquest.jpl.nasa.gov/TPF-C/tpf-C_index.cfm

2. http://www.iue.tuwien.ac.at/phd/minixhofer/node59.html

3. Cash W. 2006. Detection of Earth-like planets around nearby stars using a petal-shaped occulter. *Nature* 442: 51–53.

4. Cash W, Copi C, Heap S, Kasdin NJ, Kilston S, Kuchner M, Levine M, Lo A, Lillie C, Lyon R, Polidan R, Shaklan S, Starkman G, Traub W, Vanderbei R. 2007. External occulters for the direct study of exoplanets. Paper presented to the NASA/NSF Exoplanet Task Force, May 2007.

5. http://sco.stsci.edu/tpf_top100/

6. http://en.wikipedia.org/wiki/Terrestrial_Planet_Finder

Chapter 14

1. Des Marais DJ, Harwit MO, Jucks KW, Kasting JF, Lin DNC, et al. 2002. Remote sensing of planetary properties and biosignatures on extrasolar terrestrial planets. *Astrobiology* 2: 153–81.

2. Goode PR, Qui J, Yurchyshyn V, Hickey J, Chu MC, et al. 2001. Earthshine observations of the Earth's reflectance. *Geophys. Res. Lett.* 28: 1671–74.

3. Woolf NJ, Smith PS, Traub WA, Jucks KW. 2002. The spectrum of earthshine: a pale blue dot observed from the ground. *Astrophys. J.* 574: 430–33.

4. Owen T, Papagiannis MD. 1980. In: *Strategies for Search for Life in the Universe.* Dordrecht: Reidel.

5. Segura A, Krelove K, Kasting JF, Sommerlatt D, Meadows V, et al. 2003. Ozone concentrations and ultraviolet fluxes on Earth-like planets around other stars. *Astrobiology* 3: 689–708.

6. Sagan C, Thompson WR, Carlson R, Gurnett D, Hord C. 1993. A search for life on Earth from the Galileo spacecraft. *Nature* 365: 715–21.

7. Kiang NY, Siefert J, Govindjee, Blankenship RE. 2007. Spectral signatures of photosynthesis, I: review of Earth organisms. *Astrobiology* 7: 222–51.

8. Clark RN, Swayze GA, Wise R, Livo KE, Hoefen TM, Kokaly RF, Sutley SJ. 2003. *USGS Digital Spectral Library splib05a, Open File Report 03-395*, U.S. Geological Survey.

9. Ford EB, Seager S, Turner EL. 2001. Characterization of extrasolar terrestrialplanets from diurnal photometric variability. *Nature* 412: 885–87.

10. Wolstencroft RD, Raven JA. 2002. Photosynthesis: likelihood of occurrence and possibility of detection on earth-like planets. *Icarus* 157: 535–48.

11. Seager S, Turner EL, Schafer J, Ford EB. 2005. Vegetation's red edge: a possible spectroscopic biosignature of extraterrestrial plants. *Astrobiology* 5: 372–90.

12. Kiang NY, Segura A, Tinetti G, Govindjee, Blankenship RE, et al. 2007. Spectral signatures of photosynthesis, II: coevolution with other stars and the atmosphere on extrasolar worlds. *Astrobiology* 7: 252–74.

13. Meadows, VS. 2006. Modeling the diversity of extrasolar terrestrial planets. In: *Direct Imaging of Exoplanets: Science and Techniques*, Proceedings of IAU Coll 200, Aime C, Vakili F, eds. New York: Cambridge University Press.

14. Segura A, Meadows VS, Kasting J, Cohen M, Crisp D. 2007. Abiotic production of O_2 and O_3 in high-CO_2 terrestrial atmospheres. *Astrobiology* 7: 494–95.

15. Courtesy of R. Hanel, NASA Goddard Space Flight Center.

16. Leger A, Pirre M, Marceau FJ. 1993. Search for primitive life on a distant planet: relevance of O_2 and O_3 dectections. *Astron. Astrophys.* 277: 309–13.

17. Rubey WW, Poldervaart A. 1955. Development of the hydrosphere and atmosphere, with special reference to probable composition of the early atmosphere. In: *Crust of the Earth.* New York: Geol. Soc. Am., pp. 631–50.

18. Walker JCG. 1977. *Evolution of the Atmosphere.* New York: Macmillan.

19. Kasting JF. 1993. Earth's early atmosphere. *Science* 259: 920–26.

20. Oparin AI. 1938. *The Origin of Life.* New York: MacMillan.

21. Miller SL. 1953. A production of amino acids under possible primitive Earth conditions. *Science* 117: 528–29.

22. Hashimoto GL, Abe Y, Sugita S. 2007. The chemical composition of the early terrestrial atmosphere: formation of a reducing atmosphere from CI-like material. *J. Geophys. Res. Planets* 112: E5, E05010.

23. Pavlov AA, Kasting JF, Brown LL. 2001. UV-shielding of NH_3 and O_2 by organic hazes in the Archean atmosphere. *J. Geophys. Res.* 106: 23267–87.

24. Kharecha P, Kasting JF, Siefert JL. 2005. A coupled atmosphere-ecosystem model of the early Archean Earth. *Geobiology* 3: 53–76.

25. Schindler TL, Kasting JF. 2000. Synthetic spectra of simulated terrestrial atmospheres containing possible biomarker gases. *Icarus* 145: 262–71.

26. Kaltenegger L, Traub WA, Jucks KW. 2007. Spectral evolution of an Earth-like planet. *Astrophys. J.* 658: 598–616.

27. Kasting JF. 1997. Habitable zones around low mass stars and the search for extraterrestrial life. *Origins Life* 27: 291–307.

28. Kasting JF. 1988. Runaway and moist greenhouse atmospheres and the evolution of Earth and Venus. *Icarus* 74: 472–94.

29. Lewis JS, Prinn RG. 1984. *Planets and Their Atmospheres: Origin and Evolution.* Orlando, FL: Academic Press.

30. Beatty JK, Petersen CC, Chaikin A. 1999. *The New Solar System.* Cambridge, MA: Sky.

31. McElroy MB. 1972. Mars: evolving atmosphere. *Science* 175: 443–45.

32. Selsis F, Despois D, Parisot JP. 2002. Signature of life on exoplanets: can *Darwin* produce false positive detections? *Astron. Astrophys.* 388: 985–1003.

33. McCullough PR. 2008. Models of polarized light from oceans and atmospheres of earth-like extrasolar planets. *Astrophysical Journal*, submitted.

34. http://en.wikipedia.org/wiki/Polarization

35. Lederberg J. 1965. Signs of life: criterion-system of exobiology. *Nature* 207: 9–13.

36. Lovelock JE. 1965. A physical basis for life detection experiments. *Nature* 207: 568–70.

37. Turnbull MC, Traub WA, Jucks KW, Woolf NJ, Meyer MR, et al. 2006. Spectrum of a habitable world: Earthshine in the near-infrared. *Astrophys. J.* 644: 551–59.

38. Pavlov AA, Hurtgen MT, Kasting JF, Arthur MA. 2003. Methane-rich Proterozoic atmosphere? *Geology* 31: 87–90.

39. Segura AKJF, Cohen M, Meadows V, Crisp D, Tinetti G, Scalo J. 2005. Biosignatures from Earth-like planets around M stars. *Astrobiology* 5: 706–25.

Chapter 15

1. Labeyrie A. 1996. Resolved imaging of extra-solar planets with future 10–100 km optical interferometric arrays. *Astron. Astrophys. Suppl. Ser.* 118: 517–24.

2. Labeyrie A. 1999. Astronomy: Snapshots of alien worlds—the future of interferometry. *Science* 285: 1864–65.

3. Labeyrie A. 2002. *Hypertelescopes and exo-Earth coronography.* Proceedings of the 36th ESLAB Symposium, Noordwijk, The Netherlands.

4. http://planetquest.jpl.nasa.gov/TPF/TPFrevue/FinlReps/Boeing/Phsland2.pdf

5. Drake F. 1987. *Stars as gravitational lenses.* Presented at Bioastronomy—the Next Steps, Balaton, Hungary.

6. Eshleman VR. 1979. Gravitational lens of the Sun: its potential for observations and communications over inter-stellar distances. *Science* 205: 1133–35.

7. Einstein A. 1936. Lens-like action of a star by the deviation of light in the gravitation field. *Science* 84: 506–7.

8. Liebes S. 1964. Gravitational lenses. *Phys. Rev. B* 133: B835–&.

9. http://en.wikipedia.org/wiki/Einstein_radius

10. http://en.wikipedia.org/wiki/Voyager_1

11. Niven L, Pournelle J. 1974. *The Mote in God's Eye.* New York: Simon & Schuster, 537 pp.

12. http://www.seti.org/

13. http://www.coseti.org/introcoseti.htm

14. http://www.lswilson.ca/dewline.htm

15. http://www.seti.org/ata/

16. http://www.skatelescope.org/

17. Sagan C. 1995. *Cosmos.* Avenel, NJ: Wings Books, 365 pp.

18. Walker JCG, Kasting JF. 1992. Effects of fuel and forest conservation on predicted levels of atmospheric carbon dioxide. *Global Planet. Change* 97: 151–89.

19. Archer D. 2005. Fate of fossil fuel CO_2 in geologic time. *J. Geophys. Res. Oceans* 110: C9, C09505.

20. http://en.wikipedia.org/wiki/Milky_Way

21. http://en.wikipedia.org/wiki/Fermi_paradox

22. http://en.wikipedia.org/wiki/Zoo_hypothesis

Index

Index

Index

Index

Index